Colleto

Host Specificity, Pathology, and Host-Pathogen Interaction

Edited by

Dov Prusky
The Volcani Center
Bet Dagan, Israel

Stanley Freeman
The Volcani Center
Bet Dagan, Israel

Martin B. Dickman
University of Nebraska, Lincoln

APS PRESS

The American Phytopathological Society
St. Paul, Minnesota

Cover: Infection of bean by *Colletotrichum lindemuthianum*.
(Photo provided by R. J. O'Connell; reprinted, with permission,
from Physiological Plant Pathology 27:75-98.)

This book has been reproduced directly from copy submitted
in final form to APS Press. No editing or proofreading
has been done by the Press.

Reference in this publication to a trademark, proprietary product,
or company name by personnel of the U.S. Department of Agriculture
or anyone else is intended for explicit description only and does not
imply approval or recommendation to the exclusion of others that
may be suitable.

Library of Congress Catalog Card Number: 00-104665
International Standard Book Number: 0-89054-258-9

Second printing 2009

Printed in the United States of America on acid-free paper

The American Phytopathological Society
3340 Pilot Knob Road
St. Paul, Minnesota 55121, U.S.A.

Preface

Filamentous fungi from the genus *Colletotrichum* and its teleomorph *Glomerella* are considered major plant pathogens worldwide. They cause economically significant damage to crops in tropical, subtropical, and temperate regions. Cereals, grasses, legumes, ornamentals, vegetables, and fruit trees may be seriously affected by the pathogen. Although many cultivated fruit crops are infected by *Colletotrichum* species worldwide, the most significant economic losses are incurred when the fruiting stage is attacked. At this stage, the disease can cause a crop loss of up to 50%. *Colletotrichum* species cause typical disease symptoms known as anthracnose, that are characterized by sunken necrotic tissue where masses of orange conidia are produced. Anthracnose diseases appear in both developing and mature plant tissues. Two distinct types of diseases occur: those affecting developing fruit in the field (preharvest) and those damaging mature fruit during storage (postharvest). The ability to cause latent or quiescent infections has caused *Colletotrichum* to be ranked among the most important postharvest pathogens. Species of the pathogen appear predominantly on aboveground plant tissues; however, belowground organs, such as roots and tubers, may also be affected.

Besides having considerable economic impact, *Colletotrichum* is used as a model system for studying infection processes, fungal pathogenicity, and phytoalexins. Control of *Colletotrichum* in developed countries has proven relatively easy through the use of clean seed of resistant cultivars, but in developing countries there are considerable constraints to the adoption of control measures because of economic costs and the ubiquity of inoculum. Chemical control is practiced in high-value horticultural and tree crops, but problems of fungicide insensitivity and residues on edible produce exist.

The international workshop on Host Specificity, Pathology, and Host-Pathogen Interaction of *Colletotrichum*, supported by BARD (United States – Israel Binational Agricultural Research and Development Fund) was held in Jerusalem, Israel, from August 29 through September 1, 1998. Subjects included recent research on *Colletotrichum* in the areas of: systematics and the sexual stage, infection process, host specificity, population genetics, epidemiology, pathogenicity genes, regulation of pathogenicity and host resistance, important *Colletotrichum* diseases, mycoherbicides and their use, new strategies for the study and the control of new epidemics (i.e., *C. acutatum* in fruit crops, *C. coccodes* in vegetables, and *C. lindemuthianum* in beans). Special emphasis was placed on areas of need for further research such as subspecific variation, host range, genetics, and quantitative epidemiology. The up-to-date work which was presented by the leading scientists in these fields has contributed to the understanding of present problems dealing with *Colletotrichum* and future development in the areas of biology, pathology, and control. These important contributions have been summarized by peers in the field of research on *Colletotrichum* and appear as chapters in this book.

Robert M. Hanau, a contributor to this volume and associate professor in the Department of Botany and Plant Pathology at Purdue University, passed away 15 March 2000. Bob was born 12 November 1947 in Dallas, Texas. He received his B.A. and Ph.D. degrees from Temple University and was an NIH postdoctoral fellow with Andy Jackson at Purdue prior to his fungal research. Bob was a recognized leader in developing *Colletotrichum graminicola* as a genetically tractable experimental system and was a highly sought-after resource in the *Colletotrichum* community. Bob was a most unique individual who always appreciated a good laugh. His spirit will be missed.

Contents

HOST SPECIFICITY AND GENETIC DIVERSITY

PATHOGENICITY GENES

REGULATION OF PATHOGENICITY AND HOST RESISTANCE

MYCOHERBICIDES AND THEIR USE

BIOCONTROL

MAJOR *COLLETOTRICHUM* HOSTS

Chapter 1

Linking the Past, Present, and Future of *Colletotrichum* Systematics

Paul F. Cannon, Paul D. Bridge, and Enrique Monte

Currently available classification systems for *Colletotrichum* do not work effectively. The constituent species are inadequately defined using character suites which are inappropriate in the modern age, and they are poorly understood and frequently ignored by practicing plant pathologists. There is little usable information available on infraspecific variation, and evolution of the genus within the crop environment has hardly been considered. These are problems that are common to many groups of plant pathogens and other economically important fungi. The situation, though currently unsatisfactory, is deteriorating rapidly as the numbers of trained experts in morphological characterization are falling. A solution cannot be found overnight, but for economically important fungal groups at least, there is great potential for the augmentation of morphological classification systems with molecular data. Such data will allow the development of properly objective and even automated identification techniques.

The Past: The Importance of Historical Perspective

Fungal classification according to the best scientific methods of the day has been in existence for nearly 200 years. The system of applying names to organisms dates from the mid-eighteenth century, when binomials were introduced as shorthand versions of the descriptive phrase-names previously applied to species. The two basic units of the systematic hierarchy, species and genera, were quickly accepted, and there is still little dissent as to their value and practicality.

Fungal names are the key to all information about the organisms they represent, and communication between scientists, laymen and policymakers can only occur within the basic nomenclatural framework. It is therefore crucial that names are used by different researchers to refer to the same organism groups. Fungal names serve two purposes, as a means of reference and a summary of their relationships. The advantages of this dual system are considerable, as much more can be inferred through use of a name than through mere reference to the organism. As names are placed within a formal systematic hierarchy, attributes can be assumed without the need for repetition down the branches of the classification.

The system has some drawbacks. First, taxonomic delimitation (the range of variation included within the group of organisms given a particular name) is not fixed. The authors may then include different amounts of variation within a particular taxon, resulting in disagreements over matters such as the numbers of species within a genus, and the breadth of species concepts. Without consensus, this process leads in some cases to different names being used for organism groups. Taxa cannot practically be described initially with a full knowledge of the range of variation included within them, and a period of adjustment of taxon concepts is inevitable. Additional information may quite justifiably result in a change of circumscription of taxa. This information may result in name changes where intermediates between taxa are discovered, as a single taxonomic entity can only sensibly have one name. Further information on related taxa may also prompt name changes to maintain linkages between organism groups and the taxonomic framework.

A further problem with the nomenclatural system is that evolution is not a series of discrete steps, and the number of taxa as defined by phylogenetic branches is much greater than can be included using formal nomenclature. This problem is rapidly becoming more acute as molecular evidence is used to construct relational trees. There is no real consensus as to the extent of variation that should be included at each rank of the taxonomic hierarchy, and indeed agreement could only be achieved if all organisms (and all sections of the genome) evolved at the same steady rate. This has clearly not occurred.

If new information is to be related properly to previously published data and scientists are to be able to communicate effectively about their findings, a stable and relevant classification system is crucial for *Colletotrichum*. This system must provide for the needs of modern scientists and must allow for continuity by maintaining firm links to the valuable aspects of its taxonomic history.

A large proportion of systematic research in *Colletotrichum* must unfortunately be consigned to the garbage bin. For the first 50 years or so of this century, very large numbers of so-called species were described, based almost exclusively on assumed host specificity and all too often involving little or no comparison with previously described taxa. This is easy to criticize, but it must be recognized too that almost all modern systematic research on plant pathogens is host-limited, due to the insistence of funding agencies that taxonomic studies have direct applied relevance. The revision of *Colletotrichum* by von Arx (1) was a landmark in the classification of plant pathogens, in which around 750 "species" of *Colletotrichum* were reduced to 11 taxa based on morphological rather than host-related criteria. Species numbers have since crept upwards with more detailed studies on morphology, cultural characters, and pathogenic abilities, and currently around 40 are accepted. These are enumerated by Sutton (48), who provided an excellent summary of the history of *Colletotrichum* taxonomy.

The Present: Problems of Species Concepts

The current systematic arrangement within *Colletotrichum* is unsatisfactory for many reasons. The organism groups which are assigned species names vary widely in the degree of variation they encompass, in the characters which are used to define them, and in their perceived host specialization. In many cases also, they are not reliably distinguishable from their morphological neighbors. Infraspecific groups are hardly used, and most that are delimited are based on host virulence alone. When new taxa are recognized, there is no agreed framework which can be used to assign a taxonomic rank. This problem is not confined to *Colletotrichum*, and others (e.g., 23) have begun exploring these issues. A completely objective system for assigning ranks is not feasible, due to the complexities of evolutionary change and the need to maintain relevance of nomenclatural systems to users of classifications. However, stability is significant to users of any systematic arrangement, and measures to promote uniform and mutually agreed species concepts must be widely welcomed.

Standard species concepts are difficult to apply to many fungi (2). Traditional morphological concepts do not consider many of the aspects of variation which are important to pathologists. In the case of *Colletotrichum*, morphological concepts are difficult to define and are delimited using only small sets of characters, primarily size and shape of conidia and features of the appressoria (48). Intermediates are common, for several reasons. First, the range of variation in both size and shape of conidia is considerable for most species, with much overlap in size measurements and difficulties in defining shape variation objectively. Second, morphological features may vary considerably with environmental conditions, and comparing cultures with material observed directly from infected plant tissues is especially difficult. A related problem is that many species of *Colletotrichum* produce secondary conidia in culture directly from germinating primary spores. The secondary conidia are generally smaller and more varied in shape, especially when the culture is old. It is therefore important to compare strains using identical cultural protocols. Reference cultures have frequently been subcultured repeatedly, selecting for genotypes (or mutants) which grow more effectively in those abnormal conditions. The circumstances of their original isolation are not always recorded accurately. Reference cultures may also travel the world via different collections without their accompanying history, acquiring new reference codes at each stop. Considerable confusion arises over their identity and their links with cultures from other collections.

The biological species concept is difficult to apply to many species of *Colletotrichum* as sexual recombination (at least in experimental conditions) seems to occur infrequently or not at all. Recent research (e.g., 22 and Chapter 10 by Correll et al. in this book) is uncovering more evidence of meiotic stages of the life cycle. Even where the teleomorph is present, mating experiments are difficult and time-consuming (50). The number of species accepted in *Colletotrichum* would likely increase significantly if this approach was adopted. A further complication is that asexual forms, putatively clonal lines, are frequently very closely related to (and presumably directly descended from) meiotic taxa (4,5).

Phylogenetic species concepts in their pure sense will not be appropriately applied to *Colletotrichum* for the foreseeable future; there are too many gaps in our knowledge both of evolutionary pathways and the reproductive strategies alluded to previously. Nevertheless, the rapidly increasing amount of information being obtained from nucleic acid analyses is providing large quantities of data, with character states which are at least not directly influenced by environmental factors. Consensus trees are easily generated from sequence data, which may have a presumed phylogenetic component if outgroup taxa are included. This information currently provides the most useful structure for classification and taxon determination, but considerable further work is needed before a suitably robust system emerges for defining and naming *Colletotrichum* species.

The Need for Typification

There are clear differences in application of species names within *Colletotrichum*, which cause great confusion and seriously hinder the advancement of systematic and applied research. The most obvious of these involves *C. lindemuthianum*, for which two quite distinct morphological concepts exist. One of these places *C. lindemuthianum* as a part of the *C. gloeosporioides* complex (34), with relatively long narrow conidia illustrated and quoted as measuring 11-20 x 2.5-5.5 µm in size. The other (47,48) describes a fungus with shorter and relatively wider conidia, ranging between 9.5-11.5 µm in length and 3.5-4.5 µm in width. This places the fungus within the *C. orbiculare* complex in morphological terms, with its systematic position confirmed with ITS/28S sequence data by Sherriff et al. (42). The confusion almost certainly arose from the uncritical assumption that *Colletotrichum* species are host-specific, and by implication that any strain of *Colletotrichum* with straight conidia which occurs on legumes must be *C. lindemuthianum*. A similar situation may have occurred in *C. musae*, strains of which were reported to have significant ITS2 sequence variation by Johnston and Jones (22). The potential for error is exacerbated by the continuing lack of coordination between molecular, pathological, and morphological research.

There is a pressing need to fix the application of *Colletotrichum* names to avoid confusion in scientific communication. This process of typification is a well-established tenet of systematics, unambiguously linking published species names with individual preserved specimens. The extent of the taxon is not fixed either in morphological, genetic, or pathological terms, as there is no formal restriction on the degree of variation which can be encompassed. However, inclusion within an individual species concept of the collection designated as the type means that name can definitely be applied. The confusion over the application of the name *C. lindemuthianum* would not have happened if proper reference was made to type material.

The process of typification is often complex, especially where important names are concerned. Although the type concept is historically old in nomenclatural terms, there was no compulsion for names published before 1958 to be accompanied by citation of a type specimen. Almost all of the well-known *Colletotrichum* names were formally introduced before this date, and in many cases reference material was not formally designated, has subsequently been lost, or has deteriorated to such an extent that precludes basic characterization. Even where good material has been preserved, it is normally in the form of a dried specimen which can be studied using only restricted methods of molecular analysis. Fortunately, the nomenclatural code (16) allows for considerable flexibility in the designation of new type material for important names, and in the citation of accompanying authentic cultures which can be subjected to the full range of molecular analyses.

The process of choosing new or replacement type material must obviously be carefully conducted, in order that further confusion does not ensue. It is necessary as far as possible to select a collection which conforms to the original description and to any observation by the original author on matters such as pathogenicity and ecology, but it is equally important to maintain current usage of names where this is not ambiguous. At CABI Bioscience, we are in the process of choosing an authentic culture to represent the name *Colletotrichum gloeosporioides*, to fix its application and stabilize the classifications which are currently appearing based on molecular phylogenetic techniques. Original authentic material has not survived. We have examined a series of cultures derived from *Citrus* (the original host cited by Penzig; 39,40) from diverse geographical sources, using morphological, physiological, isoenzyme, and RAPD techniques, to assess the range of variation. We have also sequenced the ITS regions of a strain from Italy (from where the species was originally collected) in order to assess its suitability as a neotype (replacement type). It is clearly desirable that any new authentic material should be placed among the strains which are currently considered as representative of the *C. gloeosporioides* aggregate, but this is currently problematic due to inadequate understanding of the patterns of infraspecific variation. Selection of the neotype will allow objective definition of subgroups within the aggregate and provide a fixed point in the taxonomic framework.

The Future:
Towards an Integrated Classification System for *Colletotrichum*

Data derived from nucleic acid analyses currently provides the most reliable framework on which to build a classification of *Colletotrichum*, although our knowledge of molecular biology and its relevance to practical systematics is still very patchy. Current methods of obtaining genomic data are laborious, expensive, and potentially error-prone. Results are difficult to interpret and have all too often been published without adequate reference to information from other sources. Some of the pitfalls of performing molecular phylogenetic research in isolation have recently been eloquently stated by Camacho et al. (6) and Zhang et al. (53).

An essentially complete DNA sequence is known for only one fungus, *Saccharomyces cerevisiae* (14), and the function of many genes remains unclear. It is therefore difficult to relate sequence data to phenetic features such as morphology or pathogenic behaviour. Selection pressures on fungi of agricultural importance have changed beyond recognition in the last few hundred years, with the advent of monoculture, worldwide distribution of crop species, widespread use of fungicides, and the introduction of single-genotype crop varieties. These pressures must have had a considerable effect on the recent evolutionary history of *Colletotrichum* species, with reductions in crop genetic diversity almost certainly favoring the selection of clonally reproducing lines (15). This process has probably occurred through episodic selection, as discussed by Brasier (2).

Most modern molecular analyses fall into two major categories, those providing data from electrophoretic band patterns and those which result in sequence data. The first group includes RAPD and RFLP analyses, which have been widely used to compare populations and infraspecific groups of fungi. They are relatively cheap and easy to perform and can provide rapid comparative data. However, in both cases the various DNA fragments obtained are only categorized on their overall length, so that similarly sized fragments with different sequences are not separated. In addition, trivial sequence variation (in functional terms) may result in completely different band patterns, if the sequence difference occurs within a restriction site or primer amplification sequence. Data derived from these techniques, therefore, have restricted applications. They may in general be used to confirm strain relationships and identify populations, but differences in band patterns do not necessarily imply significant genetic divergence. RAPD-, RFLP-, and other band-based data, therefore, cannot be effectively used in phylogenetic studies.

Sequence data were until recently laborious and very expensive to obtain. With the advent of automated processes, at least the work involved has decreased significantly, although few laboratories can afford to obtain sequences of more than a few representative samples of the genetic diversity measurable using other techniques. Point mutations are simple to detect and analyze using sequencing techniques. Alterations such as insertions, deletions, and inversions, however, can significantly alter sequences without affecting functional capability. These alterations may cause major problems in alignment for comparative studies, especially in variable sections of the genome such as the ITS regions. The cost of sequencing will likely continue to decline, and more and more data will become available for phylogenetics and other systematic disciplines.

There is much evidence that different parts of the genome evolve at different rates, with critical functional sequences facing strong negative selection pressure and intergenic spacer regions changing in response to external mutagenic forces. Therefore, different sections of the genome may be appropriate for resolving variation at different levels of the systematic hierarchy. This has been explored by Bruns et al. (3), but the phylogenetic data available is still insufficient to confirm their findings in anything but general terms. Only a very small proportion of the total genome is currently used to infer relationships, so the risk of recreating gene trees rather than organism phylogenies is real.

In studies of *Colletotrichum*, work has concentrated on two parts of the genome, mitochondrial (mtDNA) and nuclear ribosomal (rDNA). Both are present as multiple copies within the cell, which allows analysis with less stringent experimental conditions. Sequence data are available only for rDNA, with information available from various parts of the gene cluster. Analyses have been made of the small subunit (18S), the ITS regions including the 5.8S gene, and the large subunit (28S). It is generally agreed that as ITS sequences are thought to be nonfunctional they are more variable and are useful for differentiation of taxa at and below the species level. In contrast, the 18S and 28S regions are functional and thus assumed to be more conserved, although some parts (especially the D1 and D2 domains) are known to be variable. They are therefore considered as primarily useful for phylogenetic analysis at the genus level and above. In neither case are these rules absolute; research is available which uses ITS data for phylogenetics (e.g., 33), and functional ribosomal sequences have also been used for species and infraspecific group definition, including major publications on *Colletotrichum* (e.g., 22,42). Unlinked portions of the genome such as the ß-tubulin and translation elongation factor (TEF) genes (e.g., 37) need to be analyzed further to provide independent evidence of lineages.

Species Complexes and Microspecies

The three major species aggregates currently recognized in *Colletotrichum* are based around *C. gloeosporioides*, *C. acutatum*, and *C. orbiculare*. Further assemblies will almost certainly be confirmed, including ones based around *C. capsici* and *C. graminicola*. These complexes have varied histories, but show common problems in the recognition of taxonomic subunits and links with other species.

The *C. gloeosporioides* aggregate has been acknowledged as polymorphic since the monograph of von Arx (1), but as Sutton (48) explained, uncertainty continues as to how the included nodes of variation should be treated. Recent molecular research has greatly improved the characterization of these nodes, but disagreement remains over the assigning of rank to the nodes and over details of their inter-relationships. A good example is that of *C. kahawae*. This taxon was formally introduced by Waller et al. (52) for the coffee berry disease pathogen, as distinct from strains of *C. gloeosporioides* causing other disease symptoms of coffee trees. Although *C. kahawae* could not be distinguished morphologically from other members of the *C. gloeosporioides* aggregate, it grew slower and could not utilize citrate or tartrate as sole carbon sources. This information, combined with its very distinct pathogenic behaviour, was considered by Waller and colleagues to merit recognition of *C. kahawae* as a distinct species. This importantly met the needs of plant pathologists for a distinct name for the coffee berry pathogen. The distinctiveness of the pathogen was confirmed using ITS sequence data by Sreenivasaprasad et al. (45,46), though they suggested that the differences between *C. kahawae* and the rest of the *C. gloeosporioides* aggregate were insufficient to justify its maintenance as a separate species. This conclusion was reached largely because the sequence divergence between *C. kahawae* and some strains of *C. gloeosporioides* was less than that observed among available sequences of the entire *C. gloeosporioides* aggregate. Subsequent research using SS-REP and AFLP techniques (see Chapter 10 by Correll et al. in this book) has shown the taxon to be a largely clonal population.

C. fragariae, one of the major causes of strawberry anthracnose, is another example of uncertainty over species concepts within the *C. gloeosporioides* aggregate. Traditionally, it has been considered a highly pathogenic taxon that cannot be distinguished reliably from the main body of the aggregate using morphological characters (43), although Gunnell and Gubler (18) claimed to have recognized some differential features. Strains have been assigned to *C. gloeosporioides* rather than *C. fragariae* if ascomata (referable to *Glomerella cingulata*) are produced.

Sreenivasaprasad et al. (44) examined rDNA and mtDNA of a range of strains of *C. fragariae*, comparing them with examples of *C. gloeosporioides* and *C. acutatum* isolated from strawberry. They showed close genetic similarity between the examined strains assigned to *C. fragariae* and *C. gloeosporioides,* with 97-100% homology of ITS1 sequences.

The strawberry anthracnose fungi have recently been examined in more detail by Buddie et al. (5), who studied mitochondrial and ribosomal DNA RFLPs and isoenzyme profiles. Strains of *C. fragariae* and *C. gloeosporioides* had identical rDNA RFLP band patterns, but differed in mtDNA RFLPs and isoenzyme profiles. As a strain of *C. gloeosporioides* from *Citrus* which was used as an external control had distinct rDNA as well as mtDNA RFLP patterns, they tentatively concluded that *C. fragariae* should be regarded as a holomorphic species including both the genetically similar mitotic strains previously referred to as *C. fragariae* and the meiotically reproducing isolates causing strawberry anthracnose, formerly considered as *C. gloeosporioides*. Similar conclusions were reached for a different part of the *C. gloeosporioides* aggregate by Johnson et al. (21). They used rDNA polymorphism along with morphological characterization as justification for erecting the new species *C. nupharicola*. Results from ITS1 sequencing of the *C. fragariae* cluster have recently become available (P. Martínez-Culebras, unpublished), which confirm the close relationship of the two groups of strains, but additionally suggest that non-strawberry *C. gloeosporioides* strains may cluster with them. This implies that if *C. fragariae sensu lato* is accepted as a distinct taxonomic entity (at species or infraspecific rank), it should not be restricted to strawberry isolates.

Definition of Taxa

If systematic concepts are to be applied uniformly, exploration of the range of variation within taxonomic groups must be extended. Concepts are not necessarily applicable to all fungi, as evolutionary history, selection pressures, and genetic systems vary so widely (2). This variety appears to be the case even within *Colletotrichum*, as Hodson et al. (20) demonstrated considerable variation in rDNA and mtDNA polymorphism within strain sets of *C. gloeosporioides* isolated from different tropical fruits. The rDNA and mtDNA RFLPs of geographically widely separated strains from mango were found to be almost identical, while those from avocado, banana, and papaya varied considerably. Similar results have been obtained by Freeman et al. (see Chapter 4 in this volume), who found both near-clonal and polymorphic populations of *Colletotrichum* species on various hosts in Israel. Factors such as loss of meiotic capacity and clonal reproduction may have dramatic effects on the gene pool within major evolutionary lines within the *C. gloeosporioides* complex, and the recent globalization of agriculture has greatly increased the opportunities for long-distance gene flow.

Despite these complexities, it is clearly sensible that as far as possible species and infraspecific groups should be defined in a similar way for genetically or ecologically related fungal pathogens. The first step in defining systematic concepts on an evolutionary basis is to examine the range of nucleic acid variation within the groups and subgroups currently recognized and contrast this variation with that between the taxa of concern. This topic has already been considered for *Colletotrichum* with the examination of ITS1 sequences by Sreenivasaprasad et al. (46) and ITS2/25S rDNA sequences by Sherriff et al. (42) and Johnston and Jones (22). Analysis of the sequence divergence between individual pairs of strains (Fig. 1) shows a bimodal frequency distribution, with one peak between 0 and 5% and the other between 7 and 23% divergence for the ITS1 region. Figures for the ITS2/25SD1 region (Fig. 2) are 0 and 1%, and 2.5 and 7% divergence. The first peak appears to correspond to infraspecific variation, and the second to interspecific divergence. The significance of this gap in divergence values remains unclear, but it has the potential for development as a powerful tool in assigning rank to taxonomic groups. Similar techniques have in the past been used extensively for species definition in yeast and bacterial taxa using measurements of DNA hybridization. Here, values of >70% DNA/DNA reassociation (normally at least 80-90%) are commonly regarded as indicative of infraspecific variation, and below this figure as evidence of distinction at the species

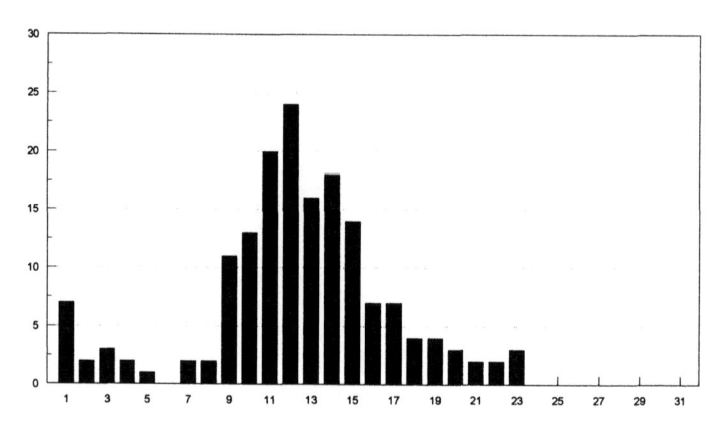

Fig. 1. Frequency of ITS1 sequence divergence values between pairs of *Colletotrichum* strains, expressed as percentages (derived from Sreenivasaprasad et al., 46).

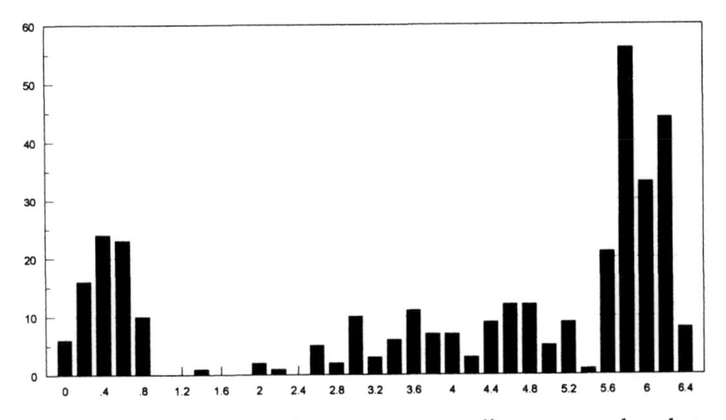

Fig. 2. Frequency of ITS2/28SD1 sequence divergence values between pairs of *Colletotrichum* strains, expressed as percentages (derived from Sherriff et al., 42).

by sequence analysis, but there seems to have been little discussion on the use of these data to assign taxonomic rank to fungal groups on a general basis. Analysis of sequences from other fungi is clearly advisable before adopting rDNA sequence divergence as a major criterion species for definition in *Colletotrichum*. Few large data sets are available, as the cost of sequencing often precludes study of more than representative strains of species. Work on *Ganoderma* by Moncalvo et al. (29,32), however, shows a similar bimodal frequency distribution of sequence divergence between strain pairs. This again appears to correspond largely to infraspecific and interspecific delimitation, although one taxon they recognized at the species level (tentatively named *G. oerstedii*) should perhaps be reassessed. Other indications of the relative values of infraspecific and interspecific sequence divergence can be found in Table 1. It should be recognized that in a number of these studies, some morphologically defined species were found to encompass a series of smaller groups definable in molecular terms, and these have been treated provisionally as independent species. Comparisons between the genera cited in the table should be approached cautiously, as the various divergence measures used give rather different results, especially because of the differing treatments of sequence gaps.

There are obstacles to the use of such data for defining species without careful interpretation, because length polymorphisms make comparison on a simple basis difficult. Assessing the importance of inversions, insertions, and deletions compared with simple base changes is also difficult. These should logically be considered as single events despite the relatively large differences in sequence which result. Little is known of the relative

Table 1. Ranges of infraspecific and intraspecific ITS sequence divergence

Fungus	Infraspecific	Interspecific	Sequence	Reference
Beauveria brongniartii	10.5%		ITS1-2	Neuvéglise et al. (34)
Colletotrichum	0-5%	7-23%	ITS1	Sreenivasaprasad et al. (46)
	0-1%	2.5-7%	ITS2	Sherriff et al. (42)[1]
Fusarium	0-2.3%	4.6-15%	ITS2	O'Donnell (36)[2]
Ganoderma	0-4%	7-18%	ITS1	Moncalvo et al. (30)
	0-5%	5-20%	ITS2	Moncalvo et al. (30)
Laccaria	1-2%	3.5-5%	ITS1-2	Gardes et al. (13)
Metarhizium anisopliae	6.3%		ITS1-2	Curran et al. (10)
Phytophthora	0-4.2%	4.2-23%	ITS1	Lee and Taylor (28)
	0-1.6%	0.5-24%	ITS2	Lee and Taylor (28)
Sclerotinia	0-0.5%		ITS1	Carbone & Kohn (8)
Seiridium	1-11.5%	11.5-18%	ITS1	Viljoen et al. (51)
Trichoderma	0-2.5%	2.0-16.7%	ITS1	Kuhls et al. (26)
	0-2%	4.4-20.3%	ITS1	Grondona et al. (17)[3]

[1] A few strains showed intermediate results
[2] Infraspecific groups within *Fusarium sambucinum*
[3] Infraspecific groups within *Trichoderma harzianum*

likelihood of these phenomena in fungal rDNA, although Swofford and Olsen (49) have considered these issues in statistical terms. There are also problems in determining the number of steps involved in a sequence change, as evolutionary processes do not necessarily follow the most parsimonious route.

Extrapolation of these data between unrelated fungal groups is clearly problematic, as Seifert et al. observed (41). However, the information may be useful as a baseline from which to explore relationships. The acceptance of rDNA sequence divergence as a valid tool for the assignment of rank to taxonomic groups within *Colletotrichum* implies that the major species aggregates centered on *C. gloeosporioides*, *C. acutatum*, and *C. orbiculare* should be treated as species units rather than superspecific taxa. Highly pathogenic taxa such as *C. kahawae* and *C. fragariae* should then be accepted as infraspecific taxa of *C. gloeosporioides* rather than as separate species, as their rDNA divergence from *C. gloeosporioides* is insufficient. However, it does not follow that such taxa should be referred to using the species name without qualification, if data specific to the infraspecific taxa are to be used and interpreted correctly.

INFRASPECIFIC NOMENCLATURE

The current nomenclatural code provides rather few formal infraspecific categories, though it does not prohibit the introduction of further levels in the systematic hierarchy for specialized applications. The ranks which are specified are subspecies, variety, and form. In addition, plant pathologists have used categories such as forma specialis, pathotype, and pathovar to denote infraspecific groups with distinct host specializations or behaviors. In very few cases are the criteria defined for deciding on rank. Of the systematic terms, subspecies has been used rarely until recently, although some workers (e.g., 7) have introduced it for biotrophic taxa with minor morphological differences linked to host preference. Variety and form have in the past been used indiscriminately for minor variants (including host-specific groups), without any real consensus as to their application. The pathological terms have never been subject to formal nomenclatural legislation, and, in a number of cases, have seemingly been introduced incidentally. It is not possible to extrapolate between organism groups by assuming that the formae speciales of unrelated species encompass similar levels of variation, especially as they are defined in terms of host response rather than pathogen characteristics.

Within *Colletotrichum*, infraspecific categories have been used sparingly, with occasional names published at the variety, forma, and forma specialis level (Table 2). As can be seen, there is no obvious link between the rank employed and the type of characters used as distinguishing factors. Bearing in mind the problems in definition of these ranks, it would be inadvisable to introduce large numbers of such infraspecific taxa without consultation. Johnston and Jones (22) considered the advisability of such trinomial systems, which at least potentially allow for the definition of host-specific, clonal taxa within the general context of the species. Complex names are awkward to use and are vulnerable to uncritical abbreviation, which may result in significant confusion where the abbreviated taxon names are used in different senses.

There is a clear problem at present, where names at species rank are used for groups of strains with widely differing internal variation. Therefore, taxa recognized as species such as *C. kahawae* must either be completely contained within another species, *C. gloeosporioides*, or the latter must be accepted as paraphyletic and defined to contain a *C. kahawae*-shaped "hole." These anomalies are in fact widespread throughout the fungal kingdom. For example, the degree of difference in molecular terms between *Beauveria bassiana* and *B. brongniartii* is no greater than that found within strains assigned to the latter taxon alone (9), and Norman and Egger (35) showed that the genus *Plicaria* as traditionally described nests within *Peziza*.

The options are to reduce the rank of *C. kahawae* (or increase that of *C. gloeosporioides*), or to accept that binomials do not necessarily refer to the same branch level of the phylogenetic tree. The latter option has serious consequences for the formal nomenclatural process if followed to completion, as the whole hierarchical structure would be compromised. Nevertheless, rankless phylogenetic classification systems are currently

Table 2. Infraspecific taxa of *Colletotrichum* reported in the *Index of Fungi* since 1940

Taxon	Stated distinguishing features
acutatum f. sp. *pineum* Dingley & Gilmour	pathogenic to *Pinus*
acutatum f. sp. *chromogenum* (Gorter) Baxter et al.	colony colour variant
capsici forma *cyamopsidicola* Desai & Prasad	pathogenic to *Cyamopsis*
dematium forma *bougainvilleae* Arya	pathogenic to *Bougainvillea*
dematium forma *circinans* (Berk.) v. Arx[11]	pathogenic to *Allium*
dematium f. sp. *clitoriicola* Pavgi & Mukhopadhyay	pathogenic to *Clitoria*
dematium forma *spinaciae* (Ellis & Halst.) v. Arx[22]	host variant
dematium forma *truncatum* (Schwein.) v. Arx[33]	smaller acervuli, host variant
euchroum forma *microsporum* Negru & Vlad	morphological variant
gloeosporioides f. sp. *aeschynomenes* Daniel et al.	biocontrol of *Aeschynomene*
gloeosporioides forma *alatae* Singh et al.	pathogenic to *Dioscorea alata*
gloeosporioides var. *aleuritidis* Saccas & Drouillon	pathogenic to *Aleurites*
gloeosporioides f. sp. *camelliae* Dickens & Cook	pathogenic to *Camellia* vars.
gloeosporioides f. sp. *clidemiae* Trujillo et al.	aggressive to *Clidemia* cults.
gloeosporioides f. sp. *cucurbitae* Menten et al.	pathogenic to *Cucumis*
gloeosporioides f. sp. *cuscutae* Zhang	pathogenic to *Cuscuta*
gloeosporioides var. *gomphrenae* Perera	pathogenic to *Gomphrena*
gloeosporioides forma *heveae* (Petch) Saccas	pathogenic to *Hevea*
gloeosporioides f. sp. *jussiaeae* O'Connell et al.	unknown
gloeosporioides f. sp. *manihotis* Chevaugeon	'adapted to *Manihot*'
gloeosporioides forma *melongenae* Fournet	differs in biometrics
gloeosporioides var. *minus* Simmonds	morphological variant
gloeosporioides f. sp. *pilosae* U.P.Singh	pathogenic to *Bidens pilosa*
gloeosporioides f. sp. *uredinicola* U.P. Singh	hyperparasite on rusts
gossypii var. *cephalosporioides* Viégas	morphological variant
graminicola f. sp. *zeae* Messaien et al.[4]	pathogenic to *Zea*
graminicola f. sp. *sacchari* Messiaen et al.[5]	pathogenic to *Saccharum*
graminicola f. sp. *secalis* Messiaen et al.	pathogenic to *Secale*
graminicola f. sp. *sorghi* Messiaen et al.[6]	pathogenic to *Sorghum*
graminicola var. *zonatum* Rajasab & Ramalingam	host variant
helianthi var. macromaculans H.C. Greene	distinguished by host response
trichellum var. *araliae* Servazzi	pathogenic to *Aralia*
vitis var. *majus* Batista & Bezerra	morphological variant

[1] Treated as *C. circinans* by Sutton (48)
[2] Treated as *C. spinaciae* by Sutton (48)
[3] Treated as *C. truncatum* by Sutton (48)
[4] Not distinguished from *C. graminicola* by Sutton (47)
[5] Treated as *C. falcatum* by Sutton (48)
[6] Treated as *C. sublineolum* by Sutton (48)

being discussed seriously (e.g., 19,31), either as alternative or additional to traditional arrangements. These define groups on a conceptual basis in terms of clades, in terms of the daughter taxa nested within a specified lineage. Such classification systems theoretically do not need types, but in practice the strains from which sequence data have been obtained function in this manner. The nomenclatural systems associated with phylogeny-based classifications have yet to be discussed fully, with some advocating the transfer of traditional taxon names by ignoring their implied rank, and others promoting a completely new, perhaps mononomial set of names (25). These revolutionary concepts have advantages in promoting nomenclatural stability, as rank-inspired name changes would become unnecessary. They do, of course, depend on the establishment of a robust and stable phylogenetic reconstruction. Even with this, they will not solve all problems. Proliferation of recognized taxa will probably be unavoidable, defined in terms of phylogenetic nodes. More seriously, circumscription of taxa will vary depending on sequences from different parts of the genome, and using different comparative methodologies. For example, the use of parsimony rather than neighbor-joining analysis results in the inclusion of *Helvella*, *Tuber*, and their hypogeous relatives within a greatly expanded phylogenetic group including *Morchella* and *Rhizina*, according to data from O'Donnell et al. (38) cited by Hibbett and Donoghue (19). The inclusion of new members will sometimes result in redefinition of existing phylogenetic taxa as topologies change.

At least until we have a comprehensive and robust phylogeny of *Colletotrichum* based on data from several parts of the genome, rankless classifications are not an option. They also present potentially serious problems in harmonizing the needs of systematists and those of the practical users of the names. We have to accept that systematic arrangements in terms of comprehensible and practically useful units can only approximate the evolutionary processes as measured by phylogenetic reconstructions. Acceptance of the paraphyly of *C. gloeosporioides* implied by the recognition of *C. kahawae* is inevitable in practical terms and answers a clear need from the pathological community for a name for the presumed basal portion of the aggregate which does not have clear host specialization. However, rank assignment is still unresolved; to recognize *C. kahawae* and similar taxa at a formal infraspecific rank would force name changes. One pragmatic, though probably temporary, solution would be to employ the informal rank of microspecies, which is used in botanical circles to describe inbreeding satellite populations (often apomictic or vegetatively propagating) within species aggregates (11,12). While we do not claim that *C. kahawae* is a direct analogue of this situation, the parallels are clear. The advantage would be that an informal infraspecific rank could be used for such populations without the need to employ trinomial nomenclature, at least until a more reliable systematic framework emerges. In time, a polynomial system for *Colletotrichum* should be developed by consensus, and in tandem with similar initiatives for other important groups of pathogenic fungi.

There is no doubt that the currently accepted *Colletotrichum* "species" are useful as terms of reference, despite the fact that they represent widely varying sizes of gene pools. Our priorities should be to define them more accurately, especially in molecular terms. We should also identify evolutionary lines within the currently recognized taxa, with particular attention paid to subpopulations which are linked to specific pathogenic traits. However, we should not lose sight of the fact that many infraspecific groups are not host-limited (see Chapter 10 by Correll et al. in this book). There is practical merit in dual nomenclatural systems, with a simplified arrangement of species and infraspecific taxa for applied mycology and a system which more closely reflects relationships for use within systematic contexts. For the time being, we have little option. However, it is important that users of names realize that their colleagues may interpret their data differently, unless the context is made clear in which the names are used. Until the internal systematic structure of species like *C. gloeosporioides* is properly understood, it is sufficient to define groups carefully without being too concerned about the nomenclatural rank at which they are introduced.

Literature Cited

1. Arx, J.A. von. 1957. Die Arten der Gattung *Colletotrichum.* Phytopathol. Z. 29:413-468.
2. Brasier, C.M. 1997. Fungal species in practice: Identifying species units in fungi. Pages 135-170 in: Species: The Units of Biodiversity. M.F. Claridge, H.A. Dawah, and M.R. Wilson, eds. Chapman and Hall, London.
3. Bruns, T.D., White, T.J., and Taylor, J.W. 1991. Fungal molecular systematics. Annu. Rev. Ecol. Systematics 22:525-564.
4. Bryson, R.J., Caten, C.E., Hollomon, D.W., and Bailey, J.A. 1992. Sexuality and genetics of *Colletotrichum*. Pages 27-46 in: *Colletotrichum*: Biology, Pathology and Control. J.A. Bailey and M.J. Jeger, eds. CAB International, Wallingford, UK.
5. Buddie, A.G., Martínez-Culebras, P., Bridge, P.D., García, M.D., Querol, A., Cannon, P.F., and Monte, E. 1999. Molecular characterization of *Colletotrichum* strains derived from strawberry. Mycol. Res. 103:385-394.
6. Camacho, F.J., Gernandt, D.S., Liston, A., Stone, J.K., and Klein, A.S. 1997. Endophytic fungal DNA, the source of contamination in spruce needle DNA. Molec. Ecol. 6:983-987.
7. Cannon, P.F. 1991. A revision of *Phyllachora* and some similar genera on the host family Leguminosae. Mycol. Papers 163. 302 pp.

8. Carbone, I. and Kohn, L.M. 1993. Ribosomal DNA sequence divergence within internal transcribed spacer 1 of the Sclerotiniaceae. Mycologia 85:415-427.
9. Couteaudier, Y., Viaud, M., and Neuvéglise, C. 1998. Combination of different independent molecular markers to understand the genetic structure of *Beauveria* populations. Pages 95-104 in: Molecular Variability of Fungal Pathogens. P.D. Bridge, Y. Couteaudier, and J.M. Clarkson, eds. CAB International, Wallingford, UK.
10. Curran, J., Driver, F., Ballard, J.W.O., and Milner, R.J. 1994. Phylogeny of *Metarhizium*: Analysis of ribosomal DNA sequence data. Mycol. Res. 98:547-552.
11. Davis, P.H. and Heywood, V.H. 1973. Principles of Angiosperm Taxonomy. Krieger, New York. 558 pp.
12. Dudman, A.A. and Richards, A.J. 1997. Dandelions of Great Britain and Ireland. BSBI Handbook, No. 9. Bot. Soc. British Isles, London. 344 pp.
13. Gardes, M., White, T.J., Fortin, J.A., Bruns, T.D., and Taylor, J.W. 1991. Identification of indigenous and introduced symbiotic fungi in ectomycorrhizae by amplification of nuclear and mitochondrial ribosomal DNA. Can. J. Bot. 69:180-190.
14. Goffeau, A. et al. [633 authors] 1997. The yeast genome directory. Nature (London) 387 (6632S). 105 pp.
15. Gordon, T.R. 1993. Genetic variation and adaptive potential in an asexual soil fungus. Pages 217-224 in: The Fungal Holomorph: Mitotic, Meiotic and Pleomorphic Speciation in Fungal Systematics. D.R. Reynolds and J.W. Taylor, eds. CAB International, Wallingford, UK.
16. Greuter, W., Barrie, F.R., Burdet, H.M., Chaloner, W.G., Demoulin, V., Hawksworth, D.L., Jorgensen, P.M., Nicolson, D.H., Silva, P.C., Trehane, P., and McNeill, J. 1994. International Code of Botanical Nomenclature (Tokyo Code). Regnum Vegetabile 131. Koeltz Scientific Books, Königstein, Germany. 389 pp.
17. Grondona, I., Hermosa, M.R., Tejada, M., Gomis, M.D., Mateos, P.F., Bridge, P.D., Monte, E., and García-Acha, I. 1997. Physiological and biochemical characterization of *Trichoderma harzianum*, a biological control agent against soilborne fungal pathogens. Appl. Environ. Microbiol. 63:3189-3198.
18. Gunnell, P.S. and Gubler, W.D. 1992. Taxonomy and morphology of *Colletotrichum* species pathogenic to strawberry. Mycologia 84:157-165.
19. Hibbett, D.S. and Donoghue, M.J. 1998. Integrating phylogenetic analysis and classification in fungi. Mycologia 90:347-356.
20. Hodson, A., Mills, P.R., and Brown, A.E. 1993. Ribosomal and mitochondrial DNA polymorphisms in *Colletotrichum gloeosporioides* isolated from tropical fruits. Mycol. Res. 97:329-335.

21. Johnson, D.A., Carris, L.M., and Rogers, J.D. 1997. Morphological and molecular characterization of *Colletotrichum nymphaeae* and *C. nupharicola* sp. nov. on water-lilies (*Nymphaea* and *Nuphar*). Mycol. Res. 101:641-649.
22. Johnston, P.R. and Jones, D. 1997. Relationships among *Colletotrichum* isolates from fruit-rots assessed using rDNA sequences. Mycologia 89:420-430.
23. Kohn, L.M. 1992. Developing new characters for fungal systematics: An experimental approach for determining the rank of resolution. Mycologia 84:139-153.
24. Krieg, N.R. 1988. Bacterial classification: An overview. Can. J. Microbiol. 34:536-540.
25. Kron, K.A. 1997. Exploring alternative systems of classification. Aliso 15:105-112.
26. Kuhls, K., Lieckfeldt, E., Samuels, G.J., Meyer, W., Kubicek, C.P., and Börner, T. 1997. Revision of *Trichoderma* sect. *Longibrachiatum* including related teleomorphs based on analysis of ribosomal DNA internal transcribed spacer regions. Mycologia 89:442-460.
27. Kurtzman, C.P. and Fell, J.W., eds. 1998. The Yeasts. A Taxonomic Study. 4th ed. Elsevier, Amsterdam. 1100 pp.
28. Lee, S.B. and Taylor, J.W. 1992. Phylogeny of five fungus-like protoctistan *Phytophthora* species, inferred from the internal transcribed spacers of ribosomal DNA. Molec. Biol. Evol. 9:636-653.
29. Moncalvo, J.-M., Wang, H.-F., and Hseu, R.-S. 1995. Phylogenetic relationships in *Ganoderma* inferred from the internal transcribed spacers and 25S ribosomal DNA sequences. Mycologia 87:223-238.
30. Moncalvo, J.-M., Wang, H.-F., and Hseu, R.-S. 1995. Gene phylogeny of the *Ganoderma lucidum* complex based on ribosomal DNA sequences. Comparison with traditional taxonomic characters. Mycol. Res. 99:1489-1499.
31. Moore, G. 1998. A comparison of traditional and phylogenetic nomenclature. Taxon 47:561-579.
32. Mordue, J.E.M. 1971. *Colletotrichum lindemuthianum*. CMI Descriptions of Pathogenic Fungi and Bacteria, No. 316. Commonwealth Mycological Institute, Kew, UK.
33. Mugnier, J. 1998. Molecular evolution and phylogenetic implications of ITS sequences in fungi and plants. Pages 253-277 in: Molecular Variability of Fungal Pathogens. P.D. Bridge, Y. Couteaudier, and J.M. Clarkson, eds. CAB International, Wallingford, UK.
34. Neuvéglise, C., Brygoo, Y., Vercambre, B., and Biba, G. 1994. Comparative analysis of molecular and biological characteristics of *Beauveria brongniartii* isolated from insects. Mycol. Res. 98:322-328.

35. Norman, J.E. and Egger, K.N. 1996. Phylogeny of the genus *Plicaria* and its relationship to *Peziza* inferred from ribosomal DNA sequence analysis. Mycologia 88:986-995.
36. O'Donnell, K. 1992. Ribosomal DNA internal transcribed spacers are highly divergent in the phytopathogenic ascomycete *Fusarium sambucinum* (*Gibberella pulicaris*). Curr. Gen. 22:213-220.
37. O'Donnell, K., Cigelnik, E., and Nirenberg, H.I. 1998. Molecular systematics and phylogeography of the *Gibberella fujikuroi* species complex. Mycologia 90:465-493.
38. O'Donnell, K., Cigelnik, E., Weber, N.S., and Trappe, J.M. 1997. Phylogenetic relationships among ascomycetous truffles and the true and false morels inferred from 18S and 28S ribosomal DNA analysis. Mycologia 89:48-65.
39. Penzig, A.G.O. 1882. Funghi agrumicoli. Contribuzione allo studio dei funghi parassiti degli agrumi. Michelia 2:385-508 + 136 pl.
40. Penzig, A.G.O. 1884. Seconde contribuzione allo studio dei funghi agrumicoli. Atti del Istituto Veneto di Scienze, Lettere ed Arti ser 6, 2:665-692.
41. Seifert, K.A., Wingfield, B.D. and Wingfield, M.J. 1995. A critique of DNA sequence analysis in the taxonomy of filamentous ascomycetes and ascomycetous anamorphs. Can. J. Bot. 73, suppl. 1:S760-S767.
42. Sherriff, C., Whelan, M.J., Arnold, G.M., Lafay, J.-F., Brygoo, Y., and Bailey, J.A. 1994. Ribosomal DNA sequence analysis reveals new species groupings in the genus *Colletotrichum*. Exp. Mycol. 18:121-138.
43. Smith, B.J. and Black, L.L. 1990. Morphological, cultural and pathogenic variation among *Colletotrichum* species isolated from strawberry. Plant Dis. 74:69-76.
44. Sreenivasaprasad, S., Brown, A.E., and Mills, P.R. 1992. DNA sequence variation and interrelationships among *Colletotrichum* species causing strawberry anthracnose. Physiol. Molec. Plant Pathol. 41:265-281.
45. Sreenivasaprasad, S., Brown, A.E., and Mills, P.R. 1993. Coffee berry disease pathogen in Africa: Genetic structure and relationship to the group species *Colletotrichum gloeosporioides*. Mycol. Res. 97:995-1000.
46. Sreenivasaprasad, S., Mills, P.R., Meehan, B.M., and Brown, A.E. 1996. Phylogeny and systematics of 18 *Colletotrichum* species based on ribosomal DNA spacer sequences. Genome 39:499-512.
47. Sutton, B.C. 1980. The Coelomycetes. Fungi Imperfecti with Pycnidia, Acervuli and Stromata. Commonwealth Mycological Institute, Kew, UK. 696 pp.
48. Sutton, B.C. 1992. The genus *Glomerella* and its anamorph *Colletotrichum*. Pages 1-26 in: *Colletotrichum*: Biology, Pathology and

Control. J.A. Bailey and M.J. Jeger, eds. CAB International, Wallingford, UK.

49. Swofford, D.L. and Olsen, G.J. 1990. Phylogeny reconstruction. Pages 411-501 in: Molecular Systematics. D.M. Hillis and C. Moritz, eds. Sinauer Associates, Sunderland, MS.
50. Turgeon, B.G., Christiansen, S.K., and Yoder, O.C. 1993. Mating type genes in ascomycetes and their imperfect relatives. Pages 199-215 in: The Fungal Holomorph: Mitotic, Meiotic and Pleomorphic Speciation in Fungal Systematics. D.R. Reynolds and J.W. Taylor, eds. CAB International, Wallingford, UK.
51. Viljoen, C.D., Wingfield, B.D., and Wingfield, M.J. 1993. Comparison of *Seiridium* isolates associated with cypress canker using sequence data. Exp. Mycol. 17:323-328.
52. Waller, J.M., Bridge, P.D., Black, R., and Hakiza, G. 1993. Characterization of the coffee berry disease pathogen, *Colletotrichum kahawae* sp. nov. Mycol. Res. 97:989-994.
53. Zhang, W., Wendel, J.F., and Clark, L.G. 1997. Bamboozled again! Inadvertent isolation of fungal rDNA sequences from bamboos (*Poaceae*: *Bambusoideae*). Molec. Phylogen. Evol. 8:205-217.

Chapter 2

The Importance of Phylogeny in Understanding Host Relationships Within *Colletotrichum*

P. R. Johnston

Among the most basic questions being asked by many research groups studying *Colletotrichum* are those relating to understanding the basis of pathogenicity. Essential to this is an understanding the basis for host-preference. Why does a particular species attack one host and not another? Why is a particular host attacked by some species but not others? Field observations suggest patterns to host relationships in *Colletotrichum*. These patterns, however, are broad, imprecise, and often overlapping. Often the patterns may be recognized only when the phylogenetic relationships within and among the *Colletotrichum* populations under study are clearly understood. Thus, an understanding of host preference within *Colletotrichum* is inextricably linked to systematics and the requirement for reliable, consistent identification of the organisms under study.

In this paper, data collected from New Zealand studies on isolates of *Colletotrichum* are used to illustrate the complexity of phylogenetic relationships amongst *Colletotrichum* isolates derived from a single host. An understanding of this complexity is essential before appropriate samples of isolates can be selected for applied studies on pathogenicity. Although molecular methods are proving to be powerful tools in helping define putative phylogenetic groups within *Colletotrichum*, the groups defined by any one of these tools are not necessarily equivalent in origin or in population structure. There are no general rules for defining relationships within *Colletotrichum*-the recognition, characterization, and the biological significance of any groups recognized must still be determined on a case-by-case basis.

Host Relationships among Fruit-Rotting *Colletotrichum* Taxa

Using over 800 freshly collected isolates, Johnston and Jones (3) studied the relationships among isolates of *Colletotrichum* associated with fruit rots in New Zealand. From morphological features (presence or absence of teleomorph and time of perithecial production, size and shape of conidia, ascospores, setae, appressoria) and appearance in culture (growth rate, mycelium production and color, pigments in agar, position of conidia, perithecia), 16 distinct groups were recognized. These morphologically defined groups included *C. orbiculare, C. musae, Glomerella miyabeana,* four *C. acutatum*-like groups, and nine *C. gloeosporioides*-like groups. The genetic distinctness of groups defined in this way has been confirmed using RAPDs (4). Most of these groups have been isolated from a range of different hosts (Fig. 1). Similarly, most hosts are associated with more than one group of *Colletotrichum* (Fig. 2).

Each of the morphologically defined groups clusters into one of four larger groups defined on the basis of rDNA sequences (Fig. 3). Despite the broad host range of these various groups, field data suggest that some host specialization is present among both the morphologically defined groups and the rDNA-defined groups. To illustrate, examples are extracted from

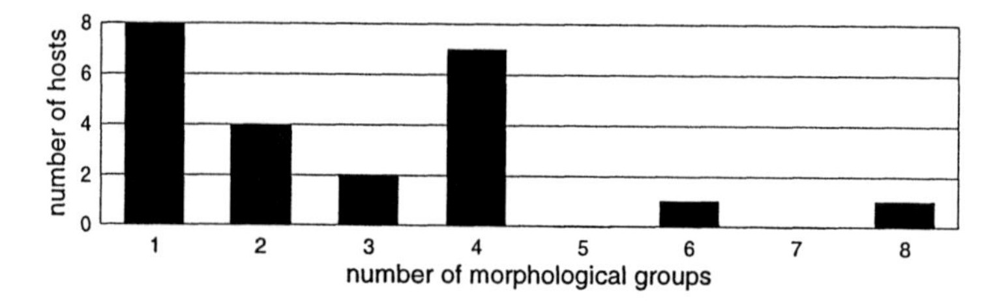

Fig. 1. Diversity of hosts associated with the *Colletotrichum* morphological groups of Johnston and Jones (3). Eight hosts have been associated with a single morphological group, but most have been associated with two or more; seven hosts, for example, have each been associated with four different morphological groups.

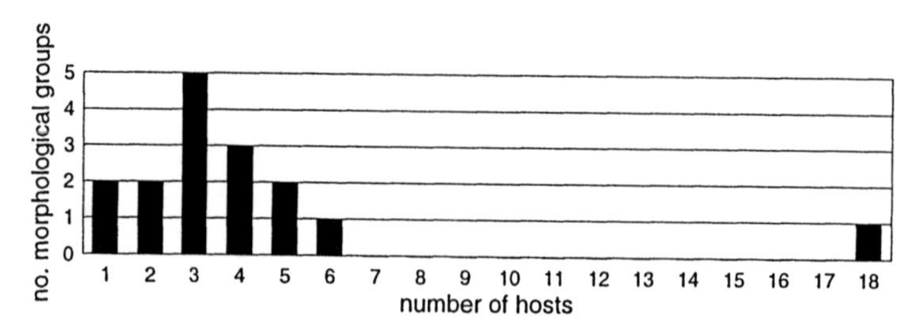

Fig. 2. Diversity of the *Colletotrichum* morphological groups of Johnston and Jones (3) associated with individual hosts. Although two of the morphological groups have been isolated from a single host, most have been isolated from two or more; five of the groups, for example, have each been isolated from three different hosts, one from six different hosts and another from 18 different hosts.

data presented by Johnston and Jones (3) (Tables 1 and 2). Table 1 compares the known host range of each of the five morphological groups within *C. gloeosporioides* rDNA Group 1. Although the host ranges between most of the five morphological groups overlap, specialization is also evident between groups. For example, morphological Group A alone is known from apple. Morphological Group B, although known from three hosts, is common only on avocado (unpublished data). Morphological Group C has been isolated from most citrus orchards sampled in New Zealand, whereas morphological Group A, also isolated from citrus, occurs only rarely on this host (unpublished data).
Some host specialization is also evident between the two *C. gloeosporioides*-like rDNA groups (Table 2). Thus, members of the *C. gloeosporioides* rDNA Group 1, commonly isolated from hosts such as apple and persimmon, have never been isolated from capsicum, tamarillo, and tomato, hosts from which members of *C. gloeosporioides* rDNA Group 2 are consistently isolated (unpublished data).

Thus, although host specialization is apparent between these various morphologically and rDNA-defined groups within *Colletotrichum*, the

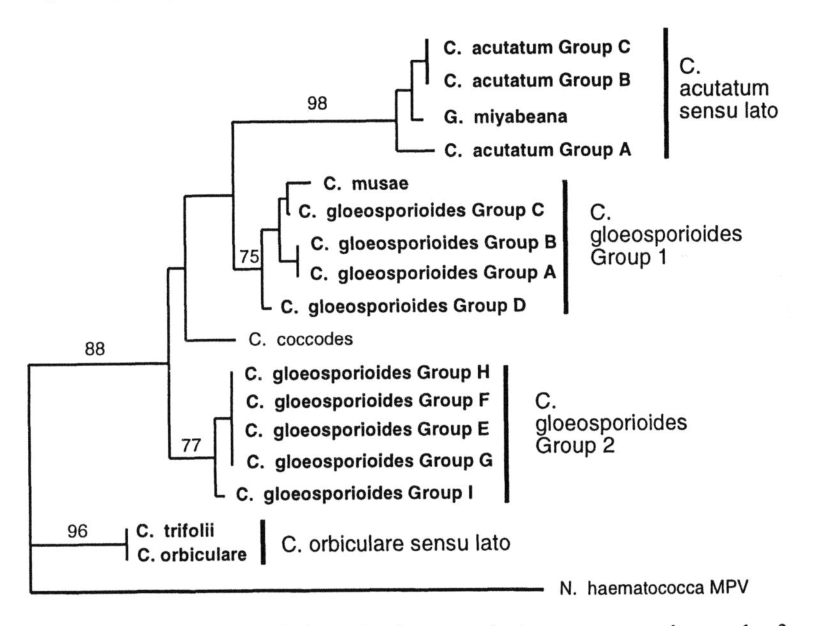

Fig. 3. Tree illustrating relationships between isolates representing each of the morphological groups of *Colletotrichum* from New Zealand fruit rots, based on sequences from the D2 domain of the large subunit rDNA. Analysis by neighbor-joining, with bootstrap values indicated where values were above 50%. (Based on Fig. 1, Johnston and Jones [3])

Table 1. Known host range (based on field isolations) of the five morphological groups recognized within *C. gloeosporioides* rDNA Group 1 (based on data from Johnston and Jones [3]).

C. gloeosporioides rDNA Group 1				
Group A	Group B	Group C	Group D	*C. musae*
apple				
avocado	avocado			
citrus		citrus		
fig	fig	fig		
pear	pear			
passionfruit				
persimmon			persimmon	
				banana

Table 2. Known host range (based on field isolations) of the two *C. gloeosporioides*-like rDNA groups recognized from fruit rots in New Zealand (data from Johnston and Jones [3]).

C. gloeosporioides rDNA Group 1	*C. gloeosporioides* rDNA Group 2
apple	
banana	
fig	
pear	
persimmon	
avocado	avocado
citrus	citrus
passionfruit	passionfruit
	capsicum
	cherimoya
	cucumber
	tamarillo
	tomato

specialization is poorly defined. *Glomerella miyabeana* provides an example to illustrate the importance of understanding the complexity of host relationships within *Colletotrichum* (Fig. 4). *G. miyabeana*, a stem pathogen of *Salix*, has been isolated from the fruit of three other plants in New Zealand. Taking one of its hosts, tomato, as an example, three

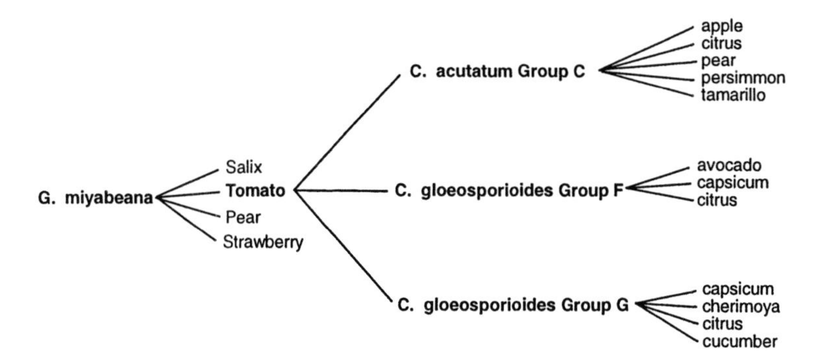

Fig. 4. Example illustrating the potential complexity of host relationships within *Colletotrichum*, based on *Glomerella miyabeana* and its four known hosts (see text for explanation).

Colletotrichum taxa (*C. acutatum* morphological Group C and *C. gloeosporioides* morphological Groups F and G) have been isolated from this host, in addition to *G. miyabeana*. Each of these morphological groups is in turn found on a range of additional hosts. Before a valid sample of isolates could be selected to study the basis of pathogenicity of *Colletotrichum* on tomato, for example, an appreciation of this complexity is required.

Genetic and Biological Relationships Between Morphological Groups within *Colletotrichum*

The rDNA groups shown in Fig. 3, based on data of *Colletotrichum* isolates from fruit rots in New Zealand, are also resolved when isolates from other parts of the world and from other substrates are included (3). To better understand the genetic relationship between the various morphological groups recognized by Johnston and Jones (3), isolates representative of each of the morphological groups within the rDNA-defined group *C. acutatum sensu lato* were compared using RAPDs and vegetative compatibility. As well as isolates from fruits, *C. acutatum* f.sp. *pineum* isolates (from New Zealand and Australia), and *C. acutatum*-like isolates from lupines from New Zealand, UK, Canada, and France were included (4).

Results from this study supported the genetic distinctness of the various morphological groups (Fig. 5) and showed that groups of isolates within *C. acutatum sensu lato* which are pathogenic towards a particular host can have independent evolutionary origins (4). For example, two genetically distinct groups are capable of causing stem and leaf blights of lupine; those

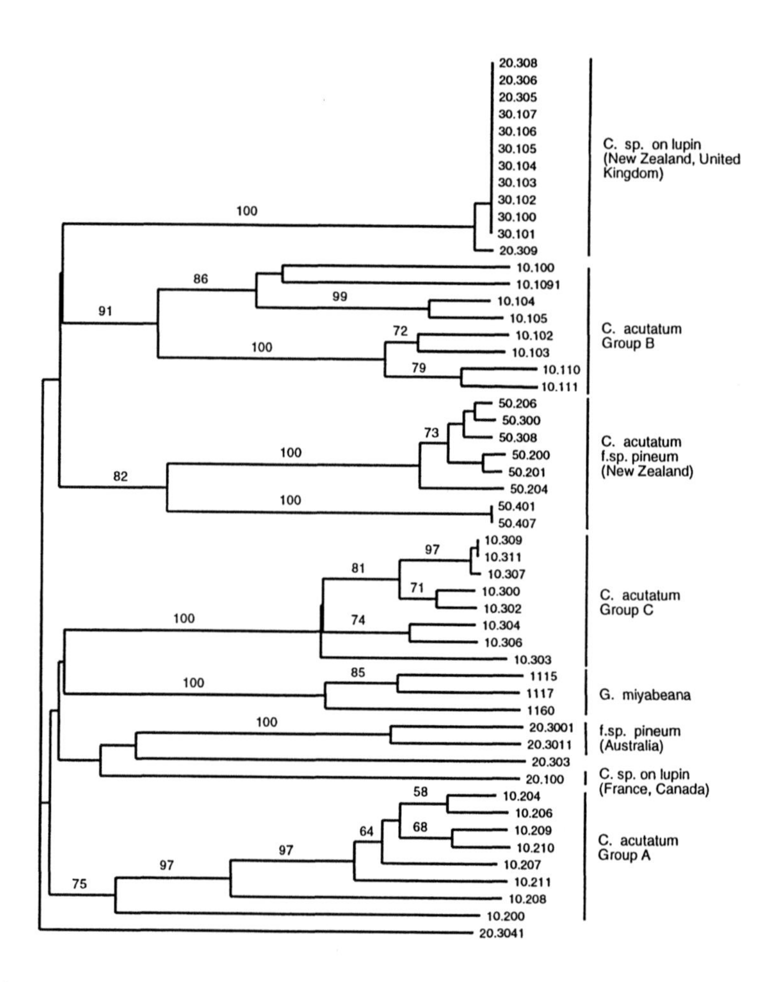

Fig. 5. Relationships among the morphological groups (indicated on right) recognized within *C. acutatum sensu lato.* Analysis by neighbor-joining, with bootstrap values indicated where values were above 50%. (Based on Fig. 17, Lardner et al. [4])

from New Zealand and the United Kingdom, and those from Canada and France (note that this group is represented by a single isolate on the tree, see [4] for explanation). Likewise, *C. acutatum* f.sp. *pineum* isolates from New Zealand are genetically distinct to those from Australia. *C. acutatum*-like taxa capable of causing terminal crook disease of pine seedlings appear to have evolved independently at least twice.

The *C. acutatum*-like pathogens of lupine and of pine from New Zealand share the following features:

- host-specialized
- highly pathogenic, causing similar crooking symptoms on young tissue
- appeared suddenly in New Zealand in historically recent times
- associated with introduced hosts with very limited genetic diversity
- putatively asexual.

Several authors have noted that it may be selectively advantageous, at least in the short term, for specialized pathogens of genetically uniform crops to be asexual, as regular sexual recombination will tend to swamp expression of the successful, highly adapted phenotype. Brasier (1,2) proposed a model called episodic selection, in which sudden and extreme ecological disturbance is invoked as a mechanism by which such highly specialized asexual organisms might arise from a less specialized, more variable, basal population. It is tempting to speculate that two separate episodic selection events may have occurred within the broad *C. acutatum* population in New Zealand, following the widespread cultivation of pine and lupine.

However, although biologically similar, the New Zealand lupine pathogen and *C. acutatum* f.sp. *pineum* populations are distinct in genetic structure. The RAPD data suggests that the *C. acutatum* f.sp. *pineum* population is genetically more variable than the New Zealand lupine pathogen population (Fig. 5) and is supported by a vegetative compatibility (VC) study using *nit*-mutants (4). All 12 isolates tested from the lupine pathogen population (from both the United Kingdom and New Zealand) belonged to a single VC group, while of the eight isolates tested from the New Zealand pine pathogen population, at least two VC groups were recognized, and more are probably present (4). Although *C. acutatum* f.sp. *pineum* has been recognized in New Zealand for over 20 years longer than the lupine pathogen, the difference in genetic variability cannot be explained as variation which has built up gradually over that time. *C. acutatum* f.sp. *pineum* isolates collected in 1966, 1967, and 1969, within the first 5 years of the disease appearing in New Zealand, show that a similar level of variation was already present at that time (4). Isolates of the lupine pathogen collected over a similar 5-year period are genetically identical. Thus, although the New Zealand lupine pathogen fits Brasier's episodic selection model well, the origin and maintenance of the genetic distinctness of *C. acutatum* f.sp. *pineum* requires a separate explanation.

Summary

It remains true that there are no general rules concerning host relationships within *Colletotrichum*. However, we can summarize the following:

- Patterns of host relationship exist in *Colletotrichum*, but they are broad, imprecise, and often overlapping.
- An understanding of this complexity, on a case-by-case basis, is essential for selecting appropriate samples of isolates for applied studies on the basis of pathogenicity within *Colletotrichum.*
- The groups so recognized cannot be assumed genetically equivalent, even when appearing to be biologically similar.
- Decisions on the taxonomic significance which should be placed on these groups require a detailed understanding of gene flow within and between them.

Literature Cited

1. Brasier, C.M. 1987. The dynamics of fungal speciation. Pages 231-260 in: Evolutionary Biology of the Fungi. A.D.M. Rayner, C.M. Brasier, and D. Moore, eds. Cambridge University Press, Cambridge.
2. Brasier, C.M. 1995. Episodic selection as a force in fungal microevolution, with special reference to clonal speciation and hybrid introgression. Can. J. Bot. 73 (Suppl. 1): S1213-S1221.
3. Johnston, P.R. and Jones, D. 1997. Relationships among *Colletotrichum* isolates from fruit-rots assessed using rDNA sequences. Mycologia 89: 420-430.
4. Lardner, R., Johnston, P.R., Plummer, K.M., and Pearson, M.N. 1999. Morphological and molecular analysis of *Colletotrichum acutatum* sensu lato. Mycol. Res. .103:275-285.

Chapter 3

Genetic Regulation of Sexual Compatibility in *Glomerella graminicola*

Lisa Vaillancourt, Juan Wang, and Robert Hanau

Fungal species that reproduce sexually can usually be classified as either self-fertile (homothallic), or self-sterile (heterothallic). However, the ascomycete genus *Glomerella* is unusual because within a single species some strains are both self-fertile and cross-fertile, while others are cross-fertile but self-sterile (44,45,49, see Chapter 4 for a review). Self-fertile isolates give rise to self-sterile progeny with a high frequency (44,49). Furthermore, while all other known heterothallic ascomycetes have only two compatible mating types, *Glomerella* has multiple mating specificities, evidenced by the fact that most isolates can mate with a majority of other isolates (44,45,49). Based on extensive studies of the genetics of mating in *Glomerella cingulata* conducted during the 1940s and 1950s, Wheeler (49) concluded that heterothallism in this species is derived from homothallism via mutations in genes controlling steps in a morphogenetic pathway necessary for self-fertility. This special type of heterothallism is called *unbalanced* heterothallism. Unbalanced heterothallism can usually be differentiated from true heterothallism because homothallic progeny will result in crosses of unbalanced heterothallic strains due to recombination between the mutant loci. In true heterothallic systems, mating-type genes occupy the same genetic locus in both specificities and self-fertile progeny are not produced.

We have investigated the genetics of mating compatibility in *G. graminicola*. Crosses between two heterothallic isolates of *G. graminicola* produced no homothallic progeny; thus, compatibility between these isolates resembled a true heterothallic mating system. However, segregation of the progeny was not consistent with a true heterothallic system. Two unlinked loci affecting mating compatibility were identified in these strains and were named *A* and *B*. The *A* gene product may be important for fertility because a particular allele of *A* must be present in at least one mate for fertility to result. The *B* locus is quite different. It has characteristics of a true heterothallic mating locus in that the two mates must have different alleles to be compatibile. Unbalanced heterothallism has been hypothesized to be a transitional state between homothallism and true heterothallism (32). A switch between the two may occur when two complementary mutant genes involved in fertility become closely linked on one chromosome. Because the genes do not recombine, self-fertile progeny are not produced. An exciting possibility is that the *B* locus in *G. graminicola* may be an example of this transitional form of "linked unbalanced heterothallism."

Mating Systems in the Fungi

The genetic variability in an interbreeding population in part determines its adaptability to environmental change. Besides mutation, sexual reproduction is a major source of variability for fungi and other eukaryotes. The contribution of sexual reproduction to genetic variability is determined by the ratio of inbreeding to outbreeding, which is, to a large extent, under the control of sexual compatibility-incompatibility systems. Fungi that reproduce sexually can be characterized as homothallic, meaning that they can undergo sexual reproduction alone, or heterothallic, meaning that they require a partner with a compatible mating type. A variety of regulatory mechanisms are associated with homothallic and heterothallic mating systems in the fungi.

Primary homothallism, of which *Emericella (Aspergillus) nidulans* is an example, involves the fusion of two genetically identical nuclei (36,54). Secondary (pseudo) homothallism is caused by the coexistence in the same thallus of nuclei with two compatible mating types. A representative of this mating system is *Neurospora tetrasperma* (29). Asci contain four spores, each with two nuclei representing opposite mating types. Thus, ascospores of *N. tetrasperma* give rise to mixed mating-type heterokaryons. In *Saccharomyces cerevisiae,* homothallism is controlled by bidirectional mating type switching (21). In this system, sexually compatible yeast cells encode opposite mating types. Homothallic sexual development occurs when some of the cells in a genetically uniform, vegetative culture switch to the opposite mating type, resulting in a mixed culture of two compatible mating types. Cells having the different mating types then fuse to produce a zygote which undergoes meiosis. Unidirectional switching due to mutation of the mating type genes results in homothallism in one of the two mating types of *Ceratocystis ulmi* (2) and *Sclerotina trifoliorum* (43).

A heterothallic phenotype can be controlled by mating type loci (true heterothallism) (37) or can be caused by a mutation in the developmental pathway for homothallism (unbalanced heterothallism) (32,49). True heterothallism includes bipolar and tetrapolar mating systems, controlled by a single locus or by two unlinked loci, respectively. All heterothallic ascomycetes that have been studied are under bipolar control involving two alternative mating-type gene complexes (11,20). These will be described in more detail. Heterothallic basidiomycetes can have either bipolar multiple allelomorphic mating systems or tetrapolar mating systems (24). Mating type loci have been cloned from a number of ascomycetes and basidiomycetes including *Saccharomyces cerevisiae* (22), *N. crassa* (19), *Cochliobolus heterostrophus* (42), *Podospora anserina* (33), *Schizophyllum commune* (40,46,47), and *Magnaporthe grisea* (23). Comparison of the nucleotide sequences of mating type loci has revealed a number of characteristic features, including extreme sequence divergence within the functional region(s), a high degree of sequence homology in the flanking regions, and genetic complexity in which more than one open reading frame is embedded within the divergent DNA in the locus (11,20). Recombination is completely suppressed in the region of these genes due to DNA heterology. Protein products of genes contained within the mating type loci physically interact in heterospecific pairs to activate various mating-specific developmental pathways. Nucleotide sequences encoding mating type loci in ascomycetes have been termed "idiomorphs," indicating that, although they are at the same position on the chromosome, they do not represent simple alleles of the same gene (30). Sexual compatibility between strains is due to the presence of different idiomorphs at the mating type loci.

The majority of ascomycetes can be readily classified as either homothallic or heterothallic. However, in three species, *Glomerella cingulata* (49), *G. graminicola* (44, 45), and *Sordaria fimicola* (31), both homothallic and heterothallic strains exist. There is substantial genetic evidence that *G. cingulata* utilizes an unbalanced heterothallic mating system resulting from mutation in the developmental pathway leading to primary homothallism (6,7,8,9,14,48,49,50,52). Strains harboring such mutations can no longer undergo sexual reproduction by themselves, and strains with complementary mutations will be sexually compatible (32). In *G. cingulata*, homothallic strains give rise to progeny with a broad range of sexual compatibilities including homothallism, heterothallism, and sterility (9,49). Homothallic strains of *G. graminicola* also give rise to both homothallic and heterothallic progeny (44).

How are the Various Mating Systems Related?

Although secondary homothallism is thought to be derived from true heterothallism and unbalanced heterothallism from primary homothallism, the relationship between primary homothallism and true heterothallism is not clear. Primary homothallism in ascomycetes has frequently been suggested to be derived from heterothallism (reviewed in 11), but true heterothallism has also been proposed to be derived from primary homothallism through the intermediate of unbalanced heterothallism (32). Of course, both may be true for different genera. Unbalanced heterothallism may confer a selective advantage because it allows a higher degree of outbreeding than homothallism. True heterothallism in turn may have certain adaptive advantages over unbalanced heterothallism. For example, in most matings involving unbalanced heterothallic strains both homothallic and sterile progeny will be produced by recombination. This will not be the case in matings involving true heterothallic strains, so true heterothallism increases fertility and outbreeding in the population. Furthermore, true heterothallism ensures sexual compatibility at an early stage of reproduction (e.g., cell-cell recognition or plasmogamy), thus reducing the possibility of an abortive mating. In an unbalanced heterothallic system, mutants could be blocked at later steps of sexual reproduction (e.g., after plasmogamy). Pairings would then be possible between certain strains with mutations in the same genes, which would result in the wasteful production of sterile fruiting structures or inviable spores.

To illustrate how an unbalanced heterothallic system could be converted to true heterothallism, we must first assume that a pair of compatible strains which display unbalanced heterothallism each have complementary mutations in one of two closely linked genes in the pathway for self-fertility. If the two genes are named *Y* and *Z*, then the genotype of the first strain can be described as *Yz*, and that of the second strain as *yZ*. If there is little or no recombination between the two genes, the gene pairs will cosegregate and the result will be a phenotype that is indistinguishable from true heterothallism. The conversion from unbalanced heterothallism to true heterothallism could be completed in this hypothetical example by deletions and/or sequence divergence in the region between the two genes, establishing true idiomorphs. Eventually, these processes could remove any resemblance between the mutant alleles and their progenitor genes. This hypothesis was supported by work with *Sordaria fimicola* (1). Two mutants, each with a mutation in one of two closely linked genes that control different steps in the reproductive pathway, were created by *in vitro* mutagenesis. The two mutants were compatible, and the segregation of sexual compatibility in the progeny was like that expected for single locus bipolar mating type control. As will be explained in more detail, our recent data suggest that we have identified a locus in *G. graminicola* which may be the first natural example of this phenomenon.

Sexual Development in *Glomerella*

The morphology and development of perithecia, asci, and ascospores has been studied in detail in *Glomerella cingulata* (26,28,51). In both selfed and mated cultures, *G. cingulata* produces specialized male and female structures. The female organ consists of an inner coil of receptive cells and an outer coil of cells that will become the perithecial wall. The male organ is a fertilization hypha which contributes a single nucleus after fusion with the receptive tip cell of the inner coil. The fertilization hypha was described as rapidly growing directly toward the receptive tip cell "as if in response to a definite stimulation" (28). Fertilization in *G. cingulata* is unlike that in *Sordaria fimicola,* in which a fertilization hypha is not observed (5). In *G. cingulata*, a single dikaryotic fusion cell proliferates to give rise to all of the asci in a single perithecium, whereas in *Sordaria fimicola,* asci of different genetic origin often occupy the same perithecium (31).

The tissue making up the perithecial wall is genetically identical to the cytoplasm of the ascospore progeny in *G. cingulata*; both are derived from the maternal parent. This is also the case in true heterothallic ascomycetes such as *N. crassa* and *Cochliobolus heterostrophus*. We have genetic evidence that suggests that perithecial development in heterothallic strains of *G. graminicola* is similar to that in *G. cingulata*. Crosses between melanin-deficient mutant strains and wild-type strains of *G. graminicola* usually produce two types of perithecia: one is the normal dark color while the other is golden brown (44). Progeny from both types of perithecia segregate 1:1 for the Mel$^-$ marker. To study the genetics and expression of sexuality in different strains of *G. graminicola*, it is important to be able to identify males and females in crosses. We anticipated that it would be possible to do this by noting the color of perithecia in matings involving a Mel$^-$ strain. We tested the hypothesis that perithecial color is an indicator of strain sexuality in crosses involving melanin-deficient mutants.

The strains of *G. graminicola* used for this study are described in Table 1. Mitochondrial DNA (mtDNA) for RFLP analysis was separated from genomic DNA in two successive cesium chloride gradients containing bisbenzimide (17,18). Two bands were observed after a single centrifugation of total DNA. The upper band was more diffuse than the lower. To identify the major components of these bands, both were recovered following enrichment in two successive cesium chloride centrifugations of DNA made from strain M1.502. DNA from each of the bands was probed with pJR70, a plasmid which contains *PYR1*, a *G. graminicola* nuclear gene encoding orotate phosphoribosyl transferase (39).

Table 1. Strains of *G. graminicola* used in this study

Strain	Origin	Phenotype
M1.001	Missouri	Wild-type
M1.502	M1.001	Melanin deficient
M2.001	North Carolina	Wild-type
M3.001	Indiana	Wild-type
M5.001	Brazil	Wild-type
M7.001	Brazil	Wild-type
M9.001	Indiana	Wild-type
M9.402	M9.001	Chlorate resistant

Only lanes containing digests of DNA from the lower band hybridized to the probe. Only DNA from the Upper band hybridized to a clone containing the mitochondrial *cox1* gene of *N. crassa* (38). Digestions of DNA from the upper band produced a smaller number of discrete fragments than did digestions of DNA from the lower band. We concluded that the upper band was enriched in mtDNA, whereas the lower band was enriched in nuclear DNA.

We prepared enriched mtDNA from a total of five progeny isolated from three golden brown perithecia and four progeny from two black perithecia from a cross of the melanin-deficient strain M1.502 and the wild-type strain M5.001. MtDNA samples from each of the nine progeny and from both parents were digested with the restriction enzyme *Hin*dIII. A restriction fragment length polymorphism (RFLP) distinguished M1.502 and M5.001 (Fig. 1A, lanes A and B). Progeny from the golden brown perithecia had the RFLP pattern of M1.502 (Fig.1A, lanes H-K), whereas progeny from the black perithecia had the pattern of M5.001 (Fig. 1A, lanes C-G) regardless of the phenotype (i.e. Mel+ or Mel$^-$) of the progeny.

Crosses between the Mel$^-$ strain M1.502 and the Mel$^+$ strain M3.001 produced only the golden brown type of perithecia. The progeny from these perithecia segregated 1:1 for the Mel phenotype. Two progeny were randomly isolated from each of three different perithecia from a cross of M1.502 and M3.001, and mtDNA was recovered from each. MtDNA samples from all of the progeny and from the two parents were digested with the restriction enzyme *Hin*dIII. Digestion with *Hin*dIII revealed an RFLP that distinguished M1.502 from M3.001 (Fig. 1B, lanes A and B). All of the progeny had the restriction pattern of M1.502, regardless of the phenotype of the progeny (Fig. 1B, lanes C-H).

Matings conducted between M1.502 and the chlorate-resistant (Chl^R) strain M9.402 produced both black and golden brown perithecia. Strain M3.001 was mated with M9.402. Since both are Mel$^+$ strains, all of the perithecia were black. Two progeny were collected from each of five perithecia, and mtDNA was prepared from each. The Chl^R phenotype segregated 1:1 among random progeny from these perithecia. MtDNA samples from each of the progeny and from the two parents were digested with the restriction enzyme *Hin*dIII. Digestion with *Hin*dIII revealed an RFLP that distinguished strains M3.001 and M9.402. All of the progeny had the restriction pattern of M9.402.

The presence of two types of perithecia in crosses involving a melanin-deficient strain indicates that the two isolates participating in the cross are hermaphrodites, the golden brown perithecia resulting when the melanin-deficient strain serves as the female parent. Similar observations have been made in studies involving melanin-deficient mutants of *Cochliobolus heterostrophus* (53). The female parent donates the mitochondria to the offspring. RFLPs in the mtDNA would be expected to be maternally

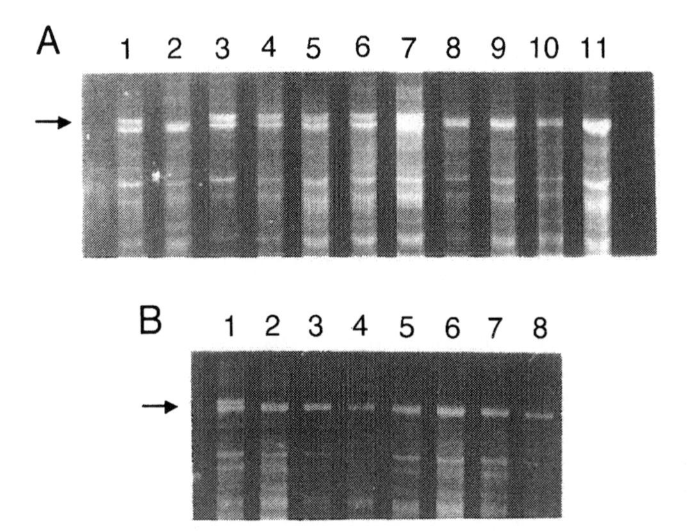

Fig. 1. MtDNA RFLPs in parent and progeny strains of *G. graminicola.* Panel A: Lanes 1 and 2 contain DNA from parent strains M5.001 and M1.502, respectively; lanes 3-5 contain DNA from black perithecia; lanes 6-9 contain DNA from golden brown perithecia. Panel B: Lanes 1 and 2 contain DNA from parent strains M3.001 and M1.502, respectively; lanes 3-5 contain DNA from the progeny, all of which have the RFLP pattern of M1.502. Panels A and B, arrows denote the position of the RFLP.

inherited. Results showed that progeny from a single perithecium of *G. graminicola* shared identical mtDNA RFLP patterns, regardless of their nuclear genotypes, and that the restriction pattern correlated with the parent that made the perithecial wall. Therefore, the color of the perithecium when a Mel$^-$ strain is included in a cross is a reliable indicator of the maternal parent of the progeny. Additional strains in our collection have been tested in matings with the Mel$^-$ strain. The majority of homothallic and heterothallic strains are hermaphrodites, but a few are female-sterile. The female-sterile strains of *G. graminicola* in our collection have never been observed to produce perithecial initials, suggesting that they are blocked early in the pathway for female fertility. Female-sterility is assumed to be a general character of the strain and not cross-dependent, since M3.001 behaved as a male in crosses with both M1.502 and M9.402. A manuscript describing these results has been been submitted.

Genes that Regulate Mating in *Glomerella cingulata*

Glomerella cingulata is a very diverse species with a wide host range. Some isolates produce only perithecia, while others produce only conidia (the latter are identified by the anamorphic name *Colletotrichum gloeosporioides*). Some produce both perithecia and conidia, though not from the same region of the thallus. A chemical substance is released by fertile perithecia of *G. cingulata* which prevents the formation of conidia nearby (50). We have also observed inhibition of conidia near perithecia in *G. graminicola* (44). Individual perithecial isolates of *G. cingulata* range from highly self-fertile to nearly self-sterile (27,49). Self-sterile conidial strains frequently can participate in crosses with perithecial strains (49). Nearly every isolate of *G. cingulata* from a particular host is fertile with nearly every other isolate from that host to some degree (27,49). However, isolates from different hosts are only rarely cross-fertile (10). In general, even highly self-fertile isolates of *G. cingulata* will mate preferentially with a different isolate rather than self-fertilize. Nonetheless, the mating system in this species is fundamentally homothallic, as demonstrated by the fact that there are self-fertile genotypes and that many self-sterile genotypes can be induced to self-fertilize under specialized conditions (12,52).

Freshly isolated perithecial cultures of *G. cingulata* generally bear their perithecia in glomerate masses, and are thus called clumped perithecial, or CP, isolates. Ascospores isolated from CP strains frequently give rise to variants which produce scattered perithecia (SP isolates). When an SP strain is paired on a plate with its CP progenitor, a prominent ridge of perithecia containing many well-formed asci and ascospores is formed at the intersection of the colonies. These perithecia arise primarily from fertilization of the SP by the CP strain. Interestingly, many of the perithecia in the ridge contain only SP progeny, in spite of the fact that SP strains are virtually self-sterile in the absence of the CP strain. This phenomenon appears to be due to induced selfing of the SP strain by a diffusible substance produced by the CP strain. Diffusible substances that induce self-fertility and increase cross-fertility have been demonstrated in *G. cingulata* (12,27), but the nature of these substances is unknown.

Edgerton (13) described the CP and SP strains of *G. cingulata* as heterothallic, comparing them with the plus and minus mating types that had recently been described in the Mucorales (3). However, there is an important difference between *G. cingulata* and the heterothallic Mucorales in that the CP and SP strains of *G. cingulata* are at least somewhat self-fertile, as well as cross-fertile. The genetic relationship between CP and SP strains was worked out over several years and reported in a series of elegant papers during the 1940s and 1950s (7,8,9,14,26,48,50,52). The switch from CP to SP is due to a mutation which occurs with a very high frequency (approximately 1 in 2000 mitotic divisions) (8). The mechanism and nature of this mutation are unknown, but its frequency is dependent on several factors. One is the age of the culture; older cultures mutate more frequently than younger ones (8). The mutation to SP occurs more frequently on artificial media than on natural substrates, perhaps explaining why the SP type has never been isolated from nature (8,41). A third factor is the genetic background of the CP strain: CP strains that are mutated for an unlinked modifier gene are stable in culture for the clustered perithecial phenotype, and only rarely produce SP progeny (48). Two different alleles of SP have been described. One allele decreases self-fertility, but a second allele does not (52). Apparently, strains containing either allele of SP mate preferentially with CP strains. The SP mutation may promote the formation of female structures, which the CP strains seem to initiate only rarely. In a CP strain, a perithecial clump originates when one of the nuclei in the thallus mutates to SP, and the region of the mycelium containing the mutant nucleus begins to differentiate protoperithecial structures. Fertile perithecia produce a diffusible substance that induces the neighboring hyphae to differentiate, leading to the clumped perithecial phenotype. The SP alleles also have an effect on conidium production, leading to scattered formation of conidiophores in conidial strains versus a clumped arrangement in CP strains. We have observed stimulation of conidia production along lines where two conidial strains meet in both *C. gloeosporioides* and *C. graminicola,* suggesting that a diffusible substance may induce conidiation in non-perithecial strains.

In addition to SP, many other mutations were isolated from wild-type isolates of *G. gloeosporioides* as ascospore cultures or as sectors (52). Most of these were fertile with their progenitor strain, and genetic analysis showed that most arose from single gene mutations which decreased self-fertility. Some isolates lost their ability to produce perithecial initials on their own, but could be induced to do so by growing them together with an ascigerous culture. These strains were conidial when grown in isolation and were identical to *C. gloeosporioides* isolates. Some mutants were completely sterile and produced neither ascospores nor conidia. Nonetheless, these cultures were able to fertilize ascigerous cultures. Analysis of these mutants led Wheeler to propose his hypothesis of unbalanced heterothallism for this species, in which mutations in the developmental pathway for self-fertility prevent self-fertility and promote cross-fertility (49).

Many puzzling questions remain regarding sexuality and fertility in this species. What are the diffusible substances that cultures of *G. cingulata* produce during perithecial formation? What is the mechanism for the high frequency of mutation from CP to SP? Why is the mutation more frequent in culture than in nature? What is the mechanism that promotes outcrossing among self-fertile isolates of *G. cingulata*? With current genetic approaches and methodologies, it may be possible to answer some of these questions at last and to build upon the solid foundation that was established by the early workers in the field. Genetic and molecular genetic techniques are better developed for *G. graminicola* than for *G. cingulata*, making *G. graminicola* the preferred system for studying mating compatibility in this genus.

Genes Regulating Heterothallic Mating in *Glomerella graminicola*

G. graminicola was originally described as a homothallic species (35). Later studies (44,45; Rollins and Hanau, unpublished) have shown that (i) field isolates can be either homothallic or heterothallic, (ii) homothallic and heterothallic strains are sexually compatible with one another, (iii) most strains are interfertile with most other strains, and (iv) homothallic strains give rise to both homothallic and heterothallic progeny. To gain an understanding of genetic factors that control mating in *G. graminicola*, we examined the inheritance of compatibility in two heterothallic strains, which are named M1.001 and M5.001. These two strains are fertile with each and both are also fertile with at least five other heterothallic strains and one homothallic strain of *G. graminicola* (44,45) (Table 2). Matings between M1.001 and M5.001 yielded no homothallic progeny. Thus, the mating system between these two strains resembles true heterothallism. However, the segregation of mating compatibility among the progeny was not consistent with true heterothallism and suggested that the parental strains differ at two genetic loci. Some of the results from our analysis of tetrads are presented in Table 3.

Only recombinant (tetratype) tetrads are included in Table 3. None of the progeny from any of the tetrads were self-fertile (column 2). None of the progeny were fertile with both parental strains (column 3). One-fourth of the progeny from each tetrad were fertile only with M1.001 (column 4), one-half were fertile only with M5.001 (column 5), and one-fourth were not fertile with either parent (column 6). Pairs of progeny that were determined to be twins by examination of molecular and other markers behaved identically in the mating tests. If the parental strains were heterothallic and bipolar, we would expect a 1:1 ratio with half the progeny fertile with M1.001 and half with M5.001. If they were heterothallic and tetrapolar, we would expect one-fourth of the progeny to mate with M1.001, one-fourth with M5.001, and one-half not to mate with either parent. Our interpretation of the unusual segregation pattern that we observed in these tetrads is that the parental strains differ at two unlinked loci which we have named *A* and *B*. Analysis of 24 random-spore progeny from crosses of M1.001 and M5.001 were consistent with the results from tetrads: 11 progeny were fertile with M5.001, five were fertile with M1.001; and eight were not fertile with either parent (χ^2 2:1:1, P> 0.5).

Table 2. Compatibility[a] of *G. graminicola* strains used in this study

Strain	1.001 Hetero thallic	5.001 Hetero thallic	2.001 Hetero thallic	3.001 Hetero thallic	7.001 Hetero thallic	9.001 Homo thallic
M1.001	-	+	+	+	+	+
M5.001	+	-	+	+	+	+
M2.001	+	+	-	+	+	+
M3.001	+	+	+	-	+	+
M7.001	+	+	+	+	-	+
M9.001	+	+	+	+	+	+

[a]A "+" indicates a compatible mating interaction, and "-" indicates incompatibility.

Table 3. Mating compatibility[a] between tetrad progeny and parental strains

Tetrad	Self Fertile	1.001+ 5.001+	1.001+ 5.001-	1.001- 5.001+	1.001- 5.001-	Viable Spores in the Tetrad
1	0	0	2	4	2	8
2	0	0	2	4	2	8
3	0	0	2	3	1	6
4	0	0	1	4	2	7

[a]Same as in legend to Table 2.

We tested our hypothesis by conducting a series of sib crosses among various progeny strains. Some representative data are presented in Table 4. In this experiment, progeny from a single tetratype tetrad were crossed in all possible combinations. Twin sister ascospores behaved identically in the mating tests, so data are presented for only one member of each pair. Results of several additional crosses were consistent with these data.

It is evident that the two loci are not equivalent in their actions. In the case of the *A* locus, fertility results if at least one of the strains contains the *A* allele. Our interpretation is that this locus is involved in a developmental pathway for fertility and that it has been mutated in M1.001 (this would be a typical example of unbalanced heterothallism). The precise nature of the *B* locus is more obscure. For fertility to result, the *B* alleles must be different in the two mates. The *B* locus may function in self-nonself recognition and resemble a typical mating-type idiomorphic locus. Alternatively, it may consist of two linked, mutually complementary genes involved in a developmental pathway for fertility. These strains may illustrate the "transitional" form of linked unbalanced heterothallism predicted by Olive's (35) hypothesis in 1958. This would be the first observed natural example of this phenomenon. Conclusive evidence regarding these possibilities will depend on molecular genetic characterization of the loci.

Conclusion

The genetics of mating compatibility in fungi is widely recognized as a valuable model system for the study of questions related to molecular evolution, gene regulation, mechanisms of self-nonself recognition, and the origins and development of sexuality. Several ascomycete fungi have been particularly well studied in this regard. These ascomycetes can be readily classified as either homothallic (self-fertile) or heterothallic (self-sterile), and all of the self-sterile species possess a bipolar mating system with two compatible mating specificities encoded by complex loci called idiomorphs. The ascomycete fungus *Glomerella* is very different. Both homothallic and heterothallic strains exist. Homothallic strains mate preferentially with heterothallic ones, and most homothallic and heterothallic strains are fertile with a majority of other isolates. Homothallic strains give rise to heterothallic progeny with a high frequency. It has been proposed that *Glomerella cingulata* utilizes a unique form of mating control called unbalanced heterothallism. In this mating system, heterothallic strains are derived from homothallic ones via mutation in the pathway for self- fertility. Strains with complementary mutations are sexually compatible. It has also been suggested that unbalanced heterothallism is an evolutionary link between true homothallism and true heterothallism. Our preliminary results suggest that *G. graminicola* may be even more versatile than *G. cingulata* and may utilize a range of mating control strategies that includes some form of true heterothallism as well as unbalanced heterothallism and true homothallism. A molecular genetic analysis of mating compatibility has never been conducted in an unbalanced heterothallic species. The coexistence of different mating control strategies in *G. graminicola*, the fact that nearly all homothallic and heterothallic strains are sexually compatible with one another, and the fact that *G. graminicola* has been fully developed as an ideal model for molecular and traditional genetic studies, presents a unique opportunity to study the relationship between these different mating strategies in a single species at a molecular level. We anticipate that a better understanding of the genetics of sexual compatibility in this species will

Table 4. Results of matings among progeny from a single recombinant tetrad[a]

Progeny	Inferred Genotypes	M1.001 *aB1*	M5.001 *AB2*	1 *AB1*	2 *aB2*	3 *aB1*	4 *AB2*
1	*AB1*	-	+	-	+	-	+
2	*aB2*	-	-	+	-	-	-
3	*aB1*	-	+	-	-	-	+
4	*AB2*	+	-	+	-	+	-

[a]Same as in legend to Table 2.

contribute valuable information for answering some of the fundamental questions that still remain about the evolution and function of mating systems in fungi.

Literature Cited

1. Ani, A.S. and Olive, L.S. 1962. The induction of balanced heterothallism in *Sordaria fimicola.* Proc. Natl. Acad. Sci. USA 48:17-19.
2. Brasier, C.M. 1984. Inter-mycelial recognition systems in *Ceratocystis ulmi*: Their physiological properties and ecological importance. Pages 451-497 in: The Ecology and Physiology of the Fungal Mycelium. D.H. Jennings and A.D.M. Rayner, eds. Cambridge University Press, Cambridge.
3. Blakeslee, A.F. 1904. Sexual reproduction in the Mucorinae. Proc. Am. Acad. Arts Sci. 40:205-319.
4. Bryson, R.J., Caten, C.E., Hollomon, D.W., and Bailey, J.A. 1992. Sexuality and genetics of *Colletotrichum*. Pages 27-46 in: *Colletotrichum*: Biology, Pathology and Control. J.A. Bailey and M.J. Jeger, eds. CAB International, Wallingford, UK.
5. Carr, A.J.H. and Olive, L.S. 1959. Genetics of *Sordaria fimicola*. III. Cross compatibility among self-fertile and self-sterile cultures. Am. J. Bot. 46:81-91.
6. Chilton, S.J.P., Lucas, G.B., and Edgerton, C.W. 1944. Genetics of *Glomerella* I. Studies on the behavior of certain strains. Am. J. Bot. 31:233-239.
7. Chilton, S.J.P., Lucas G.B., and Edgerton, C.W. 1945. Genetics of *Glomerella*. III. Crosses with a conidial strain. Am. J. Bot. 32:549-554.
8. Chilton, S.J.P. and Wheeler, H.E. 1949. Genetics of *Glomerella* VI. Linkage. Am. J. Bot. 36:270-273.
9. Chilton, S.J.P. and Wheeler, H.E. 1949. Genetics of *Glomerella* VII. Mutation and segregation in plus cultures. Am. J. Bot. 36:717-721.
10. Cisar, C.R., Speigel, F.W., TeBeest, D.O., and Trout, C. 1994. Evidence for mating between isolates of *Colletotrichum gloeosporioides* with different host specificities. Curr. Genet. 25:330-335.
11. Coppin, E., Debuchy, R., Arnaise, S., and Picard, M. 1997. Mating types and sexual development in filamentous ascomycetes. Microbiol. Mol. Biol. Rev. 61:411-428.
12. Driver, C.H. and Wheeler, H.E. 1955. A sexual hormone in *Glomerella.* Mycologia 47:311-316.

13. Edgerton, C.W. 1914. Plus and minus strains of the genus *Glomerella*. Am. J. Bot. 1:244-254.
14. Edgerton, C.W., Chilton, S.J.P., and Lucas, G.B. 1945. Genetics of *Glomerella* II. Fertilization between strains. Am. J. Bot. 32:115-118.
15. Ferreira, A.V.B., Saupe, S., and Glass, N.L. 1996. Transcriptional analysis of the *mtA* idiomorph of *Neurospora crassa* identifies two genes in addition to *mtA-1*. Mol. Gen. Genet. 250:767-774.
16. Fincham, J.R.S. and Day, P.R. 1963. Fungal Genetics. W.O. James, ed. Botanical Monographs, Volume 4. F.A. Davis Co., Philadelphia, PA.
17. Garber, R.C. and O.C. Yoder. 1983. Isolation of DNA from filamentous fungi and separation into nuclear, mitochondrial, and ribosomal components. Anal. Biochem. 135:416-422.
18. Garber, R.C. and O.C. Yoder. 1984. Mitochondrial DNA of the filamentous ascomycete *Cochliobolus heterostrophus*. Curr. Gen. 8:621-628.
19. Glass, N.L, Vollmer, S.J., Staben, C., Grotelueschen, J., Metzenberg, R.L., and Yanofsky, C. 1988. DNA of the two mating type alleles of *Neurospora crassa* are highly dissimilar. Science 241:570-573.
20. Glass, N.L. 1992. Mating type and vegetative incompatibility in filamentous fungi. Annu. Rev. Phytopathol. 30:201-224.
21. Herskowitz, I. 1988. Life cycle of the budding yeast *Saccharomyces cerevisiae*. Microbiol. Rev. 52:536-553.
22. Hicks, J., Strathern, J.N., and Klar, A.J. 1979. Transposable mating type genes in *Saccharomyces cerevisiae*. Nature 282:478-483.
23. Kang, S.C., Chumley, F.G., and Valent, B. 1994. Isolation of the mating type genes of the phytopathogenic fungus *Magnaporthe grisea* using genomic subtraction. Genetics 138:289-296.
24. Koltin Y., Stamberg, J., and Lemke, P.A. 1972. Genetic structure and evolution of the incompatibility factors in higher fungi. Bacteriol. Rev. 36:156-171.
25. Lucas, G.B., Chilton S.J.P., and Edgerton C.W. 1944. Genetics of *Glomerella* I. Studies on the behavior of certain strains. Am. J. Bot. 31:233-239.
26. Lucas, G.B. 1946. Genetics of *Glomerella* IV. Nuclear phenomena in the ascus. Am. J. Bot. 33:802-806.
27. Markert, C.L. 1949. Sexuality in the fungus, *Glomerella*. Am. Nat. 83:227-231.
28. McGahen, J.W. and Wheeler, H.E. 1951. Genetics of *Glomerella* IX. Perithecial development and plasmogamy. Am. J. Bot. 38:610-617.
29. Metzenberg, R.L. and Ahlgren, S.K. 1973. Behavior of *Neurospora tetrasperma* mating type genes introgressed into *N. crassa*. Can. J. Genet. Cytol. 15:571-576.

30. Metzenberg, R.L. and Glass, N.L. 1990. Mating type and mating strategies in *Neurospora*. BioEssays 12:53-59.
31. Olive, L.S. 1956. Genetics of *Sordaria fimicola*. Am. J. Bot. 43:97-107.
32. Olive, L.S. 1958. On the evolution of heterothallism in fungi. Am. Nat. 865:233-250.
33. Picard, M., Debuchy, R., and Coppin, E. 1991. Cloning the mating types of the heterothallic fungus *Podospora anserina*: Developmental features of haploid transformants carrying both mating types. Genetics 128:539-547.
34. Pöggeler S., Risch, S., Kück, U., and Osiewacz, H.D. 1997. Mating-type genes from the homothallic fungus *Sordaria macrospora* are functionally expressed in a heterothallic ascomycete. Genetics 147:567-580.
35. Politis, D.J. 1975. The identity and perfect state of *Colletotrichum graminicola*. Mycologia 67:58-62.
36. Pontecorvo, G., Roper, J.A., Hemmons, L.M., MacDonald, K.D., and Bufton, A.W.F. 1953. The genetics of *Aspergillus nidulans*. Adv. Genet. 5:141-238.
37. Raju, N.B. 1992. Genetic control of the sexual cycle in *Neurospora*. Mycol. Res. 96:241-262.
38. Randhir, R.J. and Hanau, R.M. 1997. Size and complexity of the nuclear genome of *Colletotrichum graminicola*. Appl. Environ. Microbiol. 63:4001-4004.
39. Rasmussen, J.B., Panaccione, D.G., Fang, G.C., and Hanau, R.M. 1992. The *PYR1* gene of the plant pathogenic fungus *Colletotrichum graminicola*: Selection by intraspecific complementation and sequence analysis. Mol. Gen. Genet. 235:74-80.
40. Specht, C.A., Stankis, M.M., Novotny, C.P., and Ullrich, R.C. 1994. Mapping the heterogeneous DNA region that determines the nine Aα mating-type specificities of *Schizophyllum commune*. Genetics 137:709-714.
41. Struble, F.B. and Keitt, G.W. 1950. Variability and inheritance in *Glomerella cingulata* Stonem. S. and V.S. from apple. Am. J. Bot. 37:563-576.
42. Turgeon, G.B., Bohlmann, H., Ciuffetti, L.M., Christiansen, S.K., Yang, G., Schafer, W., and Yoder, O.C. 1993. Cloning and analysis of the mating type genes from *Cochliobolus heterostrophus*. Mol. Gen. Genet. 238:270-284.
43. Uhm, J.Y. and Fujii, H. 1983. Heterothallism and mating type mutation in *Sclerotinia trifoliorum*. Phytopathology 73:569-572.
44. Vaillancourt, L.J. and Hanau, R.M. 1991. A method for genetic analysis of *Glomerella graminicola* from maize. Phytopathology 81(5):530-534.

45. Vaillancourt, L.J. and Hanau, R.M. 1992. Genetic and morphological comparisons of *Glomerella* (*Colletotrichum)* isolates from maize and from sorghum. Exp. Mycol. 16:219-229.
46. Vaillancourt, L.J., Raudaskoski, M., Specht, C.A., and Raper, C.A. 1997. Multiple genes encoding pheromones and a pheromone receptor define Bβ1 mating-type specificity in *Schizophyllum commune.* Genetics 146:541-551.
47. Wendland, J., Vaillancourt, L.J., Hegner, J., Lengeler, K.B., Laddison, K.J., Specht, C.A., Raper, C.A., and Köthe, E. 1995. The mating type locus Bα1 of *Schizophyllum commune* contains a pheromone receptor gene and putative pheromone genes. EMBO J. 14:5271-5278.
48. Wheeler, H.E. 1950. Genetics of *Glomerella* VIII: A genetic basis for the occurrence of minus mutants. Am. J. Bot. 37:304-312.
49. Wheeler, H.E. 1954. Genetics and evolution of heterothallism in *Glomerella.* Phytopathology 44:342-345.
50. Wheeler, H.E. 1956. Sexual versus asexual reproduction in *Glomerella.* Mycologia 48:349-353.
51. Wheeler, H.E., Olive, L.S., Ernest, C.T., and Edgerton, C.W. 1948. Genetics of *Glomerella* V. Crozier and ascus development. Am. J. Bot. 35:722-728.
52. Wheeler, H.E. and McGahen, J.W. 1952. Genetics of *Glomerella* X. Genes affecting sexual reproduction. Am. J. Bot. 39:110-119.
53. Yoder, O.C. 1988. *Cochliobolus heterostrophus*, cause of southern corn leaf blight. Pages 93-113 in: Advances in Plant Pathology, Vol. 6. G.S. Sidhu, ed. Academic Press, London.
54. Zonnenveld, B.J.M. 1988. Morphology of initials and number of nuclei initiating cleistothecia in *Aspergillus nidulans.* Trans. Br. Mycol. Soc. 90:369-373.

Chapter 4

Vegetative Compatibility in *Colletotrichum*

Talma Katan

The term vegetative compatibility refers to the ability of individual fungal strains to undergo mutual hyphal anastomosis, which results in viable fused cells containing nuclei of both parental strains in a common cytoplasm. This term is in contrast to sexual compatibility, a component of the fungal sexual cycle (Vaillancourt et al., Chapter 3 of this volume). Hyphal anastomosis is a common phenomenon in many fungi, and the genetic status of the anastomosed cell reflects the genetic relatedness of the component nuclei. When the nuclei are genetically identical (for example, due to fusion between two hyphae of the same monoconidial culture), the anastomosed cell is a *homokaryon*. When, on the other hand, the anastomosing hyphae belong to genetically different strains, the resultant anastomosed cell is a *heterokaryon*. In model ascomycetes such as *Neurospora* and *Aspergillus*, vegetative compatibility is under the control of multiple gene loci designated *het* (heterokaryon) or *vic* (vegetative incompatibility) (22,35). These genes operate to limit heterokaryosis between different strains by discriminating self from nonself within a species via an unknown mechanism. In preliminary genetic studies with limited numbers of teleomorphic (*Glomerella*) strains of *Colletotrichum acutatum, C. gloeosporioides,* and *C. graminicola*, the numbers of *vic* loci in those strains were estimated at two to seven (24,51; Correll et al., Chapter 10, this volume). Most hyphal cells of *Colletotrichum* are uninucleate (49). Microscopic examination reveals that anastomoses occur between lateral branches, which grow out of neighboring hyphae and form anastomosis bridges connecting the two hyphae (Fig. 1). The resultant fused cells are binucleate and appear not to proliferate.

Colletotrichum species are highly variable, as manifested by colony morphology, conidial shape, presence and shape of setae and appressoria, pigmentation, fungicide sensitivity, pathogenicity, and other traits. Morphology has been, and is likely to remain, the primary basis of separation between species or groups of species (48). However, because the environment influences the stability of morphological traits and intermediate forms exist, these criteria are not always adequate for reliable differentiation among *Colletotrichum* species. Non-morphological methods, which broaden the range of characters used to identify strains, should provide a means for estimating genetic diversity and understanding the population structure better, particularly below the species level.

Complex species, such as *C. acutatum* and *C. gloeosporioides,* each have a broad host range (1,7,8,14,16,17,32,34,36,37,39,44), but the spectrum for individual strains may be more limited. Considerable variation in virulence has been observed on various host plants, including colonization by strains that are opportunistic pathogens or saprophytes (38,50,52). It is not known whether phenotypically similar strains are genetically more related to one another than are dissimilar strains, whether various phenotypes can arise from one another, or whether certain biotypes constitute discrete sections within the species.

The reproduction mode in many *Colletotrichum* populations is mainly or exclusively vegetative. In the absence of a sexual stage, the only means of exchanging genetic material between two strains would be anastomosis and heterokaryosis. These processes occur between some *Colletotrichum* isolates but not others (4,6,8,51,54) and, in some cases, seem to be restricted by the existence of vegetative incompatibility. Isolates that cannot form a viable heterokaryon with each other are, in effect, genetically isolated. Isolates that can anastomose with one another and form viable heterokaryons are placed in the same vegetative-compatibility group (VCG) to indicate this fact. They may potentially share a common gene pool, and are isolated from other strains or VCGs within the species by the incompatibility mechanism.

How can vegetative compatibility be tested? When two isolates are grown together, their mycelia intermingle, and hyphal anastomoses (Fig. 1) may occur both within each isolate (intra-isolate) and between hyphae belonging to different isolates (inter-isolate). Since the mycelia are morphologically indistinguishable, direct microscopic observation is ineffective for distinguishing inter-isolate and intra-isolate anastomoses. This limitation can be overcome by applying indirect, genetic complementation tests to determine whether heterokaryosis has occurred. In this procedure, each isolate is usually represented by a recessive nutritional or pigmentation mutant. The mutants are selected to be complementary, and the normal (wild-type) allele of each nucleus in the heterokaryotic cell is expected to compensate for the mutant allele of the other nucleus, yielding a wild-type phenotype.

The experimental approach to studying vegetative compatibility in *Colletotrichum* has been adopted from similar studies in other fungi, primarily *Fusarium* (10,41) and *Verticillium* (9,30). This approach utilizes nitrate-nonutilizing (*nit*) mutants, that cannot use nitrate as the sole nitrogen source. These *nit* mutants, characterized by thin growth on nitrate-minimal medium and by resistance to chlorate, have several advantages for large-scale studies of vegetative compatibility. (i) They arise spontaneously (without mutagenic treatment) and may be selected as chlorate-resistant

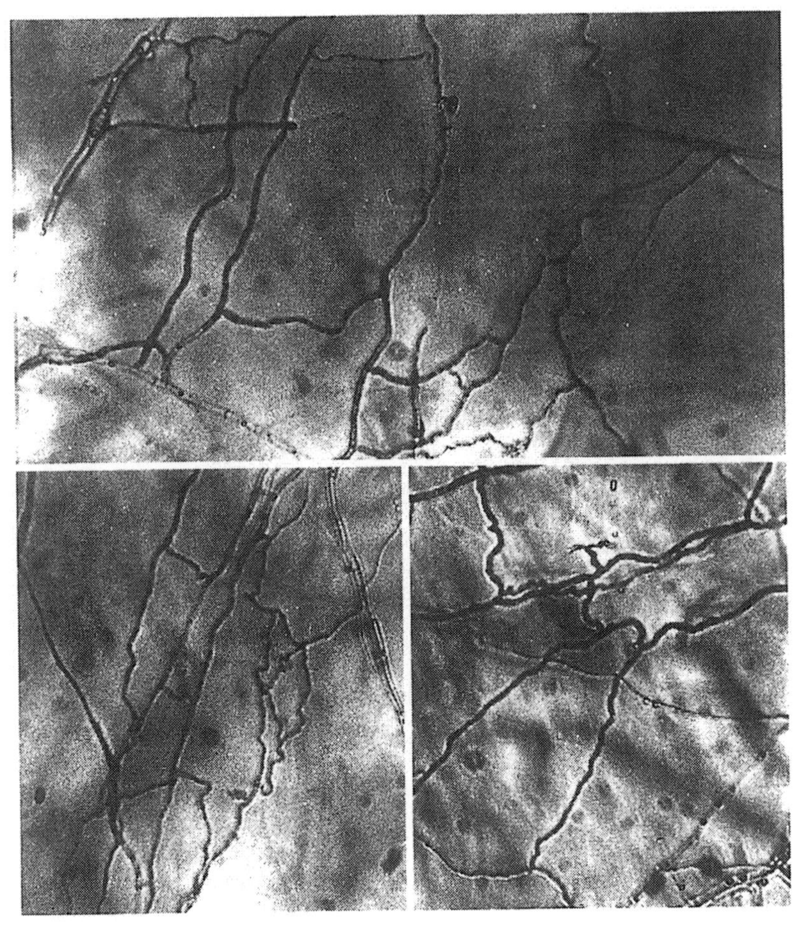

Fig. 1. Hyphal anastomosis in *Colletotrichum acutatum.*

sectors (6,41); (ii) unlike strict auxotrophs, *nit* mutants can grow on minimal medium, and clearly visible complementary heterokaryons are easily identified (Fig. 2). (iii) *nit* mutations may occur at several (at least six) loci, an essential requirement for complementation. If the mutants do not anastomose, cannot form a viable heterokaryon, or are not complementary, no wild-type growth will result from their pairing. Consequently, the results of a complementation test between mutants derived from different isolates depend on vegetative compatibility between the isolates on the one hand, and complementarity of the mutants on the other. Wild-type growth serves as visual evidence of heterokaryosis and, hence, of vegetative compatibility between the parental isolates.

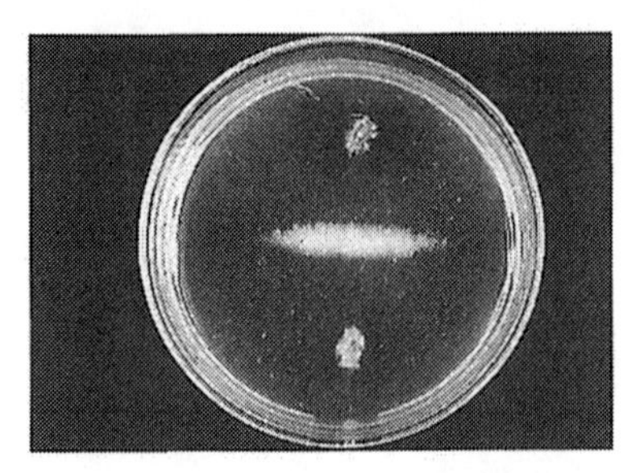

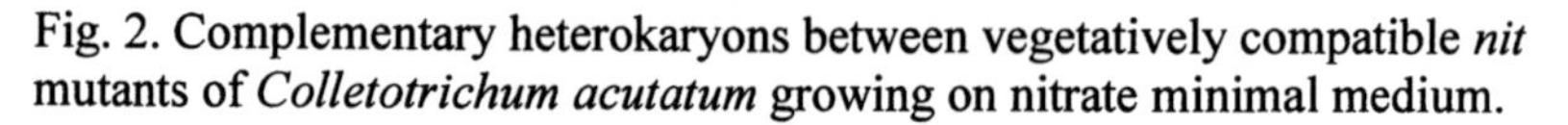

Fig. 2. Complementary heterokaryons between vegetatively compatible *nit* mutants of *Colletotrichum acutatum* growing on nitrate minimal medium.

Complementary heterokaryons of *Colletotrichum* are unstable, and their components segregate upon subculturing and conidiation (8,51; T. Katan, unpublished). In this respect, the heterokaryons of *Colletotrichum* resemble those of *Fusarium* and *Verticillium*, in which there is no nuclear migration from cell to cell along the hypha, and essentially no proliferation of the binucleate cells (5,42). In these fungi, the colony is presumably a mosaic, with a small proportion of binucleate cells and a predominance of uninucleate cells. Each heterokaryotic cell supports the growth of some adjacent uninucleate mutant cells on unsupplemented media. The intensity of heterokaryotic growth depends on the growth habit of the mutants, the frequency of lateral branching and anastomosis, the viability of the heterokaryons, and the biochemical nature of complementation. Strains defective in the capacity to form intra-isolate anastomoses, as manifested by weak or no complementation in pairings between biochemically complementary mutants derived from the same parent, generally show weak or no interaction with other strains as well.

The relative convenience of working with *nit* mutants has triggered studies of vegetative compatibility among strains and populations of *Colletotrichum* spp. (4,8,11,15,21,3132,51,54). Cumulatively, these studies reveal that the standard procedure for generating *nit* mutants is not always adequate and that modifications may be required in the chlorate media for studies of new populations (6,15,51). Vegetative-compatibility (heterokaryon) testing is based solely on self-nonself recognition. Consequently, this test alone cannot be used for taxonomic identification, unless compatibility is demonstrated with a taxonomically well-defined reference strain. Vegetative compatibility is useful for studying other aspects of *Colletotrichum* populations, such as disease etiology, population structure, host specificity, geographic distribution, and reproductive strategy. The following case studies illustrate some of these applications.

Case Studies

ANTHRACNOSE OF ALMOND

This disease has been reported from several almond (*Prunus amygdalus*)-growing regions of the world with Mediterranean climates, including Israel and California. The pathogen from Israel was identified as *C. gloeosporioides* despite its deviation from the "typical" phenotype as manifested, for example, by slow growth, low optimal growth temperature, and insensitivity to benomyl (17,32). The pathogen from California, identified as *C. acutatum*, shares some traits with strains from Israel and differs in others (1,2,16). The populations in Israel and California are both anamorphic, each constitutes a single VCG, and they are not compatible with each other (17,43). Moreover, *C. gloeosporioides* from almond in Israel is not compatible with *C. gloeosporioides* from avocado or with *C. acutatum* from anemone in the same geographic region (32). Regardless of the validity of their taxonomic identification, these pathogens of almond represent two biologically distinct entities; hence, resistance breeding (47) and control strategies targeted against one may not be useful with the other.

ANTHRACNOSE OF SPINACH

This disease is caused by *C. dematium* f.sp. *spinaciae*, a specialized form of *C. dematium* attacking *Spinacia*. Two distinct VCGs were identified among 215 isolates from North America (11). The VCGs differed in pigmentation but not in virulence. One isolate was weakly compatible with both VCGs, demonstrating cross-VCG compatibility and suggesting close relatedness between the two VCGs. The isolates from spinach were not compatible with single *C. dematium* isolates from onion or tomato, nor were the latter two compatible with each other, thus manifesting VCG diversity within the species.

ANTHRACNOSE OF CUCURBITS

This disease, caused by *C. orbiculare*, affects cucumber, watermelon, cantaloupe, and other cucurbits of the genera *Cucumis, Citrullus, Cucurbita, Lagenaria*, and *Luffa*. Three VCGs were identified among 74 isolates confirmed as virulent on differential cucurbit hosts. Cross-VCG compatibility was common between two of the VCGs (54). The cucurbit pathogens were not compatible with a VCG of *C. orbiculare* from *Xanthium* (Compositae). VCG assignment of the cucurbit pathogens corresponded with host origin, virulence phenotype (54), and RAPD (12). Along with some molecular data (12), the low VCG diversity observed among the virulent strains of *C. orbiculare* from cucurbits suggests asexual reproduction in this pathogen.

ANTHRACNOSE AND ROOT NECROSIS OF STRAWBERRY

Colletotrichum spp. can attack strawberry (*Fragaria X ananassa*) at all phases of plant growth, causing a variety of diseases in all aboveground organs. At least three species are recognized—*C. gloeosporioides*, *C. acutatum* and *C. fragariae*—and the symptoms associated with them often overlap (37). Anthracnose of strawberry was first recorded in Israel in 1995 and, using morphological criteria and species-specific PCR primers, the pathogen was identified as *C. acutatum* (15). *C. acutatum* also was isolated from rot-affected roots of chlorotic, stunted plants. The role of *C. acutatum* in the etiology of the root necrosis was confirmed by virulence tests. Isolates from all plant parts, including roots, originating from different sites were vegetatively compatible, indicating that anthracnose and root necrosis were both caused by strains from a single VCG of *C. acutatum*. *Colletotrichum* occurs on strawberry worldwide, but the relationship between genetic variation and geographic origin is not well understood. Grouping geographic populations into vegetatively compatible clonal lineages should be helpful in determining whether synonymy exists in the taxonomic identification of certain species (13,29,45); in extrapolating research results, such as for breeding programs (27) and control strategies (19,26,28), from one population to another; and in assessing epidemiological parameters of strawberry anthracnose.

LEAF CURL DISEASE OF ANEMONE

This disease, caused by *Colletotrichum* spp. in *Anemone* and also known as anthracnose, was first reported in Australia, where the pathogen was identified as *C. acutatum* (55). In Israel, the first outbreak occurred in 1978, and the pathogen was identified as *C. gloeosporioides*. Isolates from anemone in Europe, however, have been identified as *C. acutatum* and *C. gloeosporioides* (14,25). A second epidemic occurred in Israel in 1990-1992, and vegetative compatibility studies showed that isolates from the first and second epidemics belonged to the same single VCG. Representatives of this VCG were compatible with several *C. acutatum* isolates from Europe (Italy, the UK, and the Netherlands) (31), as well as a *C. acutatum* reference isolate from Australia. An additional "chromogenic" *C. acutatum* isolate from the Netherlands was not compatible with any of these isolates, suggesting that the *C. acutatum* attacking anemone in Europe consisted of representatives from at least two VCGs. By inference from their compatibility with the reference isolates from Australia and Europe, the isolates from Israel also should be identified as *C. acutatum*. Presumably, members of this VCG in Israel were imported with contaminated corms from Europe (23). About 20% of the isolates obtained from diseased anemones in Israel in recent years belong to the aforementioned VCG, while the majority of them belong to a separate, new

VCG which, unlike the previous one, is compatible with local isolates of *C. acutatum* from strawberry. The host specificity of the new VCG includes both anemone and strawberry, whereas the first VCG was found only in anemone. Using species-specific PCR primers, all the isolates were confirmed as *C. acutatum* (18,20). Both anemone and strawberry are attacked by *Colletotrichum* in Europe (13,14,26), but whether cross-infection occurs under field conditions there is unknown.

COFFEE BERRY DISEASE (CBD)

CBD, caused by *C. kahawe* (syn. *C. coffeanum*), affects green and ripe berries of *Coffea arabica* in Africa. *C. gloeosporioides* and *C. acutatum* are commonly isolated from coffee trees, but these species are considered mild pathogens or saprophytes (38). Preliminary VCG studies of *Colletotrichum* isolates from coffee in Africa gave inconclusive, somewhat contradictory results. In one study (4), vegetative compatibility could be demonstrated only among four out of 10 *C. kahawe* isolates; whereas in another study (21), 39 *C. kahawe* isolates were all assigned to a single VCG and were not compatible with single representatives of *C. gloeosporioides* or *C. acutatum* from coffee. Further VCG studies should provide information about the diversity and relatedness within and between *C. kahawe* and other *Colletotrichum* populations associated with coffee in Africa, as well as in South America, Southeast Asia and the South Pacific (38,40,46).

Conclusions and Future Outlook

The genus *Colletotrichum* encompasses multiple species, species complexes and subspecific groups, the distinction of which is often uncertain. Reproductive strategy plays a major role in determining the extent of variability within a species. A high level of genotypic variation is usually associated with sexual reproduction, as discussed elsewhere (Correll et al., Chapter 10 of this volume). Both sexually reproducing strains (*Glomerella*) and clonally reproducing anamorphic strains exist in the *C. acutatum* (15,24) and *C. gloeosporioides* (Correll et al., Chapter 10; and Manners et al., Chapter 11 of this volume) species complexes. These strains may be present in close proximity on the same or different hosts, constituting subpopulations within the species (17; Correll et al., Chapter 10, this volume). Sexual reproduction may be limited or absent in the field despite the demonstrated potential for its occurrence under laboratory conditions. Since high VCG diversity, i.e., high genotypic diversity, is thought to reflect constantly recombining alleles at multiple *vic* loci, it is a good predictor of an operative sexual cycle (17,24,51; Correll et al., Chapter 10, this volume). Low VCG diversity may indicate asexual reproduction, or limited genetic diversity at the *vic* loci. Whereas genetic relatedness between sexually competent strains can be examined by sexual crosses, this option is not available for the analysis of asexual strains. VCGs are a useful method for determining genetic relatedness among such strains. Distinct genotypic and phenotypic differences between subspecific VCGs, which are indistinguishable by species-specific PCR primers, may have important consequences for resistance breeding, chemical control, and other disease-management practices. VCG analysis also should be valuable in dissecting complex pathosystems (34,39,53), e.g., the one reported for citrus in various geographic regions, where a variety of disease symptoms have been associated with several *Colletotrichum* biotypes (3,7,33,36,50,56). Similarly, VCGs should be useful in discerning whether conspecific *Colletotrichum* isolates on the same or different hosts and/or locations are genetically related (16,17,34,39,52; Manners et al., Chapter 11, this volume).

Taxonomic identification of species by the VCG method is only applicable to anamorphic, clonally related populations that are vegetatively compatible with well-defined reference strains. Therefore, further research should concentrate on trying to correlate VCGs with molecular markers and with biological and virulence traits.

Acknowledgement

Contribution 549/99 from the ARO Institute of Plant Protection.

Literature Cited

1. Adaskaveg, J.E. and Hartin, R.J. 1997. Characterization of *Colletotrichum acutatum* isolates causing anthracnose of almond and peach in California. Phytopathology 87:979-987.
2. Adaskaveg, J.E. and Forster, H. 1998. Occurrence and management of anthracnose epidemics caused by *Colletotrichum* species on tree fruit crops in California. Phytoparasitica 26:358-359. (Abstr.)
3. Agostini, J.P. and Timmer, L.W. 1994. Population dynamics and survival of strains of *Colletotrichum gloeosporioides* on citrus in Florida. Phytopathology 84:420-425.
4. Benyon, S.M., Coddington, A., Lewis, B.G., and Varzea, V. 1995. Genetic variation in the coffee berry disease pathogen, *Colletotrichum kahawe*. Physiol. Mol. Plant Pathol. 46:457-470.
5. Bowden, R.L. and Leslie, J.F. 1992. Nitrate-nonutilizing mutants of *Gibberella zeae* (*Fusarium graminearum*) and their use in determining vegetative compatibility. Exp. Mycol. 16:308-315.
6. Brooker, N.L., Leslie, J.F., and Dickman, M.B. 1991. Nitrate non-utilizing mutants of *Colletotrichum* and their use in studies of

vegetative compatibility and genetic relatedness. Phytopathology 81:672-677.

7. Brown, A.E., Sreenivasaprasad, S., and Timmer, L.W. 1996. Molecular characterization of slow-growing orange and key lime anthracnose strains of *Colletotrichum* from citrus as *C. acutatum*. Phytopathology 86:523-527.
8. Chacko, R.J., Weidemann, G.J., TeBeest, D.O., and Correll, J.C. 1994. The use of vegetative compatibility and heterokaryosis to determine potential asexual gene exchange in *Colletotrichum gloeosporioides*. Biol. Control 4:382-389.
9. Correll, J.C., Gordon, T.R., and McCain, A.H. 1988. Vegetative compatibility and pathogenicity of *Verticillium albo-atrum*. Phytopathology 78:1017-1021.
10. Correll, J.C., Klittich, C.J.R., and Leslie, J.F. 1987. Nitrate nonutilizing mutants of *Fusarium oxysporum* and their use in vegetative compatibility tests. Phytopathology 77:1640-1646.
11. Correll, J.C., Morelock, T.E., and Guerber, J.C. 1993. Vegetative compatibility and virulence of the spinach anthracnose pathogen, *Colletotrichum dematium*. Plant Dis. 77:688-691.
12. Correll, J.C., Rhoads, D.D., and Guerber, J.C. 1993. Examination of mitochondrial DNA restriction fragment length polymorphisms, DNA fingerprints, and randomly amplified polymorphic DNA of *Colletotrichum orbiculare*. Phytopathology 83:1199-1204.
13. Denoyes, B. and Baudry, A. 1995. Species identification and pathogenicity study of French *Colletotrichum* strains from strawberry using morphological and cultural characteristics. Phytopathology 85:53-57.
14. Doornik, A.W. 1990. Hot-water to control *Colletotrichum acutatum* on corms of *Anemone coronaria*. Acta Hort. 266:491-494.
15. Freeman, S. and Katan, T. 1997. Identification of *Colletotrichum* species responsible for anthracnose and root necrosis of strawberry in Israel. Phytopathology 87:516-521.
16. Freeman, S., Katan, T., and Shabi, E. 1996. Characterization of *Colletotrichum gloeosporioides* isolates from avocado and almond fruits with molecular and pathogenicity tests. Appl. Environ. Microbiol. 62:1014-1020.
17. Freeman, S., Katan, T., and Shabi, E. 1998. Characterization of *Colletotrichum* species responsible for anthracnose disease of various fruits. Plant Dis. 82:596-605.
18. Freeman, S., Shabi, E., and Katan, T. 1998. Comparative study of *Colletotrichum* from anemone and strawberry. Phythopatology 88:S29. (Abstr.)
19. Freeman, S., Nizani, Y., Dotan, S., Even, S., and Sando, T. 1997. Control of *Colletotrichum acutatum* in strawberry under laboratory, greenhouse, and field conditions. Plant Dis. 81:749-752.

20. Freeman, S., Shabi, E., Nitzani, Y., and Katan, T. 1998. Anthracnose of anemone—a new or recurring disease? Phytoparasitica 26:155-156. (Abstr.)
21. Gichuru, E.K., Varzea, V.M.P., Rodriguez, C.J. Jr., and Masaba, D.M. 1998. Vegetative compatibility in a population of *Colletotrichum kahawe* from Kenya. (Abstr. 2.2.129) Seventh Int. Cong. Plant Pathol., Edinburgh, Scotland.
22. Glass, N.L. and Kuldau, G.A. 1992. Mating type and vegetative incompatibility in filamentous ascomycetes. Annu. Rev. Phytopathol. 30:201-224.
23. Gokkes, M., Ben-Ze'ev, I.S., Levy, E., Ben-Gal, O., and Shabtay, Y. 1993. Anemone anthracnose—two outbreaks and eradication a decade apart. Phytoparasitica 21:148. (Abstr.)
24. Guerber, J.C. and Correll, J.C. 1998. Segregation of genetic markers in *Colletotrichum acutatum*. Phytopathology 88:S34. (Abstr.)
25. Gullino, M.L., Bozzano, G., and Garibaldi, A. 1981. Saggio in vitro e in vaso di fungicidi contro l'arricciamento dell'anemone provocato de *Colletotrichum gloeosporioides*. La difessa delle piante 2:79-88.
26. Gullino, M.L., Romano, M.L., and Garibaldi, A. 1985. Identification and response to fungicides of *Colletotrichum gloeosporioides*, incitant of strawberry black rot in Italy. Plant Dis. 69:608-609.
27. Howard, C.M. and Albregts, E.E. 1979. 'Dover' a firm-fruited strawberry with resistance to anthracnose. UF/IFAS Circ. S-267, Florida.
28. Howard, C.M., Albregts, E.E., and Chandler, C.K. 1991. Evaluation of fungicides for control of gray mold and fruit anthracnose, 1990. Fungic. Nematicide Tests 46:105.
29. Howard, C.M., Maas, J.L., Chandler, C.K., and Albregts, E.E. 1992. Anthracnose of strawberry caused by the *Colletotrichum* complex in Florida. Plant Dis. 76:796-981.
30. Joaquim, T.R. and Rowe, R.C. 1990. Reassessment of vegetative compatibility relationships among strains of *Verticillium dahliae* using nitrate-nonutilizing mutants. Phytopathology 80:1160-1166.
31. Katan, T. and Shabi, E. 1994. Vegetative compatibility among *Colletotrichum* isolates from anemone. Phytoparasitica 22:90. (Abstr.)
32. Katan, T. and Shabi, E. 1996. Vegetative compatibility among isolates of *Colletotrichum gloeosporioides* from almond in Israel. Eur. J. Plant Pathol. 102:597-600.
33. Kuramae-Izioka, E.E., Lopes, C.R., Souza, N.L., and Machado, M.A. 1997. Morphological and molecular characterization of *Colletotrichum* spp. from citrus orchards affected by postbloom fruit drop in Brazil. Eur. J. Plant Pathol. 103:323-329.
34. Lenne, J.M. 1992. *Colletotrichum* diseases of legumes. Pages 134-166 in: *Colletotrichum*: Biology, Pathology and Control. J.A. Bailey and M.J. Jeger, eds. CAB International, Wallingford, UK.

35. Leslie, J.F. 1993. Fungal vegetative compatibility. Annu. Rev. Phytopathol. 31:127-150.
36. Liyanage, H.D., McMillan, R.T. Jr., and Kistler, H.C. 1992. Two genetically distinct populations of *Colletotrichum gloeosporioides* from citrus. Phytopathology 82:1371-1376.
37. Maas, J.L., ed. 1998. Compedium of Strawberry Diseases (2nd ed.). APS Press, St. Paul, Minnesota. 98 pp.
38. Masaba, D. and Waller, J.M. 1992. Coffee berry disease: The current status. Pages 237-249 in: *Colletotrichum*: Biology, Pathology and Control. J.A. Bailey and M.J. Jeger, eds. CAB International, Wallingford, UK.
39. Mills, P.R., Hodson, A., and Brown, A.E. 1992. Molecular differentiation of *Colletotrichum gloeosporioides* isolates infecting tropical fruits. Pages 269-288 in: *Colletotrichum*: Biology, Pathology and Control. J.A. Bailey and M. J. Jeger, eds. CAB International, Wallingford, UK.
40. Nandris, D., Kohler, F., Fernandez, D., Lashermes, P., Rodriguez, J., and Pellegrin, P.F. 1998. Coffee pathosystem modelling: 2. Assessment pathogen biodiversity. (Abstr. 2.1.17) Seventh Int. Cong. Plant Pathol., Edinburgh, Scotland.
41. Puhalla, J.E. 1985. Classification of strains of *Fusarium oxysporum* on the basis of vegetative compatibility. Can. J. Bot. 63:179-183.
42. Puhalla, J.E. and Mayfield, J.E. 1974. The mechanism of heterokaryotic growth in *Verticillium dahliae*. Genetics 76:411-422.
43. Shabi, E., Freeman, S., Katan, T., and Teviotdale, B. 1996. Almond anthracnose in Israel and California. Phytoparasitica 24:138-139. (Abstr.)
44. Shi, Y., Correll, J.C., Guerber, J.C., and Rom, C.R. 1996. Frequency of *Colletotrichum* species causing bitter rot of apple in the southeastern United States. Plant Dis. 80:692-696.
45. Smith, B.J. and Black, L.L. 1990. Morphological, cultural and pathogenic variation among *Colletotrichum* species isolated from strawberry. Plant Dis. 74:69-76.
46. Sreenivasaprasad, S., Brown, A.E., and Mills, P.R. 1993. Coffee berry disease pathogen in Africa: Genetic structure and relationship to the group species *Colletotrichum gloeosporioides*. Mycol. Res. 97:995-1000.
47. Striem, M.J., Spiegel-Roy, P., and Shabi, E. 1989. Evaluating susceptibility of almonds to anthracnose disease caused by *Colletotrichum gloeosporioides*. Phytoparasitica 17:107-122.
48. Sutton, B.C. 1992. The genus *Glomerella* and its anamorph *Colletotrichum*. Pages 1-26 in: *Colletotrichum*: Biology, Pathology and Control. J.A. Bailey and M.J. Jeger, eds. CAB International, Wallingford, UK.

49. TeBeest, D.O., Shilling, C.W., Riley, L.H., and Weidemann, G.J. 1989. The number of nuclei in spores of three species of *Colletotrichum*. Mycologia 81:147-149.
50. Timmer, L.W. and Brown, G.E. 1998. Biology and control of anthracnose diseases of citrus. (Abstr.) Phytoparasitica 26:358.
51. Vaillancourt, L.J. and Hanau, R.M. 1994. Nitrate-nonutilizing mutants used to study heterokaryosis and vegetative compatibility in *Glomerella graminicola* (*Colletotrichum graminicola*). Exp. Mycol. 18:311-319.
52. Waller, J.M. 1992. *Colletotrichum* diseases of perennial and other cash crops. Pages 167-185 in: *Colletotrichum*: Biology, Pathology and Control. J.A. Bailey and M.J. Jeger, eds. CAB International, Wallingford, UK.
53. Waller, J.M. 1998. Recent advances in the understanding of *Colletotrichum* diseases of some tropical crops. Phytoparasitica 26:359. (Abstr.)
54. Wasilwa, L.A., Correll, J.C., Morelock, T.E., and McNew, R.E. 1993. Reexamination of races of the cucurbit anthracnose pathogen *Colletotrichum orbiculare.* Phytopathology 83:1190-1198.
55. Woodcock, T. and Washington, W.S. 1979. *Colletotrichum acutatum* on *Anemone* and *Ranunculus*. Australas. Plant Pathol. 8:10.
56. Zulfiqar, M., Brlansky, R.H., and Timmer, L.W. 1996. Infection of flower and vegetative tissues of citrus by *Colletotrichum acutatum* and *C. gloeosporioides.* Mycologia 88:121-128.

Chapter 5

Dissecting the Cell Biology of *Colletotrichum* Infection Processes

R. O'Connell, S. Perfect, B. Hughes, R. Carzaniga,
J. Bailey, and J. Green

Species of *Colletotrichum* employ diverse strategies for invading host tissue, ranging from intracellular hemibiotrophy to subcuticular/intramural necrotrophy. These pathogens also develop a series of specialized infection structures, e.g., germ tubes, appressoria, penetration pegs, infection vesicles, primary hyphae, and secondary hyphae. *Colletotrichum* thus provides excellent models for studying the cell biology of fungal pathogenesis. We are using the bean anthracnose fungus, *C. lindemuthianum*, as a model system to study the development and adhesion of infection structures, biotrophy, and necrotrophy. Our approach has been to raise monoclonal antibodies (MAbs) to infection structures isolated from host tissue to identify fungal components expressed at different stages of the infection process (38,43). In this way, we have obtained a panel of MAbs recognizing glycoproteins specific to the extracellular matrices surrounding spores, germ tubes, and appressoria (44,45), the plasma membrane of appressoria (46) and the biotrophic interface formed between intracellular hyphae and living host cells (47). In this chapter we will review recent progress in the characterization of these glycoproteins and the successful use of one antibody to clone a biotrophy-related gene. To put this work in context, we will first briefly describe the cytology of infection of bean tissue by *C. lindemuthianum* and contrast this with the infection processes of other members of the genus.

Infection Strategies of *Colletotrichum* Species

The early stages of fungal development on the plant surface are essentially the same for all *Colletotrichum* species. Conidia adhere to the cuticle and germinate to produce germ tubes, which in turn differentiate melanized appressoria. Narrow penetration pegs emerging from the base of the appressorium then penetrate the cuticle directly, although there are a few reports of indirect penetration through stomata or wounds without appressoria (33,53,58). Major differences only become apparent after penetration, when two main types of infection strategy can be distinguished: intracellular hemibiotrophy and subcuticular, intramural necrotrophy (4,56).

Intracellular Hemibiotrophy

Many *Colletotrichum* species initially establish infection through a brief biotrophic phase, associated with large intracellular primary hyphae. They later switch to a destructive, necrotrophic phase, associated with narrower secondary hyphae which ramify throughout the host tissue. Among the species adopting this strategy, the initial biotrophic phase can vary in duration from less than 24 hours to over 3 days. The intracellular primary hyphae formed by these fungi can vary greatly in morphology. At least two *Colletotrichum* species produce intracellular mycelia that are entirely confined to the initially infected epidermal cell. In the case of *C. destructivum* infecting cowpea (Fig. 1A) or alfalfa, the intracellular hyphae are enormously swollen, lobed structures that eventually expand to fill the host cell (27,28). Similarly, the primary hyphae of *C. truncatum* extensively colonize pea epidermal cells (Fig. 1B), but these hyphae are smaller in diameter than those of *C. destructivum* and become highly branched and convoluted (41). In *C. orbiculare* on cucumber (Fig. 1C), *C. malvarum* on *Sida* (Fig. 1D), *C. lindemuthianum* on bean (Fig. 1E) and *C. trifolii* on alfalfa, initial penetration of epidermal cells is immediately followed by the formation of a large, spherical infection vesicle. One or more large-diameter, filamentous primary hyphae subsequently grow out from the infection vesicle and go on to colonize many other host cells (2,34,37). The intracellular hyphae of these four pathogens appear identical, which is consistent with molecular and serological data suggesting they are all *formae speciales* of a single aggregate species, namely *C. orbiculare* (2,54). In some other pathogens producing intracellular hyphae, e.g., *C. sublineolum* and *C. acutatum*, the morphological distinction between infection vesicles and primary hyphae is less clear in than in *C. orbiculare* (35,62).

The best-studied example of intracellular hemibiotrophy is the *C. lindemuthianum*-bean interaction (36,37). In susceptible bean cultivars, penetrated cells initially remain alive, as shown by the normal ultrastructure of their cytoplasm and their continued ability to plasmolyze and accumulate vital stains (37). During this biotrophic phase, the host plasma membrane expands and invaginates around the developing infection vesicles and primary hyphae, and an interfacial matrix layer is deposited between the fungal cell wall and host plasma membrane. However, around 24 hours after penetration, the host plasma membrane loses its functional integrity and host cells start to degenerate and die (37). Although the intracellular hyphae

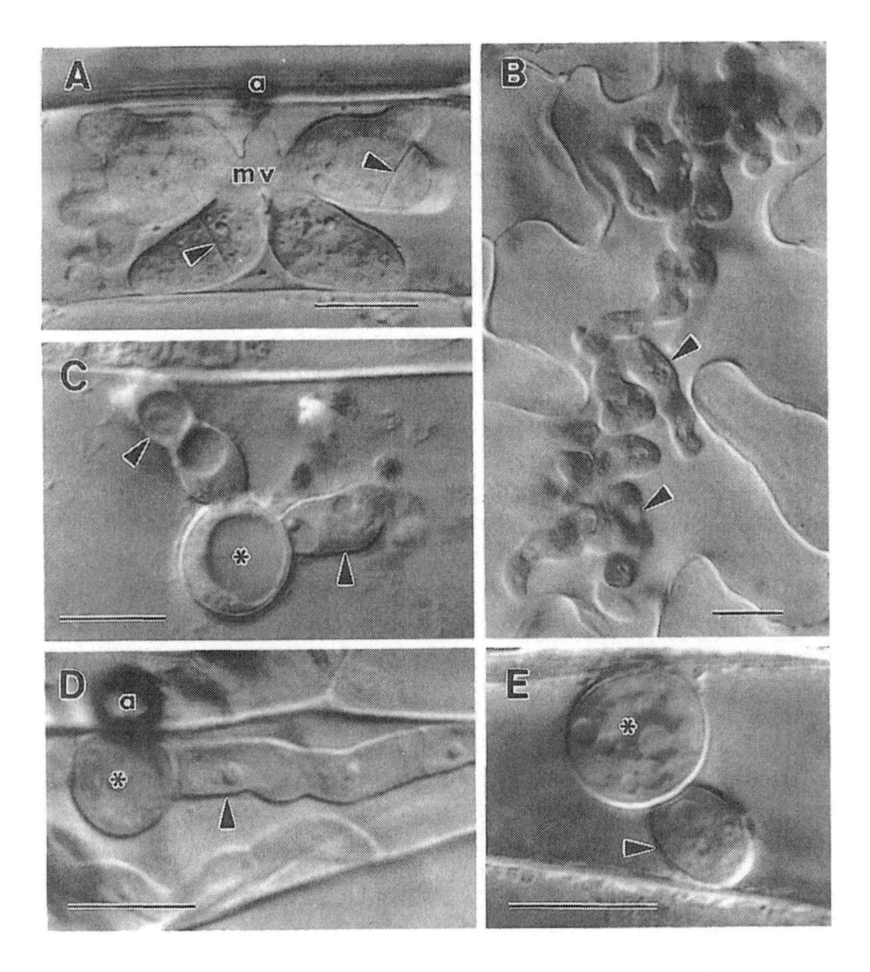

Fig. 1. Morphology of biotrophic intracellular hyphae, viewed with interference contrast microscopy. Bars = 10 µm, a = appressorium. A: *C. destructivum* on cowpea hypocotyl (fresh tissue, 72 h). A swollen, multilobed infection vesicle (MV) fills the epidermal cell. Arrowheads indicate septa. B: *C. truncatum* on pea leaf (cleared tissue, 72 h). Convoluted, branched primary hyphae (arrowheads) fill the epidermal cell. C: *C. orbiculare* on cucumber hypocotyl (fresh tissue, 48 h). Large primary hyphae (arrowheads) arise from a distinct, spherical infection vesicle (asterisk). D: *C. malvarum* on *Sida spinosa* leaf (cleared tissue, 96 h). Morphology is identical to *C. orbiculare*. E: *C. lindemuthianum* on bean hypocotyl (fresh tissue, 72 h). Morphology is identical to *C. orbiculare*. D and E reprinted, by permission of the publishers, from Bailey et al. (2) and O'Connell and Bailey (36).

presumably obtain nutrients from the dead host cells, this is a benign form of necrotrophy: cell death is gradual and confined to infected cells and is not associated with extensive dissolution of host cell walls. Many host cells die, but active host defense mechanisms are not induced (1), and there are no macroscopic symptoms since the dead tissue does not become brown. Moreover, as the primary hyphae colonize further host cells, a biotrophic relationship is re-established in each newly infected cell. Hence, biotrophy and necrotrophy may occur simultaneously in different regions of the primary mycelium (36).

A more destructive form of necrotrophy commences with the appearance of narrow secondary hyphae, which ramify through host tissues both inter- and intracellularly (37). At this stage, host cells are killed rapidly in advance of infection and host cell walls are extensively degraded by fungal depolymerizing enzymes (63). The typical brown anthracnose symptoms now appear. Host defense compounds are synthesized at this stage (1), but fungal growth is not arrested because the host cells producing these compounds are killed before inhibitory concentrations can accumulate.

The infection of sorghum tissues by *C. sublineolum* follows a very similar pattern to the *C. lindemuthianum*-bean interaction, with biotrophy progressing from cell to cell (62). However, the sequence of events described previously is not typical of all intracellular hemibiotrophic species. In *C. destructivum* and *C. truncatum*, the biotrophic phase of interaction is, like the primary mycelium, entirely confined to the first infected epidermal cell (3,41). By the time narrow secondary hyphae radiate out from the epidermal cell, host cells are being killed and cell walls dissolved ahead of infection. Some species producing intracellular hyphae, e.g., *C. trifolii* and *C. graminicola*, appear to have no detectable biotrophic phase, with infections resembling the benign necrotrophic phase of *C. lindemuthianum* (34,49). However, the viability of host cells infected by these pathogens has not been critically assessed by plasmolysis or vital staining, and the possible occurrence of a brief biotrophic phase cannot be excluded.

SUBCUTICULAR, INTRAMURAL NECROTROPHY

The infection of cowpea by *C. capsici* illustrates a second type of infection strategy. Following penetration of the cuticle, this pathogen does not immediately enter into the cell lumen, but instead develops beneath the cuticle, within the periclinal and anticlinal walls of epidermal cells (50). From the earliest stage of penetration, intramural development is associated with extensive swelling and dissolution of host cell walls. It is unclear whether the underlying host cells remain alive, but symptoms do not appear until 24 hours after penetration, so a brief biotrophic phase or a benign necrotrophic phase may occur. As with the intracellular hemibiotrophs, it appears that host defense mechanisms are not induced, and the fungus continues to ramify intramurally, only later penetrating into host cells. Subsequently, this fungus spreads rapidly throughout the tissue both intra- and intercellularly, killing host cells and dissolving cell walls ahead of infection. This type of infection strategy is not associated with production of morphologically distinct primary and secondary mycelia. Further examples of subcuticular, intramural pathogens include *C. capsici* on cotton (51), *C. circinans* on onion (60) and *C. gloeosporioides* on papaya (10) and *Stylosanthes* (20). Other *Colletotrichum* species combine both types of infection strategy, producing intracellular hyphae and subcuticular hyphae within the same tissue. *C. magna* on watermelon and forms of *C. gloeosporioides* attacking avocado, citrus, and rubber all behave in this way (7,11,67; R. O'Connell, unpublished).

Spore Adhesion: The Role of Cell Surface Glycoproteins

The surface of *Colletotrichum* conidia differs from that of all other fungal cell types in possessing a brush-like outer layer of short fibrils arranged perpendicular to the cell wall (39,59). This outer layer, termed the spore coat, could not be visualized with TEM until the application of cryopreparation techniques such as freeze-fracture (Fig. 2A) or freeze-substitution (Fig. 2B), which preserve mucilaginous materials better than conventional methods (17). Interestingly, the ordered arrays of rodlets characteristic of hydrophobin layers in other fungi (23) were not visible in freeze-fracture replicas of *C. lindemuthianum* spores (Fig. 2A). The spore coat has a spongy, porous structure (39), similar to the mannoprotein surface layers reported in the yeast form of *Candida albicans* (6). In *C. lindemuthianum*, the spore coat also appears to be composed largely of glycoproteins, since it reacts strongly with carbohydrate-specific stains but is completely removed from the cell wall by digestion with a proteolytic enzyme (B. Hughes, unpublished). Moreover, the spore coat appears to lack the major cell wall polysaccharides such as chitin (Fig. 2B) and ß-1,3-glucans (Fig. 2C).

To investigate the composition of glycoproteins on the surface of *C. lindemuthianum* spores, we have used a lectin from *Bauhinia purpurea* (BPA), specific for *N*-acetylgalactosamine, and a monoclonal antibody UB20, raised against germinating conidia (39,45). TEM-immunogold labeling of frozen sections showed that the glycoproteins recognized by UB20 are organized into two distinct layers, namely the outer surface of the spore coat and a thin layer of the cell wall and cytoplasm adjoining the plasma membrane (Fig. 2D). In yeast (*Saccharomyces cerevisiae*) there is evidence that some cell wall glycoproteins are released from glycosyl phosphatidylinositol lipid

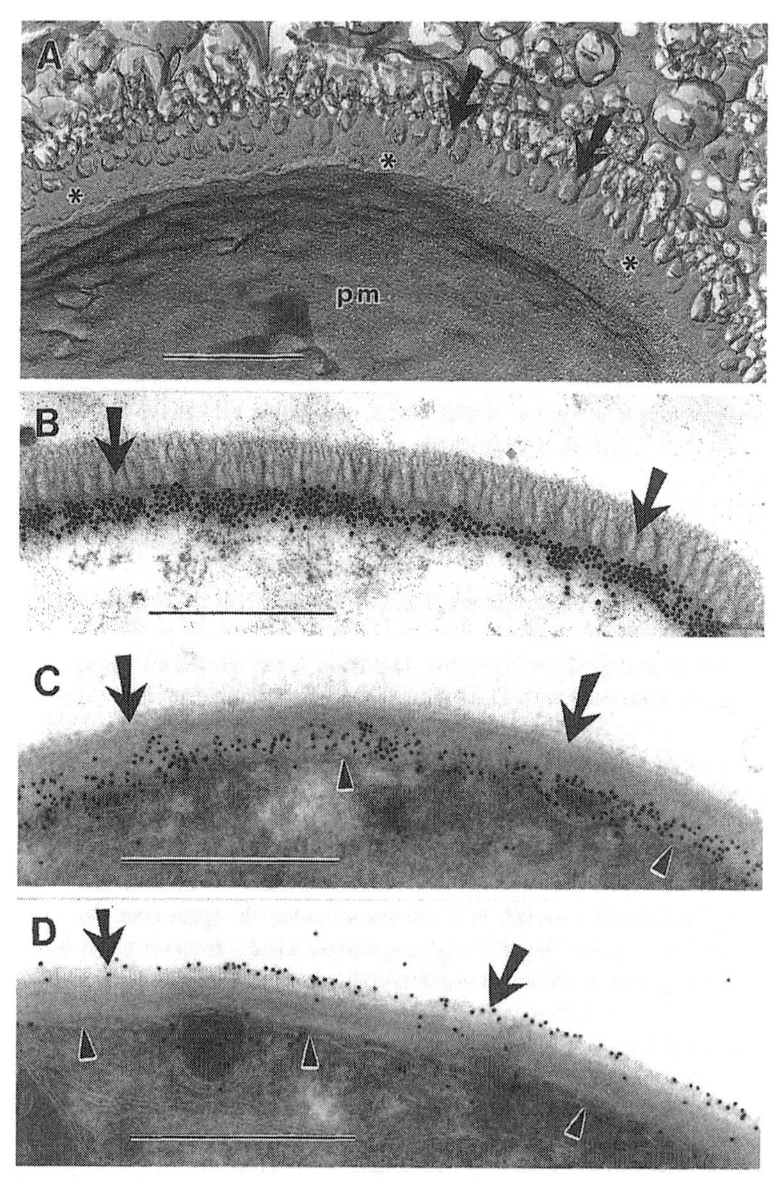

Fig. 2. The cell surface of *C. lindemuthianum* conidia. Bars = 0.5 µm. A: Freeze-etch replica showing porous spore coat (arrows), overlying cell wall (asterisks) and plasma membrane (pm). B: Freeze-substituted spore gold-labeled with wheat germ lectin. Cell wall is intensely labeled while spore coat (arrows) is unlabeled. C: Immunogold labeling of β-1,3-glucans in ungerminated spore (frozen section). Cell wall is intensely labeled but spore coat (arrows) is unlabeled. D: Immunogold labeling of UB20 glycoproteins in ungerminated spore (frozen section). Outer surface of spore coat (arrows) and cell wall/cytoplasm close to plasma membrane (arrowheads) are labeled.

anchors that link these glycoproteins initially to the plasma membrane (55). In *C. lindemuthianum*, the possibility that some of the glycoproteins recognized by UB20 at the cell wall/plasma membrane interface are also linked to membrane lipids deserves further study.

In recent work, we have shown that the spore coat can be completely removed from intact spores simply by extraction with hot (90°C) water or 0.2% SDS. Similar treatments have been used to extract surface components of yeast cells and are likely to solubilize cell surface glycoproteins that are not covalently linked to wall polysaccharides (8). Western blotting with UB20 and BPA showed that a small number of glycoproteins are extracted by these procedures, including major bands at 110 and >200 kDa and minor bands at 45 and 155 kDa. UB20 recognizes a carbohydrate epitope on these glycoproteins that is carried on *N*-linked side-chains of the complex type.

Biotinylation of proteins and glycoproteins can be used to identify components that are exposed at the cell surface (9). Biotinylation *via* either primary amine groups or carbohydrate moieties confirmed that the glycoproteins labeled by UB20 and BPA in Western blots are exposed on the surface of *C. lindemuthianum* spores and that they represent the main glycoprotein components present in the spore surface extracts (B. Hughes, unpublished). The water-soluble mucilage that surrounds spores in acervuli did not contain significant amounts of these glycoproteins.

In *Candida albicans*, mannoprotein surface layers resembling the spore coat have been implicated in the regulation of cell surface hydrophobicity and adhesion to host cells (15). Evidence from several *Colletotrichum* species indicates that the initial rapid attachment of ungerminated conidia also involves hydrophobic interactions (31,52,66). As a preformed structural feature of the dormant spore, the spore coat is present at the correct time and location to function in this attachment. Such a role is supported by the observation that removal of the spore coat by hot water extraction abolishes the ability of conidia to rapidly attach to polystyrene (R. Carzaniga, unpublished data). In *C. musae* and *C. graminicola*, enzymatic removal of surface proteins also prevented attachment of ungerminated conidia, suggesting that cell surface proteins mediate this process (31,52). In *C. graminicola*, mannose-containing glycoprotein(s) appear to be involved, since incubation of conidia with the mannose-specific lectin ConA completely blocked attachment (31).

To investigate whether the surface glycoproteins extracted from *C. lindemuthianum* spores have adhesive properties, the following approach was used (B. Hughes, unpublished). A hot water spore extract was separated by SDS-PAGE, blotted onto nitrocellulose and incubated with colored polystyrene microspheres. The hydrophobic polystyrene beads bound selectively to the 110 kDa glycoprotein. This glycoprotein may therefore contribute to the

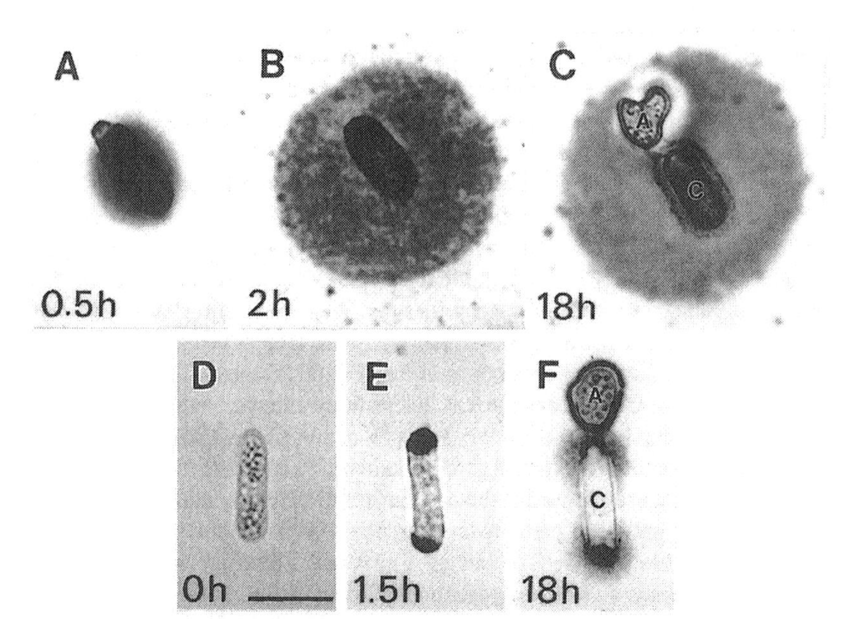

Fig. 3. Release of protein exudates by germinating conidia of *C. gloeosporioides* (A-C) and *C. magna* (D-F) after incubation on glass slides for varying times. Proteins were detected with anionic colloidal gold followed by silver enhancement. Bar = 20 μm. In *C. gloeosporioides*, proteins are released from the entire spore surface as early as 0.5 hr (A). By 2 h, prior to germ tube emergence, a large circular film of protein surrounds the spore (B), reaching its maximum extent by 18 hr (C). In *C. magna*, protein release is not detectable until 1.5 hr (E) and is confined to the spore apices, producing a dumb-bell-shaped deposit over the glass by 18 hr (F).

hydrophobicity of the spore surface and could be involved in the initial rapid attachment of conidia to the plant surface.

In addition to passive hydrophobic interactions, spore adhesion in *C. lindemuthianum, C. musae*, and *C. graminicola* appears to require active metabolism, including protein synthesis (31,52,66). In studies on these three species, adhesion was assayed after incubating spores with various test substrata for between 30 min and several hours, i.e., a considerable time after initial attachment, which occurs within seconds of contact. These active processes appear to be associated with a second phase of adhesion, which may serve to consolidate the initial hydrophobic attachment. In some species, e.g., *C. gloeosporioides, C. graminicola*, and *C. magna*, this second phase of adhesion is correlated with the release of a protein exudate, which spreads outwards from the spore as a thin film over the substratum or leaf surface (21,30). Staining with silver-enhanced colloidal gold revealed that the timing and location of protein release varies between different species (Fig. 3). In *C.*

gloeosporioides, proteins were released from the entire surface of the spore within 30 min of initial attachment, forming a circular halo that extends up to 20 μm from the spore after 2 hr (Fig. 3A-C). In *C. magna*, protein release was not detectable until 1.5 hr after initial attachment and was confined to the spore apices, producing a dumb-bell-shaped deposit over the glass surface at 18 hr (Fig. 3D-F). Some species, including *C. lindemuthianum*, do not appear to release any protein exudates (44). It is therefore unclear whether secretion of these materials contributes to spore adhesion.

Differentiation of Appressoria: Evidence for Specialization of the Plasma Membrane

In most, though by no means all, *Colletotrichum* species, differentiation of an appressorium is essential for host penetration. Appressoria develop on an inductive surface when apical growth of the conidial germ tube stops and the tip swells and becomes delimited by a septum. Maturation of the appressorium involves formation of a penetration pore in the base of the cell, deposition of new wall layers, and secretion of extracellular matrix materials (Fig. 4A). The latter probably have a key role in firmly anchoring the appressorium to host surfaces (44). Subsequently, melanin is deposited in a layer of the cell wall close to the plasma membrane (26), although this polymer probably also occurs throughout the wall and extracellular matrix. In some species, the penetration pore becomes surrounded by a funnel-shaped elaboration of the inner wall layer, termed the appressorial cone (Fig. 4A). This structure does not contain chitin or melanin and is continuous with the wall of the penetration peg (40). In other species, appressorial cones are not present but the cell wall instead forms a thickened ring around the penetration pore, similar to the pore wall overlay of *Magnaporthe grisea* (19,28,41,49). Apical growth resumes with the emergence of the penetration peg through the pore. Subsequent penetration of the plant cuticle and cell wall probably involves a combination of mechanical force, in the form of high turgor pressure, and enzymic degradation (19,29). In *M. grisea* appressoria, high turgor is generated through the accumulation of molar concentrations of glycerol, which is retained within the cell because melanin is impermeable to glycerol (12,32). A similar mechanism is likely to operate in *Colletotrichum* appressoria.

The mature appressorium is clearly a highly asymmetric, polarized cell, with an upper, domed region and a flattened basal region containing the complex pore apparatus. Other examples of polarized cells include mammalian epithelial and sperm cells, *Fucus* zygotes, and the growing bud and mating projection in *Saccharomyces cerevisiae* (57). One expression of this polarity is that the plasma membrane of these cells is divided into discrete domains associated with specific proteins. Using a monoclonal antibody, UB27, we have demonstrated that the plasma membrane of *C. lindemuthianum* appressoria is also differentiated into two domains. UB27 recognizes a 48-kDa glycoprotein, designated CLA1 (*Colletotrichum lindemuthianum* Appressorium 1), which has large *O*-linked carbohydrate side-chains (46). After solubilization and phase-separation of appressorial proteins in Triton X-114, CLA1 partitioned into the detergent phase, suggesting that it is a hydrophobic protein, probably an integral membrane component (46). Immunogold labeling showed that CLA1 is only present in the plasma membrane of appressoria and not other fungal cell types. Within appressoria, the glycoprotein is abundant in the plasma membrane lining the upper, domed region of the cell, but appears to be absent from a circular region surrounding the basal penetration pore (Fig. 4B).

Plasmolysis experiments provide further evidence that the plasma membrane is specialized into two domains. *C. lindemuthianum* appressoria, which had penetrated bean epidermal cells to form infection vesicles, were plasmolyzed in 0.85 M potassium nitrate and examined with TEM (Fig. 4C). In previous studies of *C. lindemuthianum* and *Magnaporthe grisea*, the melanized cell wall collapsed (cytorrhysis) when appressoria were placed into hyperosmotic solutions because melanin acts as a semi-permeable barrier that excludes even small solute molecules (18,32,65). We did not observe cytorrhysis, probably because the solute was able to diffuse through the non-melanized walls of the infection vesicle and penetration peg. When appressoria were plasmolyzed, the plasma membrane retracted away from the wall in the basal region of the cell, producing a concave area in the protoplast directly above the penetration pore (Fig. 4C). In contrast, the plasma membrane remained in continuous close contact with the cell wall in the upper, domed region of the cell. This contact suggests that the membrane domain containing the CLA1 glycoprotein adheres more tightly to the cell

Fig. 4. Ultrastructure of *Colletotrichum* appressoria. Bars = 1 µm. A: Mature appressorium of *C. orbiculare* on cucumber, showing appressorial cone (arrowheads) surrounding penetration pore (asterisk), and extracellular matrix (arrows) encrusted with electron-opaque melanin granules. n = nucleus, m = mitochondrion, l = lipid body. B: Basal region of *C. lindemuthianum* appressorium on a polycarbonate membrane, showing immunogold labeling with MAb UB27. The plasma membrane is strongly labeled, but labeling stops abruptly at point indicated by arrows, leaving region around penetration pore (asterisk) and appressorial cone (arrowheads) unlabeled. Reprinted, by permission of the publishers, from Pain et al. (46). C: Asymmetrical plasmolysis of a *C. lindemuthianum* appressorium. Bean epidermal strips were plasmolyzed for 30 min in 0.85 M KNO_3 and processed for TEM as in O'Connell et al. (37) except the fixative solutions were supplemented with 0.7 M KNO_3. Note the plasma membrane has only retracted from the appressorial wall in a small region around the penetration pore. The penetration peg (pp) had formed a vesicle in the underlying epidermal cell.

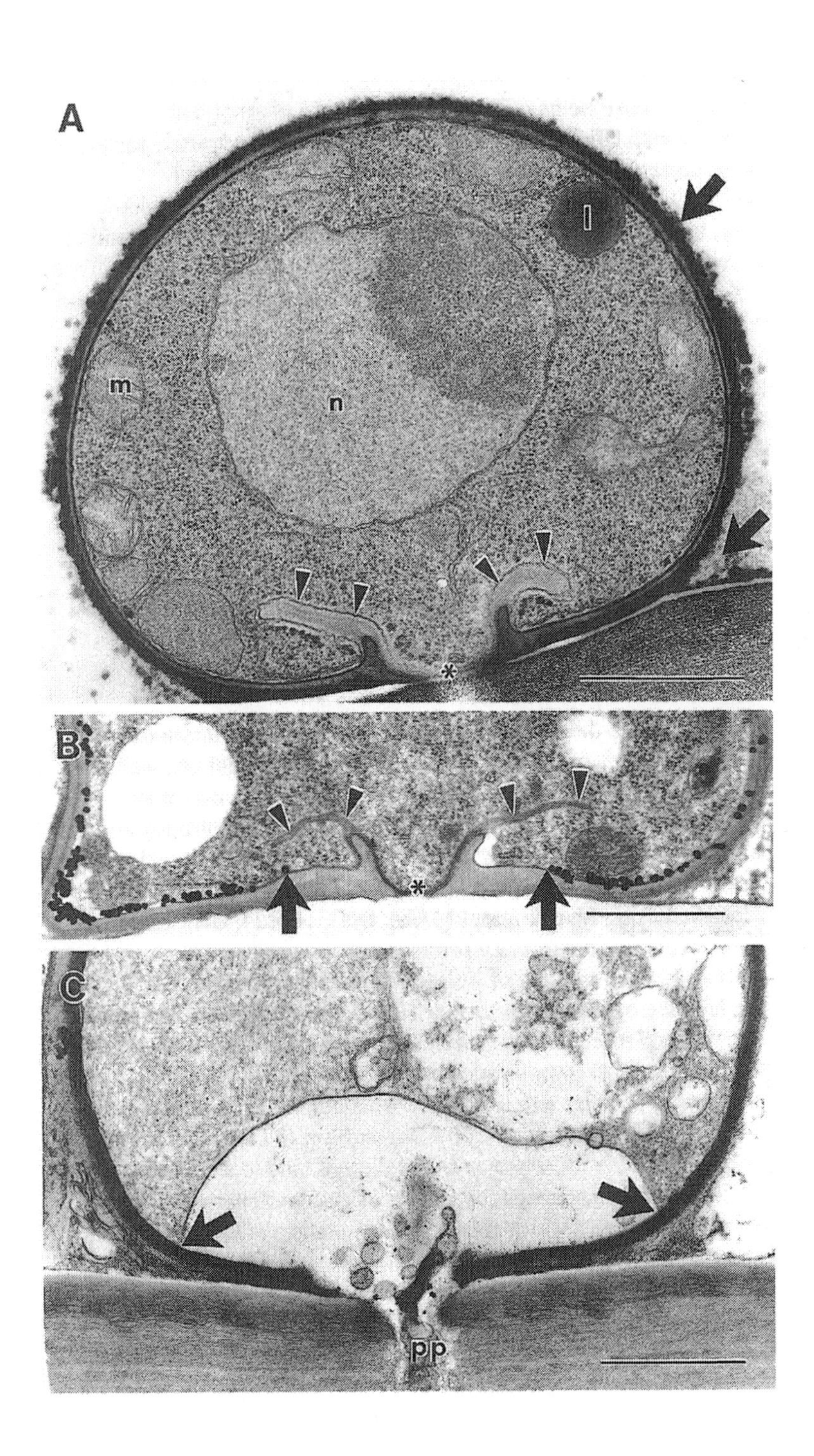
A
l
m
n
B
C
pp

wall. However, since we have not immunolabeled a plasmolyzed appressorium with UB27, it remains unclear whether the retracted portion of membrane corresponds precisely with the domain lacking CLA1.

The CLA1 glycoprotein may possibly function as an integrin-like molecule, binding the plasma membrane to the cell wall (22). The finding that a proportion of CLA1 cannot be solubilized from appressoria by Triton X-114 suggests there may be physical links between this glycoprotein and components of the cell wall and/or cytoskeleton (46). Such interactions could be important for maintaining separation between the two membrane domains and would effectively immobilize CLA1 within the domed part of the cell. The domain lacking CLA1 may represent newly synthesized membrane associated with the emergence of the penetration peg. A change in membrane composition could be an important preparation for subsequent penetration and the establishment of biotrophy.

The Biotrophic Interface: Identification and Sequencing of a Biotrophy-Related Gene

A feature in common between the haustoria formed by obligate biotrophic fungi and the intracellular hyphae of *C. lindemuthianum* is the presence of an interfacial matrix layer separating the fungal cell wall from the invaginated plant plasma membrane (14). This region may play important roles in the establishment and maintenance of biotrophy and the avoidance or suppression of host defense responses. Several novel proteins specifically associated with *C. lindemuthianum* intracellular hyphae have been identified using four different MAbs: UB23, UB24, UB25, and UB50 (42,47). The localization of these proteins was determined by immunofluorescence labeling of isolated intracellular hyphae. UB23 gave punctate labeling of the hyphal surface (Fig. 5A), whereas UB24 and UB25 gave stronger and more uniform surface labeling (Figs 5B,C). In contrast, UB50 appears to label cytoplasmic components of the hyphae (Fig. 5D). In Western blots of proteins extracted from isolated intracellular hyphae, UB23 and UB50 labeled high (>200 kDa) and low (27 kDa) M_r components, respectively, whereas UB24 did not bind to any proteins (42). UB25 recognized a protein epitope in a set of glycoproteins with *N*-linked carbohydrate side-chains which appear to be multimers of a 40.5-kDa subunit (47). TEM-immunogold labeling of infected bean tissue showed that UB23, UB24, and UB25 specifically label the walls of intracellular hyphae and the surrounding interfacial matrix (47; R. O'Connell, unpublished). In the case of UB25, labeling was specific to the biotrophic phase of growth inside living host cells. This specificity was confirmed by immunofluorescence labeling of infected epidermal strips prepared from bean leaves, in which

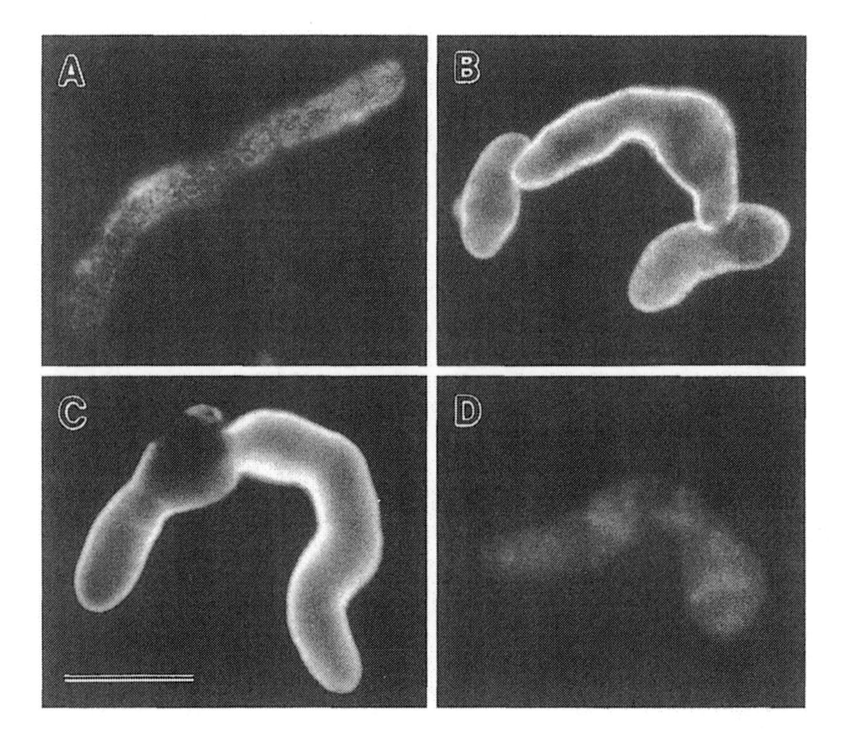

Fig. 5. Immunofluorescence labeling of *C. lindemuthianum* intracellular hyphae (isolated from bean leaves) using four different monoclonal antibodies. Bar = 10 µm. UB23 gives punctate labeling of hyphal surfaces (A). UB24 (B) and UB25 (C) label hyphal surfaces uniformly. UB50 labels intracellular components (D).

infection vesicles and primary hyphae were labeled by UB25, but secondary necrotrophic hyphae were not (48). The specificity of UB25 for intracellular hyphae has allowed these structures to be purified by immunomagnetic separation (43).

MAb UB25 was used to immunoscreen a cDNA library prepared from infected bean tissue, allowing the sequence of the corresponding antigen to be elucidated (48). The full-length cDNA isolated has been designated *CIH1 (Colletotrichum* Intracellular Hypha 1). *In vitro* transcription and translation of *CIH1* and transfection of mammalian COS cells showed that UB25 recognized the expressed product in both procedures, confirming that the clone was a true positive. Southern analysis of bean and *C. lindemuthianum* genomic DNA showed that the CIH1 glycoprotein is fungally encoded, while Northern analysis confirmed *CIH1* was only expressed *in planta* (48).

The predicted open reading frame for *CIH1* contains a putative signal peptide at the N-terminus, which is likely to be removed from the mature protein during processing. This is consistent with CIH1 being a secreted product that is present in the fungal wall and interfacial matrix. Although there is no overall homology with other protein sequences in databases, the predicted amino acid sequence appears to be made up of two domains. The N-terminal domain is very rich in proline residues, which account for 30% of the amino acids, and within this proline-rich region there is a series of short repetitive motifs of three types (LPEP, YKPK, and VEGP). The C-terminal domain contains two potential sites for *N*-glycosylation, which corroborates evidence from Western blotting studies (47), together with six cysteine residues, which are potential sites for the formation of disulphide bridges. The C-terminal domain also contains two long, partial repeats of 43 amino acids. These repeats show strong sequence similarity to repeats found in lytic enzymes from Gram-positive bacteria and bacteriophage, which are thought to be involved in the binding of these enzymes to bacterial cell wall components (5). An intriguing possibility is that these regions of the CIH1 glycoprotein are involved in interactions with other fungal wall components or with plant cell surface polysaccharides or glycoproteins.

Thus, the CIH1 glycoprotein recognized by UB25 is a fungal proline-rich glycoprotein, which forms a multimeric structure in the interfacial matrix between the intracellular hyphae and the plant plasma membrane. CIH1 has some similarities to proline-rich proteins (PRPs) and hydroxyproline-rich glycoproteins (HRGPs) that are important structural components of plant cell walls. These characteristically have repetitive motifs, although they differ from those found in the CIH1 polypeptide. In addition, although the HRGPs are extensively *O*-glycosylated at hydroxyproline and serine residues, they lack *N*-linked carbohydrate side chains, which are present in the CIH1 glycoprotein. PRPs and HRGPs can also form cross-linked structures, mediated by oxidative cross-linking. The predicted sequence of CIH1 contains three tyr-lys motifs, which have been implicated in the formation of these cross-links (24). Oxidative cross-linking could thus have a role in the formation of the multimers of the CIH1 glycoprotein which are present in infected plants.

Since CIH1 is present at the biotrophic interface between intracellular hyphae and the invaginated host plasma membrane, it may have a role in the establishment and maintenance of biotrophy. CIH1 may function to prevent the plant recognizing the fungus, perhaps by mimicking plant PRPs and HRGPs and so presenting a 'pseudo'-plant cell wall to the host. The CIH1 glycoprotein could also act as a barrier to host defense molecules or restrict the movement of fungal elicitors, enzymes, or toxins. The avoidance, or suppression, of host defense responses such as hypersensitive cell death and callose deposition is a key feature of intracellular biotrophic interactions

(16), and CIH1 could contribute to the ability of *C. lindemuthianum* to escape such responses in bean tissues.

New Approaches for the Study of *Colletotrichum* Infection Processes

By raising MAbs to the isolated infection structures of *C. lindemuthianum*, we have been able to identify a series of stage-specific glycoproteins in spores, germ tubes, appressoria, and intracellular hyphae. These results led to the isolation of a fungal gene that could play an important role in the pathogenicity of *C. lindemuthianum* and other *Colletotrichum* species. In the future, this cell biological approach to the isolation of fungal pathogenicity genes could be extended to include the use of phage display antibodies. Phage antibodies are libraries of human antibody fragments displayed on bacteriophage, in which the antibody protein has been mutagenized *in vitro* to produce a very large repertoire of monoclonal antibodies (64). Phage antibodies eliminate the need for animals and cell culture, do not suffer from problems of immunodominance, and can be raised to virtually any type of molecule, including those with low immunogenicity.

Recent advances in light microscopy, combined with the introduction of new fluorescent probes, have greatly enhanced our ability to visualize fungal infection processes and host responses. In conventional fluorescence microscopy of thick specimens, out-of-focus blur seriously degrades the image. However, in confocal laser scanning microscopy (CLSM), light from above or below the plane of focus is excluded, resulting in improved image contrast and resolution (13). Using conventional light microscopy techniques, it is often difficult to distinguish whether fungal infection hyphae are intracellular, intramural, subcuticular, or growing on the plant surface. The ability of CLSM to collect optical sections through intact tissues can rapidly resolve this question, without the need to fix, embed in wax or resin and cut microtome sections (61). One limitation to the use of CLSM is that optical sections cannot be obtained more than two to three cell layers into plant tissue without the image becoming degraded. The new technique of two-photon microscopy offers much better depth penetration than CLSM (13), allowing fungal infection hyphae to be imaged inside thick tissue samples.

Optical sectioning is non-invasive. It can, therefore, be used to monitor dynamic processes in living cells, using fluorescent vital stains specific for different cell components and organelles. Two-photon microscopy is particularly suitable for live-cell imaging because it causes less phototoxicity and dye bleaching than CLSM. The green fluorescent protein (GFP) is now widely used as a transgenic reporter molecule to observe temporo-spatial patterns of gene expression and protein targeting in living cells (25). Using GFP under the control of a constitutive promoter, the extent and time-course of fungal colonization of host tissue can be observed directly, without any staining. These major advances in imaging technology promise to provide many new insights into *Colletotrichum* infection processes in the years ahead.

Acknowledgements

The authors would like to thank Naomi Pain, Caroline Nash, Richard Pring, Olu Latunde-Dada, Gwyneth Jones, Phillip Wharton, and Katie Hutchison for their many contributions to our understanding of *Colletotrichum* infection processes. IACR-Long Ashton Research Station receives grant-aided support from the Biotechnology and Biological Sciences Research Council of the U.K.

Literature Cited

1. Bailey, J.A. and Deverall, B.J. 1971. Formation and activity of phaseollin in the interaction between bean hypocotyls (*Phaseolus vulgaris*) and physiological races of *Colletotrichum lindemuthianum*. Physiol. Plant Pathol. 1:435-449.
2. Bailey, J.A., Nash, C., Morgan, L.W., and O'Connell, R.J. 1996. Molecular taxonomy of *Colletotrichum* species causing anthracnose on the Malvaceae. Phytopathology 86:1076-1083.
3. Bailey, J.A., Nash, C., O'Connell, R.J., and Skipp, R.A. 1990. Infection process, host specificity and taxonomic relationships of a *Colletotrichum* species causing anthracnose disease of cowpea, *Vigna unguiculata*. Mycol. Res. 94:810-814.
4. Bailey, J.A., O'Connell, R.J., Pring, R.J., and Nash, C. 1992. Infection strategies of *Colletotrichum* species. Pages 88-120 in: *Colletotrichum*: Biology, Pathology and Control. J.A. Bailey and M.J. Jeger, eds. CAB International, Wallingford.
5. Birkeland, N.-K. 1994. Cloning, molecular characterization, and expression of the genes encoding the lytic functions of lactococcal bacteriophage φLC3: Dual lysis system of modular design. Can. J. Microbiol. 40:658-665.
6. Bobichon, H., Gache, D., and Bouchet, P. 1994. Ultrarapid cryofixation of *Candida albicans*: evidence for a fibrillar reticulated external layer and mannan channels within the cell wall. Cryo Lett. 15:161-172.
7. Brown, G.E. 1977. Ultrastructure of penetration of ethylene-degreened Robinson tangerines by *Colletotrichum gloeosporioides*. Phytopathology 67:315-320.

8. Casanova, M. and Chaffin, W. L. 1991. Cell wall glycoproteins of *Candida albicans* as released by different methods. J. Gen. Microbiol. 137:1045-1051.
9. Casanova, M., Lopez-Ribot, J. L., Martinez, J. P., and Sentandreu, R. 1992. Characterization of cell wall proteins from yeast and mycelial cells of *Candida albicans* by labeling with biotin: Comparison with other techniques. Infect. Immun. 60:4898-4906.
10. Chau, K.F. and Alvarez, A.M. 1983. A histological study of anthracnose on *Carica papaya*. Phytopathology 73:1113-1116.
11. Coates, L.M., Muirhead, I.F., Irwin, J.A., and Gowanlock, D.H. 1993. Initial infection processes by *Colletotrichum gloeosporioides* on avocado fruit. Mycol. Res. 97:1363-1370.
12. De Jong, J.C., McCormack, B.J., Smirnoff, N., and Talbot, N.J. 1997. Glycerol generates turgor in rice blast. Nature 389:244-245.
13. Gilroy, S. 1997. Fluorescence microscopy of living plant cells. Annu. Rev. Plant Physiol. Plant Mol. Biol. 48:165-190.
14. Green, J.R., Pain, N.A., Cannell, M.E., Jones, G.L., Leckie, C.P., McCready, S., Mendgen, K., Mitchell, A.J., Callow, J.A., and O'Connell, R.J. 1995. Analysis of differentiation and development of the specialized infection structures formed by biotrophic fungal plant pathogens using monoclonal antibodies. Can. J. Bot. 73:S408-S417.
15. Hazen, K.C. and Hazen, B.W. 1992. Hydrophobic surface protein masking by the opportunistic fungal pathogen *Candida albicans*. Infect. Immunol. 60:1499-1508.
16. Heath, M.C. and Skalamera, D. 1997. Cellular interactions between plants and biotrophic fungal parasites. Adv. Bot. Res. 24: 195-225.
17. Hoch, H.C. 1986. Freeze-substitution in fungi. Pages 183-212 in: Electron Microscopy of Microorganisms. H.C. Aldrich and W.J. Todd, eds. Plenum, New York.
18. Howard, R.J, and Ferrari, M.A. 1989. Role of melanin in appressorium function. Exp. Mycol. 13:403-418.
19. Howard, R.J., and Valent, B. 1996. Breaking and entering: host penetration by the fungal rice blast pathogen *Magnaporthe grisea*. Annu. Rev. Microbiol. 50:491-512.
20. Irwin, J.A.G., Trevorrow, P.R., and Cameron, D.F. 1984. Histopathology of compatible interactions involving biotypes of *Colletotrichum gloeosporioides* that cause anthracnose of *Stylosanthes* spp. Aust. J. Bot. 32:631-640.
21. Jones, G.L., Bailey, J.A., and O'Connell, R.J. 1995. Sensitive staining of fungal extracellular matrices using colloidal gold. Mycol. Res. 99:567-573.
22. Kaminskyj, S.G.W. and Heath, I.B. 1995. Integrin and spectrin homologs, and cytoplasm-wall adhesion in tip growth. J. Cell Sci. 108:849-856.

23. Kershaw, M.J. and Talbot, N.J. 1998. Hydrophobins and repellants: Proteins with fundamental roles in fungal morphogenesis. Fungal Genet. Biol. 23:18-33.
24. Kieliszewski, M.J. and Lamport, D.T.A. 1994. Extensin: Repetitive motifs, functional sites, post-translational codes, and phylogeny. Plant J. 5:157-172.
25. Köhler, R.H. 1998. GFP for *in vivo* imaging of subcellular structures in plant cells. Trends Plant Sci. 3:317-320.
26. Kubo, Y. and Furusawa, I. 1986. Localization of melanin in appressoria of *Colletotrichum lagenarium*. Can. J. Microbiol. 32:280-282.
27. Latunde-Dada, A.O., Bailey, J.A., and Lucas, J.A. 1997. Infection process of *Colletotrichum destructivum* O'Gara from lucerne (*Medicago sativa* L). Eur. J. Plant Pathol. 103:35-41.
28. Latunde-Dada, A.O., O'Connell, R.J., Nash, C., Pring, R.J., Lucas, J.A., and Bailey, J.A. 1996. Infection process and identity of the hemibiotrophic anthracnose fungus (*Colletotrichum destructivum* O'Gara) from cowpea (*Vigna unguiculata* (L.) Walp.). Mycol. Res. 100:1133-1141.
29. Mendgen, K. and Deising, H. 1993. Infection structures of fungal plant pathogens–a cytological and physiological evaluation. New Phytol. 124:193-213.
30. Mercure, E.W., Kunoh, H., and Nicholson, R.L. 1995. Visualization of materials released from adhered, ungerminated conidia of *Colletotrichum graminicola*. Physiol. Mol. Plant Pathol. 46:121-135.
31. Mercure, E.W., Leite, B., and Nicholson, R.L. 1994. Adhesion of ungerminated conidia of *Colletotrichum graminicola* to artificial hydrophobic surfaces. Physiol. Mol. Plant Pathol. 45:421-440.
32. Money, N.P. 1997. Mechanism linking cellular pigmentation and pathogenicity in rice blast disease. Fungal Genet. Biol. 22:151-152.
33. Morin, L., Derby, J.-A.L., and Kokko, E.G. 1996. Infection process of *Colletotrichum gloeosporioides* f. sp. *malvae* on Malvaceae weeds. Mycol. Res. 100:165-172.
34. Mould, M.J.R., Boland, G.J., and Robb, J. 1991. Ultrastructure of the *Colletotrichum trifolii-Medicago sativa* pathosystem. II. Post-penetration events. Physiol. Mol. Plant Pathol. 38:195-210.
35. Nair, J. and Corbin, J.B. 1981. Histopathology of *Pinus radiata* seedlings infected by *Colletotrichum acutatum* f. sp. *pinea*. Phytopathology 71:777-783.
36. O'Connell, R.J. and Bailey, J.A. 1991. Hemibiotrophy in *Colletotrichum lindemuthianum*. Pages 211-222 in: Electron Microscopy of Plant Pathogens. K. Mendgen and D.-E. Lesemann, eds. Springer-Verlag, Berlin.

37. O'Connell, R.J., Bailey, J.A., and Richmond, D.V. 1985. Cytology and physiology of infection of *Phaseolus vulgaris* by *Colletotrichum lindemuthianum*. Physiol. Plant Pathol. 27:75-98.
38. O'Connell, R.J., Pain, N.A., Bailey, J.A., Mendgen, K., and Green, J.R. 1996. Use of monoclonal antibodies to study differentiation of *Colletotrichum* infection structures. Pages 79-97 in: Histology, Ultrastructure and Molecular Cytology of Plant-Microorganism Interactions. M. Nicole and V. Gianinazzi-Pearson, eds. Kluwer, Dordrecht.
39. O'Connell, R.J., Pain, N.A., Hutchison, K.A., Jones, G.L., and Green, J.R. 1996. Ultrastructure and composition of the cell surfaces of infection structures formed by the fungal plant pathogen *Colletotrichum lindemuthianum*. J. Microsc. 181:204-212.
40. O'Connell, R.J. and Ride, J.P. 1990. Chemical detection and ultrastructural localization of chitin in cell walls of *Colletotrichum lindemuthianum*. Physiol. Mol. Plant Pathol. 37:39-53.
41. O'Connell, R.J., Uronu, A.B., Waksman, G., Nash, C., Keon, J.P.R., and Bailey, J.A. 1993. Hemibiotrophic infection of *Pisum sativum* by *Colletotrichum truncatum*. Plant Pathol. 42:774-783.
42. Pain, N.A. 1994. Immunological analysis of infection structures and the biotrophic interface formed in the *Colletotrichum*-bean interaction. Ph.D. Thesis, University of Birmingham.
43. Pain, N.A., Green, J.R., Gammie, F., and O'Connell, R.J. 1994. Immunomagnetic isolation of viable intracellular hyphae of *Colletotrichum lindemuthianum* (Sacc. & Magn.) Briosi & Cav. from infected bean leaves using a monoclonal antibody. New Phytol. 127:223-232.
44. Pain, N.A., Green, J.R., Jones, G.L., and O'Connell, R.J. 1996. Composition and organisation of extracellular matrices around germ-tubes and appressoria of *Colletotrichum lindemuthianum*. Protoplasma 190:119-130.
45. Pain, N.A., O'Connell, R.J., Bailey, J.A., and Green, J.R. 1992. Monoclonal antibodies which show restricted binding to four *Colletotrichum* species: *C. lindemuthianum, C. malvarum, C. orbiculare* and *C. trifolii*. Physiol. Mol. Plant Pathol. 41:111-126.
46. Pain, N.A., O'Connell, R.J., and Green, J.R. 1995. A plasma membrane-associated protein is a marker for differentiation and polarisation of *Colletotrichum lindemuthianum* appressoria. Protoplasma 188:1-11.
47. Pain, N.A., O'Connell, R.J., Mendgen, K., and Green, J.R. 1994. Identification of glycoproteins specific to biotrophic intracellular hyphae formed in the *Colletotrichum lindemuthianum*-bean interaction. New Phytol. 127:233-242.

48. Perfect, S.E., O'Connell, R.J., Green, E.F., Doering-Saad, C., and Green, J.R. 1998. Expression cloning of a fungal proline-rich glycoprotein specific to the biotrophic interface formed in the *Colletotrichum*-bean interaction. Plant J. 15:273-279.
49. Politis, D.J. and Wheeler, H. 1973. Ultrastructural study of penetration of maize leaves by *Colletotrichum graminicola*. Physiol. Plant Pathol. 3:465-471.
50. Pring, R.J., Nash, C., Zakaria, M., and Bailey, J.A. 1995. Infection process and host range of *Colletotrichum capsici*. Physiol. Mol. Plant Pathol. 46:137-152.
51. Roberts, R.G. and Snow, J.P. 1984. Histopathology of cotton boll rot caused by *Colletotrichum capsici*. Phytopathology 74:390-397.
52. Sela-Buurlage, M.B., Epstein, L., and Rodriguez, R.J. 1991. Adhesion of ungerminated *Colletotrichum musae* conidia. Physiol. Mol. Plant Pathol. 39:345-352.
53. Sénéchal, Y., Sanier, C., Gohet, E., and D'Auzac, J. 1987. Différents modes de pénétration du *Colletotrichum gloeosporioides* dans les feuilles d'*Hevea brasiliensis*. C.R. Acad. Sci. Paris 305:537-542.
54. Sherriff, C., Whelan, M.J., Arnold, G.M., Lafay, J-F., Brygoo, Y., and Bailey, J.A. 1994. Ribosomal DNA sequence analysis reveals new species groupings in the genus *Colletotrichum*. Exp. Mycol. 18:121-138.
55. Shimoi, H., Iimura, Y., and Obata, T. 1995. Molecular cloning of CWP1: A gene encoding a *Saccharomyces cerevisiae* cell wall protein solubilised with *Rorobacter faecitabidus* protease I. J. Biochem. 118:302-311.
56. Skipp, R.A., Beever, R.E., Sharrock, K.R. Rikkerink, E.H.A., and Templeton, M.D. 1995. *Colletotrichum*. Pages 119-143 in: Pathogens and Host Parasite Specificity in Plant Disease: Histopathological, Genetic, Biochemical and Molecular Basis. U.S. Singh, K. Kohmoto and R.P. Singh, eds. Elsevier Science Ltd, Oxford.
57. Stafford, C.J., Green, J.R., and Callow, J.A. 1992. Organisation of glycoproteins into plasma membrane domains on *Fucus serratus* eggs. J. Cell Sci. 101:437-448.
58. Van der Bruggen, P. and Maraite, H. 1987. Histopathology of cassava anthracnose disease caused by *Colletotrichum gloeosporioides* f. sp. *manihotis*. Parasitica 43:3-21.
59. Van Dyke, C.G. and Mims, C.W. 1991. Ultrastructure of conidia, conidium germination, and appressorium development in the plant pathogenic fungus *Colletotrichum truncatum*. Can. J. Bot. 69:2455-2467.
60. Walker, J.C. 1921. Onion smudge. J. Agric. Res. 20:685-721.

61. Wei, Y.D., Byer, K.N., and Goodwin, P.H. 1997. Hemibiotrophic infection of round-leaved mallow by *Colletotrichum gloeosporioides* f. sp. *malvae* in relation to leaf senescence and reducing reagents. Mycol. Res. 101:357-364.
62. Wharton, P.S. and Julian, A.M. 1996. A cytological study of compatible and incompatible interactions between *Sorghum bicolor* and *Colletotrichum sublineolum*. New Phytol. 134:25-34.
63. Wijesundera, R.L.C., Bailey, J.A., and Byrde, R.J.W. 1984. Production of pectin lyase by *Colletotrichum lindemuthianum* in culture and in infected bean *(Phaseolus vulgaris)* tissue. J. Gen. Microbiol. 130:285-290.
64. Winter, G., Griffiths, A.D., Hawkins, R.E., and Hoogenboom, H.R. 1994. Making antibodies by phage display technology. Annu. Rev. Immunol. 12:433-455.
65. Wolkow, P.M., Sisler, H.D., and Vigil, E.L. 1983. Effect of inhibitors of melanin biosynthesis on structure and function of appressoria of *Colletotrichum lindemuthianum*. Physiol. Plant Pathol. 22:55-71.
66. Young, D.H. and Kauss, H. 1984. Adhesion of *Colletotrichum lindemuthianum* spores to *Phaseolus vulgaris* hypocotyls and to polystyrene. Appl. Environ. Microbiol. 47:616-619.
67. Zakaria, M. 1995. *Colletotrichum* diseases of forest tree nurseries in Malaysia. Ph.D. Thesis. University of Bristol.

Chapter 6

Early Molecular Communication Between *Colletotrichum gloeosporioides* and Its Host

P.E. Kolattukudy, Yeon-Ki Kim, Daoxin Li, Zhi-Mei Liu, and Linda Rogers

Molecular communication begins as soon as a fungal conidium lands on a plant surface. Since this contact is often on the plant cuticle, cuticular components are ideally located to play a major role in plant-fungus interactions. The consequences of this early interaction can be crucial for the survival of both the pathogen and the host. In this chapter we examine the molecular details of the different phases of the early communication between *Colletotrichum gloeosporioides* and its hosts. We present a working hypothesis that proposes discrete stages in this interaction that involve sequential activation of transcription of different sets of genes that ultimately leads to penetration of the pathogen into its host. Experimental evidence in support of the proposed stages comes not only from studies on *Colletotrichum* but also on other fungi, and the suggestion that similar events also occur during *Colletotrichum* interactions with plants need further experimental verification. Nevertheless, this hypothesis provides a framework for elucidating the molecular details that characterize each stage and thus understanding the early events in the plant-fungus interaction. As a first approximation, the stages (Fig. 1) include relief of self-inhibition by the diffusion of self-inhibitors into the cuticle, followed by the host surface contact-induced events that include expression of the early genes. These events prime the conidia to respond to host signals that induce germination and appressoria formation by the transcriptional activation of another set of genes. Appressoria then form penetration pegs that penetrate the cuticular and cell wall barriers of the host with the assistance of extracellular hydrolases produced as a result of transcriptional activation of their genes. In the following sections we provide a glimpse of some of the molecular events involved in these early processes in the interaction of *Colletotrichum* with its host.

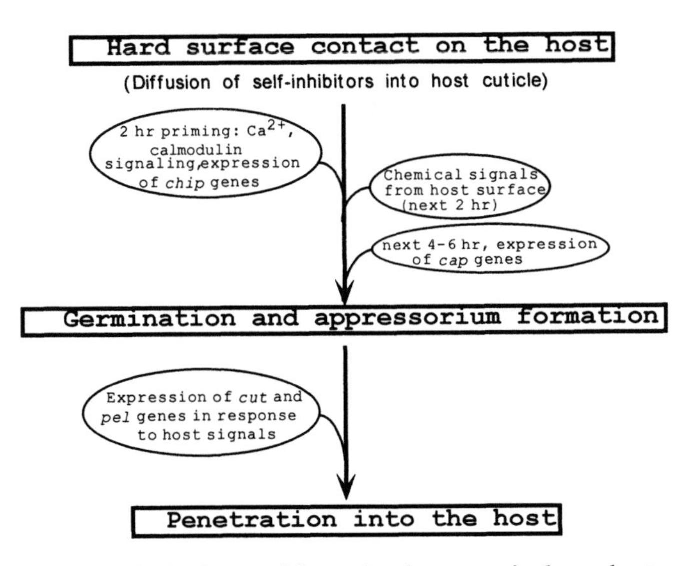

Fig. 1. Hypothetical scheme of the molecular events in the early stages of fungal infection.

Relief of Self-Inhibition

Conidia of many, if not most, fungal species have chemicals that prevent them from undergoing germination and appressorium formation until the conidia are well dispersed in a favorable environment for colonization of their hosts. *Colletotrichum gloeosporioides* was reported to have an alkaloid-like self-inhibitor (37). Later, chromatographic isolation and chemical identification led to a proposed cyclic lipid structure (42). However, total synthesis of the proposed molecule yielded a biologically inactive material. Further studies on the self-inhibitor isolated from *C. gloeosporioides* showed that it was a mixture of indole derivatives that are soluble in organic solvents (59). Thus, such self-inhibitors of *C. gloeosporioides* could diffuse into the plant cuticle and thus relieve self-inhibition, as recently indicated for *Magnaporthe grisea* (21).

Priming by Hard Surface Contact

Conidia of many fungi are known to require contact with a hard surface before they can respond to the host signals (12). *C. gloeosporioides* can germinate on hydrophilic soft agar surfaces or hydrophobic soft petrolatum, but the germ tube does not differentiate to form an appressorium on either of these soft surfaces. When the chemical signals from the host were found to induce appressorium formation, contact with a hard surface for about 2 hr

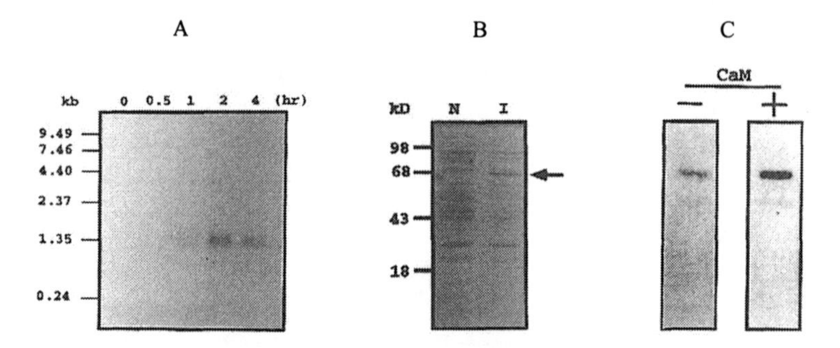

Fig. 2. A: RNA blot showing induction of *cam* gene by hard surface contact; B: SDS-PAGE showing expression of the putative calmodulin-dependent kinase (CgCMK) from *C. gloeosporioides* in *E. coli*; C: in-gel kinase assay demonstrating calmodulin-dependent kinase activity assayed as described (28). N, non-induced; I, induced.

was required before the conidia were capable of responding to the chemicals (14,25). Two approaches were taken to probe into the molecular events during this priming period: one tested whether a plausible signaling occurred and whether candidate genes were induced during this period (29), and the other examined the nature of the genes uniquely expressed during the priming period (38).

Ca^{+2}-Calmodulin Signaling During Hard Surface Contact

Hard surface contact may be analogous to the touch response in higher plants where transcriptional activation of calmodulin (*cam*)-like genes were triggered by touch stimuli (1). Therefore, we tested whether hard surface contact triggers *cam* gene expression. First, we cloned *cam* from *C. gloeosporioides* by PCR using primers designed from *cam* of other filamentous fungi such as *Neurospora crassa* and *Aspergillus oryzae* (41,64). The single product obtained revealed a 450-bp open reading frame encoding a protein of 149 amino acids. At the nucleotide level the *cam* from *C. gloeosporioides* showed 87% and 88% identity with those of *N. crassa* (41) and *C. trifolii* (GenBank accession no. U15993), respectively. Southern analysis indicated the presence of a single *cam* gene in the *C. gloeosporioides* genome. RNA blot analysis showed a single band that hybridized with *cam* cDNA at 1300 bp (29). The *cam* transcript level increased 11-fold reaching a maximum after a 2-hr contact with a hard surface and subsequently decreased (Fig. 2A). Ethephon enhanced the transcript level to a maximum of 13-fold after a 2-hr contact with a hard surface. That calmodulin (CaM) is involved in conidial germination and appressorium formation was indicated by the finding that a CaM antagonist, compound 48/80, severely inhibited both germination and appressorium formation (Fig. 3), as also seen with the *cam* transcript level and inhibition by the antagonist in *C. trifolii* (3,8).

If CaM is involved in conidial germination and appressorium formation, CaM-dependent kinase (CaMK) may also be involved in this process. To test for such a possibility, we first used an RT-PCR approach to isolate a DNA fragment encoding a segment of CaMK. RT-PCR using primers based on the conserved DLKPEN from domain VI and DIWSIG from domain IX of fungal CaMKs (32,47) and RNA from hard surface-treated conidia yielded a 190-bp product (29). Cloning and sequencing revealed that one of the clones showed homology to CaMKs. Screening of a cDNA library prepared from hard surface-treated conidia with the 190-bp CaMK probe gave a clone representing *C. gloeosporioides* CaMK, designated CgCMK. This clone revealed an open reading frame that would encode a protein of 420 amino acids. Alignment of CgCMK to optimize homology with the predicted amino acid sequences of CaMK from *Aspergillus*, yeast (YCMK1 and YCMKII), and *Metarrhizium* showed that CgCMK shared 31, 28, and 32% identity, respectively. CgCMK does contain a CaM-binding domain and the 11 conserved kinase domains. That the cloned cDNA represents CaMK was shown by expression of the protein in *E. coli* as a thio-fusion protein (Fig. 2B) and in-gel kinase assay of the protein separated by SDS-PAGE, showing CaM-dependent phosphorylation (Fig. 2C).

If CaM and CgCMK are involved in germination and appressorium formation, selective inhibitors of CaMK should block these processes. In fact, germination and appressorium formation were severely inhibited by KN93, a selective inhibitor of CaMK (Fig. 3). To test at which stage KN93 of these proteins increased during the early phase (2 hr) of hard surface contact. KN93 inhibited the phosphorylation of these proteins, strongly supporting the notion that CaMK is involved in the phosphorylation of these proteins, a process that is important for germination and appressorium formation (29).

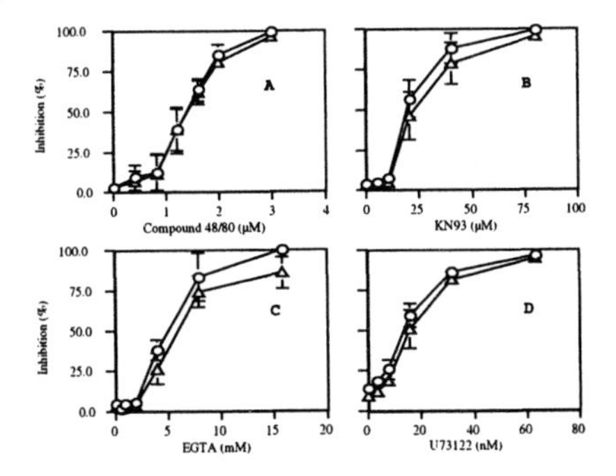

Fig. 3. Inhibition of germination (Δ) and appressorium formation (O) by compound 48/80, KN93, EGTA, and U73122 (7). Conidia were incubated in 10 μm ethephon on glass plates.

Inhibition of CaMK by KN93 not only decreased the frequency of appressorium formation, but also affected the structure of the appressoria that formed (Fig. 4). In controls, well-rounded, melanized dark brown appressoria were found. With about 20 μM KN93, <50% of the conidia formed appressorium-like structures, which showed little melanization. In the absence of the limiting structural component, melanin, the poorly rounded appressorium-like structures produced long germ tube-like structures and appressorium-like structures arising from the original appressorium, as previously observed with melanin-deficient mutants (6,33) and with the protein phosphatase inhibitor, calyculin A (15). That KN93 inhibited steps prior to the formation of scytalone, a key intermediate in melanin biosynthesis, was shown by the observation that in the presence of both scytalone and KN93, melanization was restored. Exogenous scytanone also prevented the formation of the secondary structures from the primary appressorium, supporting the hypothesis that such secondary structures are produced because of the absence of the confining melanin structural barrier. However, the inhibition of germination and appressorium formation by KN93 was not reversed by scytalone. Thus, the target of KN93 involved in the germination and appressorium formation is probably different from the target in the melanin synthesis pathway. It is likely that CaMK is involved in the regulation of the polyketide synthase step, the likely rate-limiting step in the melanin synthesis pathway. CaM involvement in the regulation of aflatoxin production was recently reported (54). Involvement of CaM in the regulation of biosynthesis of secondary metabolites may be more widespread than currently realized.

If CaMK is involved in the early phase of interaction of *C. gloeosporioides* with its host, CaMK inhibitors should inhibit any protein phosphorylation during the hard surface contact. When the nucleotide pools in the conidia were labeled with ^{32}P before placing the conidia on glass plates, and the resulting protein phosphorylation was examined by SDS-PAGE, two proteins of 18 kDa and 43 kDa were more selectively labeled after treatment with hard surface and ethephon. Phosphorylation of these proteins increased during the early phase (2 hr) of hard surface contact.

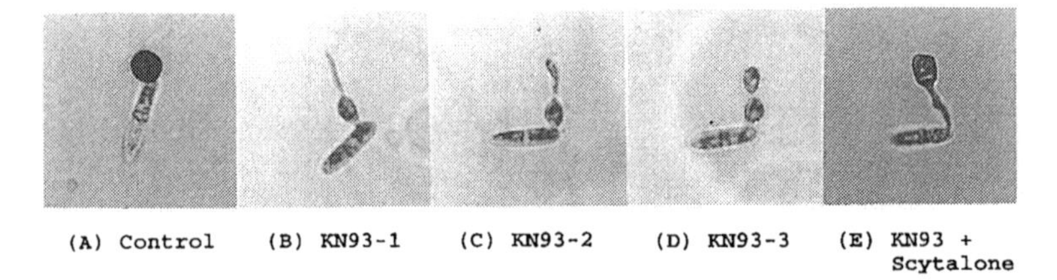

Fig. 4. Effect of KN93 treatment on the morphology of appressoria and recovery of melanization by scytalone.

KN93 inhibited the phosphorylation of these proteins, strongly supporting the notion that CaMK is involved in the phosphorylation of these proteins, a process that is important for germination and appressorium formation (29).

If CaM and CaMK signaling are involved in triggering germination and appressorium formation in *C. gloeosporioides*, involvement of Ca^{+2} is to be expected. Prevention of participation of exogenous Ca^{+2} by inclusion of a chelator (EGTA) in the medium severely inhibited germination and appressorium formation (Fig. 3). Release of internal Ca^{+2} sources by IP_3 generated by phospholipase C may also be involved in germination and appressorium formation. In fact, an inhibitor of phospholipase C, U73122, at nanomolar concentrations severely inhibited both. Dioctanoylglycerol did not induce appressorium formation by conidia of *C. gloeosporioides*, unlike that reported for *M. grisea* (58), suggesting that the effect of the phospholipase C inhibitor probably involves IP_3 rather than diacylglycerol, the other product of phospholipase C. Both EGTA and U73122 inhibited germination and appressorium formation only if they were added during the early (a few hrs) phase of hard surface contact, and appressorium formation was more sensitive to inhibition than germination. Thus, Ca^{+2}, CaM, and CaMK signaling probably play a highly significant role in the early phase of interaction between *C. gloeosporioides* and its host.

Colletotrichum Hard Surface-Induced Protein (Chip) Genes

One way to explore the molecular events involved in the early phase of contact of the conidium with a hard surface is to examine gene expression triggered by the hard surface contact. A differential display approach was used to detect the unique transcripts that appeared in the conidia of *C. gloeosporioides* during a 2-hr contact with hard surface (38) (Fig. 5A). DNA from the differential display representing the up-regulated genes were sequenced and used for northern blot analysis to confirm the induction by hard surface contact. Such DNA fragments were used to identify clones in a cDNA library made with RNA from hard surface-treated conidia. The isolated cDNA clones representing genes whose expression was induced by hard surface contact were sequenced. This approach yielded seven unique clones representing genes designated *chip* genes. *Chip1* showed an open reading frame that would encode a 147-amino acid protein of 16.2 kDa. This protein shows a very high homology to ubiquitin-conjugating enzymes from a variety of organisms from yeast to human (Fig. 5B); therefore, it is designated $UBC1_{cg}$ Northern blot analysis showed that the $ubc1_{cg}$ transcript of 1 kb was detectable within 2 hr of contact of the conidia with the hard surface; the level increased up to 6 hr and then decreased (Fig. 5C).

Ethylene, the ripening hormone of the host, that is known to induce germination and appressorium formation as indicated later, enhanced $ubc1_{cg}$ expression. The maximal level reached with this hormone on hard surface was much higher (six-fold) than that reached on hard surface without

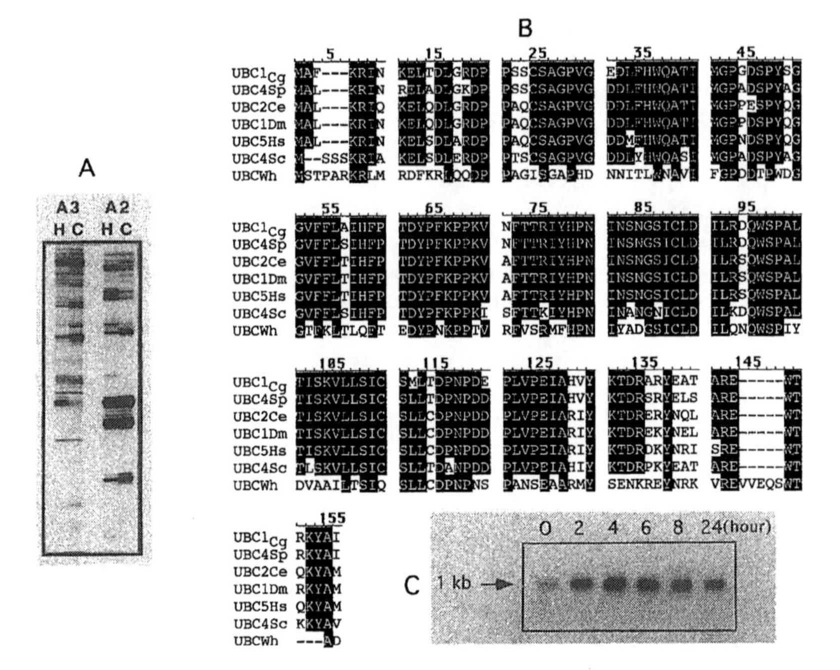

Fig. 5. A: Area of a differential display gel showing the amplified products obtained with primer combinations of oligo (dT) primer HT11A and arbitrary 5' decamers H-AP2 (A2) or H-AP3 (A3) using as templates cDNAs derived from conidia on hard surface for 2 hr (H) and control (C, nontreated conidia). B: Homology comparison of UBC1$_{cg}$ with UBC4Sp of *S. pombe*, UBC2Ce of *C. elegans*, UBCDm of *Drosophilia*, UBCHs of human, UBC4Sc of *Saccharomyces cerevisiae*, and UBCWh of wheat. Homologous residues are shaded. C: Northern blots showing time course of induction of *ubc1*$_{Cg}$ transcript by hard surface contact in *C. gloeosporioides* conidia.

ethylene (two-fold). Southern blot analysis showed that the *C. gloeosporioides* genome contains one copy of *ubc1*$_{cg}$ That the sequence similarity of UBC1$_{cg}$ to yeast UBC4 is also reflected in its function was shown by complementation of the *ubc4$^-$ubc5$^-$* mutant of *S. cerevisae*. When the yeast mutant cells were transformed with plasmids with *ubc1*$_{cg}$ inserted in the proper orientation, growth deficiency and heat sensitivity of the mutant were overcome, demonstrating that the UBC1$_{cg}$ is a ubiquitinylating enzyme (38). Selective protein degradation by the ubiquitin-proteosome system plays a critical role in many situations involving reprogramming of protein synthesis (60). UBC1$_{cg}$ is probably involved in the reprogramming of protein synthesis needed for conidia germination and differentiation into appressoria.

The identity and function of the other *chip* gene products remain to be established as they do not show significant homology to proteins currently in the data base. *chip*2 appears to be a DNA binding protein, and *chip*3 appears to be a seven transmembrane receptor-type protein (Z.M. Liu, Y. Kim, P.E. Kolattukudy, unpublished). Functional analyses are currently in progress.

Response of *C. gloeosporioides* to Host Signals

Avocado fruit surface wax induces germination and appressorium formation of *C. gloeosporioides* conidia (48,50). This induction is highly specific to the host wax as wax from other plants do not elicit this response, unlike the relief of self-inhibition in *M. grisea* conidia where any wax can remove the lipophylic self-inhibitors and thus stimulate germination and appressorium formation. The biological specificity of host-wax stimulation was also demonstrated by the observation that conidia of *Colletotrichum* species that infect other hosts are not induced to germinate and differentiate by the avocado wax (48). Bioassays of thin-layer chromatographic fractions of avocado wax showed that the fatty alcohol fraction was most active in inducing appressorium formation. In fact, synthetic *n*-primary alcohols of 24 or more carbon atoms induced appressorium formation, whereas shorter ones showed much less activity. Gas chromatographic examination of primary alcohol fractions from waxes of other plants showed that they contained very long chain alcohols (unpublished). However, such waxes showed very little appressorium-inducing activity. Further examination revealed that surface waxes contain components that stimulate and others that inhibit appressorium formation (L. Rogers, P.E. Kolattukudy, unpublished). For example, in avocado wax there are terpenoid materials that migrate in the chromatograms close to the primary alcohol fraction; these materials promote appressorium formation. Terpenoids can show synergistic effect with primary alcohols. For example, ursolic acid at very low levels stimulates appressorium formation and shows a synergistic effect with primary alcohols (Table 1). The inhibitory effect of non-host waxes on appressorium formation can be seen when they are tested in combination with the host wax (Table 2). The presence of potent inducers and inhibitors of conidia germination and appressorium formation in plant surface waxes and their relative amounts and potency can either stimulate or inhibit pathogenesis. This is an area that has received little attention. Metabolic engineering may be used to enhance the levels of inhibitors to prevent germination of fungal conidia on the host surface and thus protect plants against fungal diseases.

The nature of host signals that prompt the attack of the latent anthracnose fungi at the time of host ripening has been a mystery. An ideal candidate for such a signaling function would be a volatile host component produced only at ripening. We suspected that ethylene, the fruit ripening

Table 1. Induction of appressorium formation by *C. gloeosporioides* the primary alcohol fraction of various waxes with and without the addition of ursolic acid.

	% Appressorium formation		
		ursolic acid (μg)	
Source of Primary Alcohol	0	10^{-7}	10^{-8}
Avocado	78.5	86.1	81.4
Broccoli	1.5	70.1	69.8
Cabbage	5.7	72.1	65.8
Jade	0.7	75.5	61.3
Sweet potato tuber	7.5	64.9	63.4
Senecio odoris	3.6	67.3	71.2
C_{28} alcohol	33.1	73.4	75.3
Ursolic acid		36.0	19.1

In each assay, the TLC-fractionated primary alcohol, equivalent to 2.5 μg total wax, was used, except for C_{28} alcohol (4 ng) and ursolic acid as noted (L. Rogers and P.E. Kolattukudy, unpublished).

Table 2. Effect of addition of other plant waxes on appressorium formation induction in *C. gloeosporioides* by avocado wax (48).

Addition	Amount μg % of Control	Appressorial Induction
Broccoli wax	0.5	14
	2.0	19
Jade wax	0.5	38
	2.0	18
Senecio odoris wax	2.0	27

hormone, could fit the bill extremely well. In fact, volatiles produced by climacteric tomato, avocado, and banana fruits or ethylene at ≤ 1 μl/liter induced germination and appressorium formation by conidia of *C. gloeosporioides* and *C. musae* (14). Interestingly, the hormone caused branching of germ tubes and differentiation of the ends of branches into appressoria, resulting in the formation of up to six appressoria from a single conidium (Fig. 6). This ethylene effect on the fungal conidium was like the well-studied ethylene response of higher plants. For example, an analogue (propylene), but not other hydrocarbons like methane, could induce appressorium formation, and the conidial response was inhibited by silver ion and competitively by 2,5-norbornadiene. Although these observations suggest behavioral similarities between the ethylene receptor in the plant and that of the fungus, all of our attempts to clone the fungal ethylene receptor by hybridization with the cloned plant receptors and by RT-PCR using a variety of degenerate primers failed (D. Li and P.E. Kolattukudy, unpublished).

Ethylene induction of germination and appressorium formation appears to be unique to *Colletotrichum* species that infect climacteric fruits, but not to other *Colletotrichum* species that infect other plants. For example, *C. gloeosporioides* and *C. musae* responded to ethylene, whereas *C. pisi*, *C. orbiculare, C. trifolii*, and *C. lindemuthianum* did not respond. There is evidence that ethylene induction of multiple appressorium formation is relevant to postharvest infection. *C. gloeosporioides* conidia formed multiple appressoria on normally ripening tomato fruits that produce ethylene and infected such fruits. On the other hand, on transgenic tomato fruits that did not produce ethylene because of the presence of antisense 1-aminocyclopropane-1-carboxylic acid synthase gene, *C. gloeosporioides* conidia neither germinated nor differentiated into appressoria and formed no lesions. Exogenous ethylene induced germination and multiple appressoria formation leading to lesion formation. On orange fruits that do not produce ethylene, *C. gloeosporioides* conidia germinated and formed single appressoria within a day. Subsequent exposure to ethylene resulted in branching of germ tubes and formation of multiple appressoria leading to lesion formation. Commercially used citrus degreening by ethylene

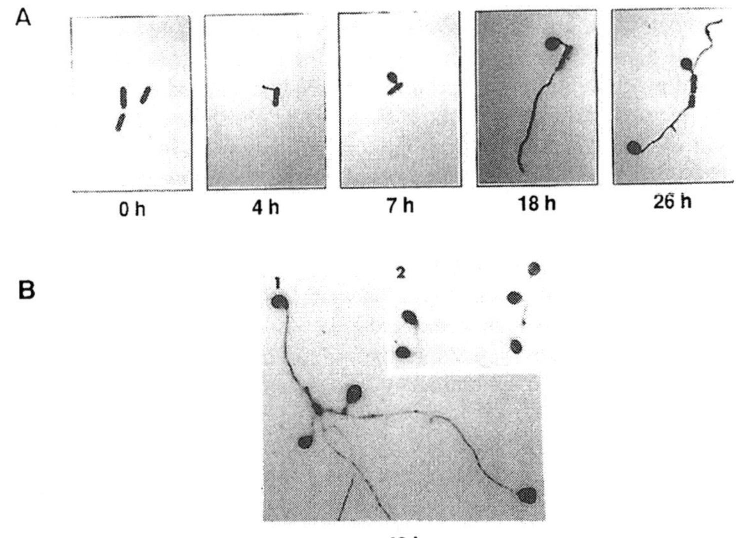

Fig. 6. A: Successive stages of germination and appressorium formation of a typical spore of *C. musae* exposed to 7 μM ethephon. B: Multiple appressorium production from a single spore of *C. musae* (1) and *C. gloeosporioides* (2) exposed for 48 h to ethylene (x1200.)

treatment, to improve fruit color by removing chlorophyll from the peel, is known to cause an increase in anthracnose (2).

Involvement of Protein Phosphorylation in Appressorium formation

Labeling of the nucleotide pool by incubation of the conidia with $^{32}PO_4^{-3}$ for 3 hr followed by exposure of the conidia to appressorium-inducing host signals, host wax or ethylene caused a selective increase in the phosphoproteins could be detected after 0.5 to 1.0 hr of ethylene or wax treatment and reached maximal levels after about 3 hr. That this phosphorylation is involved in appressorium formation was suggested not only by its time-course, but also by the finding that the protein kinase inhibitors that inhibited the phosphorylation of these proteins also inhibited appressorium formation. For example, ethylene-induced protein phosphorylation and appressorium formation were inhibited by H-7, that preferentially inhibits serine/threonine kinase, and by genistein, that preferentially inhibits tyrosine phosphorylation. On the other hand, wax-induced phosphorylation and appressorium formation were inhibited only by genistein but not by H-7. Thus, the signal transduction inhibited by H-7 is probably involved preferentially in the ethylene signaling but not in wax signaling. Norbornadiene uniquely inhibited appressorium formation induced by ethylene but not that induced by host wax, demonstrating that both signals do not share all of the signal transduction components.

Calyculin A, an inhibitor of protein phosphatase, caused accumulation of the same phosphoproteins as those found to be phosphorylated by treatment with wax and ethylene. It induced appressorium formation, even in the presence of yeast extract that normally does not allow appressorium formation. In the presence of calyculin A, secondary growth of long germ tubes and additional appressoria from the primary appressoria were observed. In the presence of slightly higher (>1.0 μm) concentrations of calyculin A, abnormal looking structures were formed. In any case, it is clear that the phosphoproteins generated either by activation of protein kinases caused by the host signals, or by inhibition of phosphatase, are involved in appressorium formation.

Protein Kinases in *C. gloeosporioides*

Protein phosphorylation has been found to be involved in several stages of the interaction of *C. gloeosporioides* with its host, from the early phase of contact to the penetration into the host. Therefore, we used an RT-PCR approach to examine the kinases using RNA purified from hard surface-treated conidia. Degenerate oligonucleotides coding for the conserved kinase domains VI and IX (62) yielded the expected size (190 bp) product which was cloned. When individual clones were sequenced, four of them showed high levels of homology to other known protein kinases (Fig. 7) of which one was identified to be a segment of CgMK as indicated in a previous section. Another, designated CgRAN1, showed 64% and 80% identities to *ran*1 of *Schizosaccharomyces pombe* and *ran*1-like kinase of *F. solani* f.sp. *pisi* T8, respectively. The *ran*1 gene was reported to encode a protein kinase that functions to prevent activation of the meiotic pathway under conditions that normally promote vegetative growth (40). *ran*1 mutants conjugate without the requirement of nutritional starvation and sporulate from the haploid state without the requirement for heterozygosity (26,43). Although the function of CgRAN1 is not known at the moment, it might be involved in controlling the conidia germination and appressorium formation. A third one, designated CgMK30, showed about 50% identity to other serine/threonine protein kinases such as barley sucrose nonfermenting-1 (SNF1)-related protein kinase BKIN12 (20) and yeast GIN4 (GenBank accession no. U33057) which is involved in cytokinesis. A fourth one, designated CgMK77, was highly homologous to other serine/threonine protein kinases such as NIMA which is involved in the control of the cell cycle of *Aspergillus nidulans* (44) and KSP1 which is not essential for vegetative growth of yeast (16). PCR with primers designed for domain VI and VIII and cDNA from hard surface-treated conidia of *C. gloeosporioides* as template yielded a 204-bp product (Z.M. Liu, P.E. Kolattukudy, unpublished). 3'-RACE yielded a 1.2-kb clone that showed homology to domains VI to X of ser/thr-kinase of yeast-79% and 71% identity with the kinases from *S. pombe* (GenBank accession no. Z97210) and *S. cerevasicae* (GenBank accession no. S76380), respectively. It also showed 84% identity with domain VI to IX of a ser/thr protein kinase from *Fusarium* designated FSK (GenBank accession no. U61839). Early stages of infection by *C. trifolii* has been suggested to involve protein kinase C and a protein kinase with homology to the *N. crassa* kinase required for hyphal elongation (3,4,8). *C. gloeosporioides*, like other phytopathogens, must sense

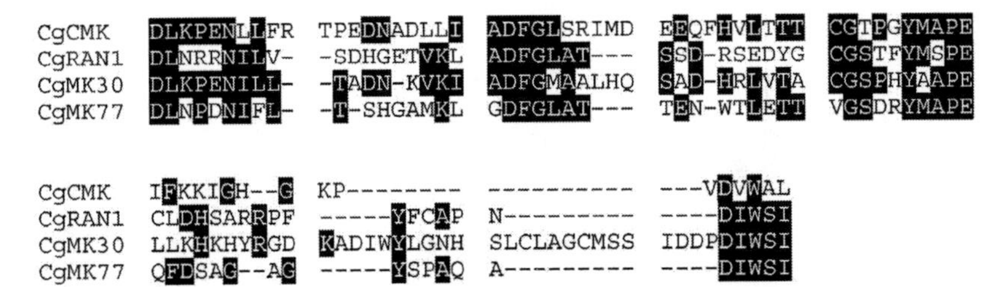

Fig. 7. Amino acid sequence comparison in domains from VI to IX of several protein kinases which were isolated by RT-PCR approach with RNA extracted from hard surface-treated conidia of *C. gloeosporioides*. The full CgCMK sequence is in GenBank (AF034963).

environmental factors and mount responses. The two component signal transduction system may be used for such purposes in which the sensor component autophosphorylates a histidine residue in response to an environmental stimulus and then transfers the phosphoryl group to an aspartate residue of the response regulator. Ethylene receptor ETR1 from *Arabidopsis thaliana* (5) and osmosensor Sln1 from *Saccharomyces cerevisiae* (45) are examples of such sensing in the eukaryotes.

Our search for an ETR-1 like receptor in *C. gloeosporioides,* which also responds to ethylene, led to cloning of a two-component kinase from this organism (D. Li, P.E. Kolattukudy, unpublished). PCR with degenerate primers designed based on the conserved H (HEL/I/V/MRTP) and N (NAV/L/IKFT/S) boxes of prokaryotic histidine kinases and cDNA from the mycelia of *C. gloeosporioides* as template generated a 369-bp product. The sequence of this clone showed that it could encode 123 amino acids with sequence homology to prokaryotic histidine kinases. Screening of a *C. gloeosporioides* cDNA library with this clone as a probe yielded a two-component histidine kinase designated CgHK1. The *C. gloeosporioides* gene for this kinase has an open reading frame of 2238 bp interrupted by one intron. The 82.2-kDa protein that would be encoded by this gene shows homology to the histidine kinase domain and response regulator of the prokaryotic two-component histidine kinase family (24,46) (Fig. 8).

Penetration Through the Cuticle

The appressorium generates an infection peg that penetrates into the host. Highly melanized appressoria can exert tremendous turgor pressure that can play a major role in the cuticular penetration process (23). However, there is evidence for assistance by extracellular cutinase secreted by the fungus. Cutinase has been purified from *C. gloeosporioides* cultured in the presence of cutin (10). Targeted secretion of cutinase by the infection peg arising from the appressorium was demonstrated using immunofluores-cence techniques (49). Antibodies prepared against the enzyme protected against lesion development on papaya fruits (10). A cutinase-deficient mutant of *C. gloeosporioides* could not produce lesions except when exogenous cutinase was added (9). Potent inhibitors of cutinase protected papaya fruits from lesion formation by *C. gloeosporioides* in laboratory bioassays (11), and spraying an inhibitor in the field periodically over the entire growing season protected the fruits from anthracnose (Fig. 9).

Cutinase genes from *C. gloeosporioides* and *C. capsici* have been cloned and sequenced, and the transcription initiation site has been identified (13). Comparison of these sequences with the *Fusarium* cutinase previously cloned identified similarities not only in the intron position, but also in the conserved active site residues, although only 43% of the amino acid residues were conserved among all of the three enzymes. Although the *cis*-elements involved in the regulation of expression of the cutinase gene in *C.*

METNIRGIHQALSRNNSSSSSSVYEMPTADLTELARKGIIKVTTDSNGRLSYTTEVRLRS 60
RNGEYRWHLIRCVEIDNIDFGNGASSYFGSATDINDHKLLEAKLKEAMESKSRFLSNMS**H** 120
EIRTPLIGISGMVSFLQDTTLNEEQRDYTNTIQTSANSLLMIINDILDLSKVDAGMMKLK 180
YEWFHTRSLIEDVNELVSTMAIAKRLELNYIVEEDVPAWVKGDKVRIRQVLLNVIG**NAIK** 240
FTAEGEVFSRCRIFTNTKSALSNDEIMLEFAVIDTGRGFTKEEADLIFKPFSQIDGSSTR 300
QHGGSGLGLVISRQLVELHGGKMEGTAVPGKGSTFTFTAKFALPTSADHPDGPIGPEAFK 360
PVVQAPEEVTVLRPAVKASPAAKATSGASPSTDAEFVSPALASSGSSDPSIRSNISHVTE 420
RTSVSSINVGLVHFSEAARASGQDLSEMKLELPLSESSPGTTPTPETSKPAKLEDFRPPM 480
YSILVICPQTHSREATAQHIETTLPKDVPHQITALDSVDKAQSLLGGDESVNFTHIVLNL 540
PSPEEIIGLMERITKSMTISNTTILILSDSVQRQAVMKLATETKYEQLVSENLVTFIYKP 600
VKPSRFAVIFDPDKVRDLSIDRNRSTAQRMVETQKASYQEIEKRMGNKGYKVLLVEDNPV 660
NQKVLKKYLKKVGVEVEVVADGAECTETVFSRSHSYFSLILCDLHMPRKDGYQACREIRQ 720
WEKERGFKKLPIIALSANVMSDVSG 745

Fig. 8. CHK-1 of *C. gloeosporioides*. The deduced amino acid sequence contains the conserved H and N boxes (in *boldface*) and the putative site for ATP binding (*underlined*).

gloeosporioides have not been established, detailed studies on the regulation of the cutinase gene from *Fusarium solani* f.sp. *pisi* suggest the following. A low level of constitutively expressed cutinase, present on the conidia that land on the plant surface, generates small amounts of cutin monomers from the host cuticle. The unique monomers then trigger transcriptional activation of the inducible cutinase gene. The transcription factors from *F. solani* f.sp. *pisi* have been cloned and studied (31,35,36), but similar studies have not been done with *C. gloeosporioides*.

Breaking Carbohydrate Barriers

Once the infection peg penetrates through the cuticle, it must penetrate through the underlying carbohydrate barriers. Pectin degradation by the enzymes secreted by the fungus is thought to be important in this phase of the infection process. A pectate lyase induced by pectin in *F. solani* f.sp. *pisi* was suggested to be important in the infection of its host because antibodies prepared against the lyase protected the host against infection (7). Similarly, *C. gloeosporioides* pectate lyase was recently found to be

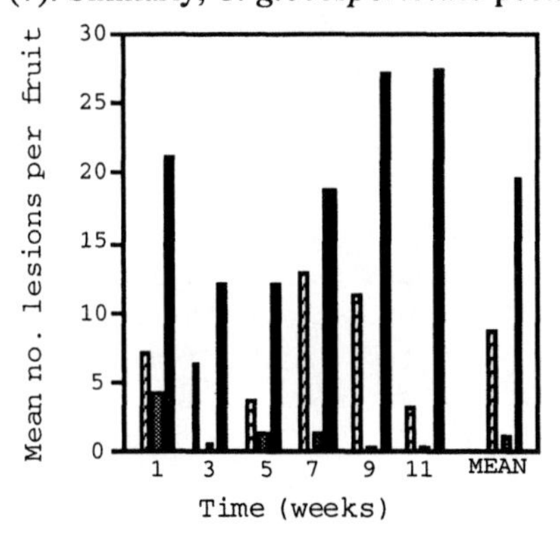

Fig. 9 Protection of papaya fruits in the field from *C. gloeosporioides* by a cutinase inhibitor. Chlorpyrifos, an organic phosphate inhibitor of cutinase (▨), a fungicide (Mancozeb) (■), or water (■), were sprayed on the growing fruits on a biweekly interval and after the fruits reached maturity the lesions observed were recorded (W. Nishijima, M.B. Dickman, S. Patil, and P.E. Kolattukudy, unpublished).

important to pathogenesis because antibodies prepared against this lyase protected avocado fruits (61). Knocking out single genes that encode pectin degrading enzymes in some fungi have failed to show a significant decrease in fungal virulence; therefore, such enzymes have been suggested to be unimportant for pathogenesis. However, pathogenic fungi usually have multiple genes that encode enzymes that degrade the physical barriers in the plant and animal hosts. For example, *Candida* has about 10 genes that encode aspartic proteinase (56) and the *Aspergillus* species that cause invasive aspergillosis in immunocompromised patients make proteinases that use different catalytic mechanisms including serine proteinases (52,55), metalloproteinases (39,53), and aspartic proteinases (34). Knocking out one proteinase gene was found to enhance the production of other proteinases (51). In view of such facts, single-gene knock-out would not be an appropriate approach to explore the role of extracellular proteinases in pathogenesis, and such knock-outs did not reveal a decrease in virulence (27,57). The classical mutagenesis approach did show that proteinase-deficient mutants were less virulent than the wild type (30). Recent examination of such a mutant of *A. fumigatus* (30) showed that it produces drastically decreased levels of serine proteinase, metalloproteinase, and aspartic proteinase, suggesting that the mutation affected a gene that regulates proteinase production (M.V. Ramesh and P.E. Kolattukudy, unpublished). Knock-out of *areA,* a regulatory gene that affects synthesis of all of the proteinases, has been shown to decrease virulence (22).

Multiple pectin-degrading enzymes are used by fungi to help invade through the pectinaceous barriers of the host. A pectate lyase induced by pectin in *F. solani* f.sp. *pisi* was purified, and antibodies prepared against this protein were found to protect the host against infection (7). However, knock-out of the gene encoding this lyase did not significantly reduce virulence, yet antibodies prepared against the inducible lyase protected the host against infection by the gene-disrupted mutant (W. Guo, L. Rogers, P.E. Kolattukudy, unpublished). Thus, the fungus was suspected to produce multiple pectate lyases that are immunologically similar. In fact, multiple genes were cloned, characterized, and found to be differentially regulated. For example, *pelA* is induced in culture by pectin and *pelB* is expressed constitutively at very low levels probably to produce oligomers from the pectin in the environment and thus trigger induction of *pelA* that produces high levels of lyase (17). *p*elC is an intracellular enzyme that may be involved in catabolizing the oligomers taken up into the fungal cells (18). *pelD* was found to be produced only *in planta* and not in culture with any substrates so far tested (19). All four *pel* gene products have been expressed in *Pichia pastoris* and characterized as *endo* pectate lyases, and they were all found to be immunologically cross reactive; antibodies against any of them inhibited the lyase activity of each. *pelA* knock-out did not decrease virulence substantially, and *pelD* was found to be produced by the *pelA*-disrupted mutant (W. Guo, L. Rogers, P.E. Kolattukudy, unpublished). The *pelD* knock-out mutant also did not show a significant decrease in virulence.

Knock-out of both *pelA* and *pelD* caused a drastic decrease in virulence (unpublished). These results demonstrated that pectate lyases are important virulence factors in the pathogenesis by *F. solani* f.sp. *pisi*. These results also illustrate how single gene knock-out can lead to misleading conclusions if the role of multiple genes that encode polymer-degrading enzymes is not carefully considered. The observation that antibodies prepared against a pectate lyase from *C. gloeosporioides* protected its host (61) suggests that pectin-degrading enzymes are also virulence factors in anthracnose.

Conclusion

Even though plant-fungus communication clearly starts as soon as the conidia land on the host, exploration of the molecular basis of this communication is only in its beginning stages. Elucidation of the early processes involved in the interaction between *C. gloeosporioides* and its host may lead to methods to intervene in this process and thus may provide new methods to protect plants against fungal infection.

Literature Cited

1. Braam, J. and Davis, R.W. 1990. Rain-, wind-, and touch-induced expression of calmodulin and calmodulin-related genes in *Arabidopsis*. Cell 60:357-364.
2. Brown, G.E. 1975. Factors affecting postharvest development of *Colletotrichum gloeosporioides* in citrus fruits. Phytopathology 65:404-409.
3. Buhr, T.L. and Dickman, M.B. 1997. Gene expression analysis during conidial germ tube and appressorium development in *Colletotrichum trifolii*. Appl. Environ. Microbiol. 63:2378-2383.
4. Buhr, T.L., Oved, S., Truesdell, G.M., Huang, C.X., Yarden, O. and Dickman, M.B. 1996. A kinase-encoding gene from *Colletotrichum trifolii* complements a colonial growth mutant of *Neurospora crassa*. Mol. Gen. Genet. 251:565-572.
5. Chang, C, Kwok, S.F, Bleecker, A.B, and Meyerowitz, E.M. 1993. *Arabidopsis* ethylene-response gene ETR1: Similarity of product to two-component regulators. Science 262:539-544.
6. Chumley, F.G., and Valent, B. 1990. Genetic analysis of melanin-deficient, nonpathogenic mutants of *Magnaporthe grisea*. Mol. Plant-Microbe Interact. 3:135-143.
7. Crawford, M.S. and Kolattukudy, P.E. 1987. Pectate lyase from *Fusarium solani* f. sp. *pisi*: purification, characterization, *in vitro* translation of the mRNA, and involvement in pathogenicity. Arch. Biochem. Biophys. 258:196-205.

8. Dickman, M.B., Buhr, T.L., Warwar, V., Truesdell, G.M. and Huang, C.X. 1995. Molecular signals during the early stages of alfalfa anthracnose. Can J. Bot. 73 (suppl. 1):S1169-S1177.
9. Dickman, M.B. and Patil, S.S. 1986. Cutinase deficient mutants of *Colletotrichum gloeosporioides* are non-pathogenic to papaya fruit. Physiol. Mol. Plant Pathol. 28:235-242.
10. Dickman, M.B., Patil, S.S. and Kolattukudy, P.E. 1982. Purification, characterization and role in infection of an extracellular cutinolytic enzyme from *Colletotrichum gloeosporioides* Penz. on *Carica papaya* L. Physiol. Plant Pathol. 20:333-347.
11. Dickman, M.B., Patil, S.S. and Kolattukudy, P.E. 1983. Effects of organophosphorus pesticides on cutinase activity and infection of papayas by *Colletotrichum gloeosporioides.* Phytopathology 73:1209-1214.
12. Emmett, R.W. and Parbery, D.G. 1975. Appressoria. Annu. Rev. Phytopathol. 13:147-167.
13. Ettinger, W.F., Thukral, S.K. and Kolattukudy, P.E. 1987. Structure of cutinase gene, cDNA, and the derived amino acid sequence from phytopathogenic fungi. Biochemistry 26:7883-7892.
14. Flaishman, M.A. and Kolattukudy, P.E. 1994. Timing of fungal invasion using host's ripening hormone as a signal. Proc. Natl. Acad. Sci. USA 91:6579-6583.
15. Flaishman, M.A., Hwang, C.-S. and Kolattukudy, P.E. 1995. Involvement of protein phosphorylation in the induction of appressorium formation in *Colletotrichum gloeosporioides* by its host surface wax and ethylene. Physiol. Mol. Plant Pathol. 47:103-117.
16. Fleischmann, M., Stagljar, I., and Aebi, M. 1996. Allele-specific suppression of a *Saccharomyces cerevisiae* prp20 mutation by overexpression of a nuclear serine/threonine protein kinase. Mol. Gen. Genet. 250:614-625.
17. Guo, W., Gonzalez-Candelas, L. and Kolattukudy, P.E. 1995. Cloning of a novel constitutively expressed pectate lyase gene *pelB* from *Fusarium solani* f. sp. *pisi* (*Nectria haematococca*, mating type VI) and characterization of the gene product expressed in *Pichia pastoris.* J. Bacteriol. 177:7070-7077.
18. Guo, W., Gonzalez-Candelas, L. and Kolattukudy, P.E. 1995. Cloning of a new pectate lyase gene *pelC* from *Fusarium solani* f. sp. *pisi* (*Nectria haematococca*, mating type VI) and characterization of the gene product expressed in *Pichia pastoris*. Arch. Biochem. Biophys. 323: 352-360.
19. Guo, W., Gonzalez-Candelas, L. and Kolattukudy, P.E. 1996. Identification of a novel *pelD* gene expressed uniquely in planta by *Fusarium solani* f. sp. *pisi* (*Nectria haematococca* , mating type VI) and characterization of its protein product as an endo-pectate lyase. Arch. Biochem. Biophys. 332:305-312.

20. Halford, N.G., Viconte-Carbajosa, J., Sabelli, P.A., Shewry, P.R., Hannappel, U. and Keiis, M. 1992. Molecular analysis of a barley multigene family homologous to the yeast protein kinase gene *SNF*1. Plant J. 2:791-797.
21. Hegde, Y. and Kolattukudy, P.E. 1997. Cuticular waxes relieve self-inhibition of germination and appressorium formation by the conidia of *Magnaporthe grisea*. Physiol. Mol. Plant Pathol. 51: 75-84.
22. Hensel, M., Arst, H.N. Jr., Aufauvre-Brown, A. and Holden, D.W. 1998. The role of the *Aspergillus fumigatus areA* gene in invasive pulmonary aspergillosis. Mol. Gen. Genet. 258: 553-557.
23. Howard, R.J. and Valent, B. 1996. Breaking and entering: Host penetration by the fungal rice blast pathogen *Magnaporthe grisea*. Annu. Rev. Microbiol. 50:491-512.
24. Hrabak, E.M. and Willis, D.K. 1992. The *lem*A gene required for pathogenicity of *Pseudomonas syringae pv. syringae* on bean is a member of a family of two-component regulators. J. Bacteriol. 174: 3011-3020.
25. Hwang, C.-S. and Kolattukudy, P.E. 1995. Isolation and characterization of genes expressed uniquely during appressorium formation by *Colletotrichum gloeosporioides* conidia induced by the host surface wax. Mol. Gen. Genet. 247:282-294.
26. Iino, Y. and Yamamoto, M. 1985. Mutants of *Schizosaccharomyces pombe* which sporulate in the haploid state. Mol. Gen. Genet. 198:416-421.
27. Jaton-Ogay, K., Paris, S., Huerre, M., Quadroni, M., Falchetto, R., Togni, G., Latge, J.-P. and Monod, M. 1994. Cloning and disruption of the gene encoding an extracellular metalloprotease of *Aspergillus fumigatus*. Mol. Microbiol. 14: 917-928.
28. Kameshita, I. and Fusisawa H. 1989. A sensitive method for detection of calmodulin-dependent protein kinase II activity in sodium dodecyl sulfate-polyacrylamide gel. Anal. Biochem. 183:139-143.
29. Kim, Y.-K., Li, D., and Kolattukudy, P.E. 1998. Induction of Ca^{2+}-calmodulin signaling by hard surface contact primes *Colletotrichum gloeosporioides* conidium to germinate and form appressorium. J. Bacteriol. 180:5144-5150.
30. Kolattukudy, P.E., Lee, J.D., Rogers, L.M., Zimmerman, P., Ceselski, S., Fox, B., Stein, B. and Copelan, E.A. 1993. Evidence for possible involvement of an elastolytic serine protease in aspergillosis. Infect. Immun. 61:2357-2368.
31. Kolattukudy, P.E., Rogers, L.M., Li, D., Hwang, C.-S. and Flaishman, M.A. 1995. Surface signaling in pathogenesis. Proc. Natl. Acad. Sci. USA 92:4080-4087.
32. Kornstein, L.B., Gaiso, M.L., Hammell, R.L., and Bartelt, D.C. 1992. Cloning and sequence determination of a cDNA encoding *Aspergillus nidulans* calmodulin-dependent multifunctional protein kinase. Gene 113:75-82.

33. Kubo, Y., Suzuki, K., Furusawa, I., and Yamamoto, M. 1983. Scytalone as a natural intermediate of melanin biosynthesis in appressoria of *Colletotrichum lagenarium*. Exp. Mycol. 7:208-215.
34. Lee, J.D. and Kolattukudy, P.E. 1995. Molecular cloning of the cDNA and gene for an elastinolytic aspartic proteinase from *Aspergillus fumigatus* and evidence of its secretion by the fungus during invasion of the host lung. Infect. Immun. 63:3796-3803.
35. Li, D. and Kolattukudy, P.E. 1995. Cloning and expression of cDNA encoding a protein that binds a palindromic promoter element essential for induction of fungal cutinase by plant cutin. J. Biol. Chem. 270:11753-11756.
36. Li, D. and Kolattukudy, P.E. 1997. Cloning of cutinase transcription factor 1, a transactivating protein containing Cys_6Zn_2 binuclear cluster DNA-binding motif. J. Biol. Chem. 272:12462-12467.
37. Lingappa, B.T. and Lingappa, Y. 1967. Alkaloids as self-inhibitors of fungi. Nature 214:516-517.
38. Liu, Z.-M. and Kolattukudy, P.E. 1998. Identification of a gene product induced by hard-surface contact of *Colletotrichum gloeosporioides* conidia as a ubiquitin-conjugating enzyme by yeast complementation. J. Bacteriol. 180:3592-3597.
39. Markaryan, A., Morozova, I. ,Yu, H., and Kolattukudy, P.E. 1994. Purification and characterization of an elastinolytic metalloprotease from *Aspergillus fumigatus* and immunoelectron microscopic evidence of secretion of this enzyme by the fungus invading the murine lung. Infect. Immun. 62:2149-2157.
40. McLeod, M. and Beach, D. 1996. Homology between the *ran*1 gene of fission yeast and protein kinases. EMBO J. 5:3665-3671.
41. Melnick, M.B., Melnick, C., Lee, M., and Woodward, D.O. 1993. Structure and sequence of the calmodulin gene from *Neurospora crassa*. Biochim. Biophys. Acta. 1171:334-336.
42. Meyer, W.L., Lax, A.R., Templeton, G.E., and Brannon, M.J. 1983. The structure of gloeosporone, a novel germination self-inhibitor from conidia of *Colletotrichum gloeosporioides*. Tetrahedron Lett. 24:5059-5062.
43. Nurse, P. 1985. Mutants of the fission yeast *Schizosaccharomyces pombe* which alter the shift between cell proliferation and sporulation. Mol. Gen. Genet.198:497-502.
44. Osmani, S.A., May, G.S., and Morris, N.R. 1987. Regulation of the mRNA levels of *nim*A, a gene required for the G2-M transition in *Aspergillus nidulans*. J. Cell. Biol. 104:1495-1504.
45. Ota, I.M. and Varshavsky, A. 1993. A yeast protein similar to bacterial two-component regulators. Science 262:566-569.
46. Parkinson, J.S. and Kofoid, E.C. 1992. Communication modules in bacterial signaling proteins. Annu. Rev. Genet. 26:71-112.

47. Pausch, M.H., Kaim, D., Kunisawa, R., Admon, A., and Thorner, J. 1991. Multiple Ca^{2+}/calmodulin-dependent kinase genes in a unicellular eukaryote. EMBO J. 10:1511-1522.
48. Podila, G.K., Rogers, L.M., and Kolattukudy, P.E. 1993. Chemical signals from avocado surface wax trigger germination and appressorium formation in *Colletotrichum gloeosporioides*. Plant Physiol. 103:267-272.
49. Podila, G.K., Rosen, E., San Francisco, M.J.D., and Kolattukudy, P.E. 1995. Targeted secretion of cutinase in *Fusarium solani* f. sp. *pisi* and *Colletotrichum gloeosporioides*. Phytopathology 85:238-242.
50. Prusky, D. and Saka, H. 1989. The role of epicuticular wax of avocado fruit in appressoria formation of *Colletotrichum gloeosporioides* Phytoparasitica 17: 140.
51. Ramesh, M.V. and Kolattukudy, P.E. 1996. Disruption of the serine protease gene (*sep*) in *Aspergillus flavus* leads to a compensatory increase in the expression of a metalloprotease gene (*mep20*). J. Bacteriol. 178: 3899-3907.
52. Ramesh, M.V., Sirakova, T., and Kolattukudy, P.E. 1994. Isolation, characterization, and cloning of cDNA and the gene for an elastinolytic serine protease from *Aspergillus flavus*. Infect. Immun. 62: 79-85.
53. Ramesh, M.V., Sirakova, T.D., and Kolattukudy, P.E. 1995. Cloning and characterization of the cDNAs and genes (*mep20*) encoding homologous metalloproteinases from *Aspergillus flavus* and *A. fumigatus*. Gene 165:121-125.
54. Rao, J.P. and Subramanyam, C. 1998. A putative role for calmodulin in aflatoxin production. Sixth International Mycological Congress, 23-28 Aug. 1998, Jerusalem, Israel. Abstract, p. 148.
55. Reichard, U., Buttner, S., Eiffert, H., Staib, F., and Ruchel, R. 1990. Purification and characterisation of an extracellular serine proteinase from *Aspergillus fumigatus* and its detection in tissue. J. Med. Microbiol. 33: 243-251.
56. Ruchel, R., De Bernadis, F., Ray, T.L., Sullivan, P.A., and Cole, G.T. 1992. *Candida* acid proteinases. J. Med.Vet. Mycol. 30 (Suppl. 1):123-132.
57. Tang, C.M., Cohen, J., Krausz, T., van Boorden, S., and Holden, D.W. 1993. The alkaline protease of *Aspergillus fumigatus* is not a virulence determinant in two murine models of invasive pulmonary aspergillosis. Infect. Immun. 61:1650-1656.
58. Thines, E., Eilbert F., Sterner, O., and Anke, H. 1997. Signal transduction leading to appressorium formation in germinating conidia of *Magnaporthe grisea*: Effects of second messengers diacylglycerols, ceramides and sphingomyelin. FEMS Microbiol. Lett. 156:91-94.
59. Tsurushima, T., Ueno, T., Fukami, H., Irie, H., and Inoue, M. 1995. Germination self-inhibitors from *Colletotrichum gloeosporioides* f. sp. *jussiaea.* Mol. Plant-Microbe Interact. 8: 652-657.

60. Varshavsky, A. 1997. The ubiquitin system. Trends Biochem. Sci. 22:383-387.
61. Wattad, C., Kobiler, D., Dinoor, A., and Prusky, D. 1997. Pectate lyase of *Colletotrichum gloeosporioides* attacking avocado fruits: cDNA cloning and involvement in pathogenicity. Physiol. Mol. Plant Pathol. 50:197-212.
62. Wilks, A.F. 1991. Cloning members of protein-tyrosine kinase family using polymerase chain reaction. Meth. Enzymol. 200:533-546.
63. Yasui, K., Kitamoto, K., Gomi, K., Kumagai, C., Ohya, Y., and Tamura, G. 1995. Cloning and nucleotide sequence of the calmodulin-encoding gene (cmdA) from *Aspergillus oryzae*. Biosci. Biotechnol. Biochem. 59:1444-1449.

Chapter 7

Regulation of Melanin Biosynthesis Genes During Appressorium Formation by *Colletotrichum lagenarium*

Y. Kubo, Y. Takano, G. Tsuji, O. Horino, and I. Furusawa

A prerequisite for invasion of host plants by some plant pathogenic fungi is the differentiation of an infection structure called the appressorium (8). The appressorium, which arises as an apical swelling of the germ tube, promotes adhesion to the host surface and provides the mechanical force and enzymes required for initial penetration. Species of *Colletotrichum* which cause anthracnose diseases and *Magnaporthe* which cause blast diseases produce appressoria with thickened cell walls that are darkly pigmented with melanin. Melanization of fungal cells has been assumed to contribute to increased endurance under adverse conditions, such as irradiation, lysis by other microorganisms, oxygen radicals, or high temperature (5).

Studies with melanin-deficient mutants and melanin biosynthesis inhibitors demonstrated that melanin biosynthesis of *Colletotrichum lagenarium* (Pass.) Ellis et Halsted (syn. *Colletotrichum orbiculare* (Berk. Et Mont.) von Arx) and *Magnaporthe grisea* (Hebert) Barr. is essential for appressorial penetration of their host plants (15). In *M. grisea*, the melanized appressorium wall functions as a semi-permeable membrane, mediating the osmotic generation of large turgor pressures which provide the driving force for mechanical penetration (12). Recently, accumulation of glycerol in appressoria was found to generate high osmotic pressure (6). Cell wall melanin acts essentially as an impermeable barrier to glycerol which passes through the plasma membrane (25). In *C. lagenarium* , the increase in structural rigidity of appressorial cell walls generated by melanin is envisaged to resist the high turgor pressure within the appressorium and to direct the force vertically downwards through the non-melanized penetration pore (15).

The biosynthetic pathway for melanin in *C. lagenarium* has been elucidated with melanin-deficient mutants and melanin biosynthesis inhibitors. The pathway starts from polyketide synthesis and proceeds to form 1,3,6,8-tetrahydroxynaphthalene (1,3,6,8-THN). The subsequent metabolic conversion consists of two dehydration and reduction steps, ie., reduction of 1,3,6,8-THN to scytalone, dehydration of scytalone to 1,3,8 trihydroxynaphthalene (1,3,8-THN), reduction of 1,3,8-THN to vermelone and dehydration of vermelone to 1,8-dihydroxynaphthalene (1,8-DHN). The final step of the pathway is oxidative polymerization of 1,8-DHN to form a melanin polymer. We cloned and sequenced three structural genes and one regulatory gene involved in melanin biosynthesis in *C. lagenarium*. A dehydratase and a reductase involved in melanin biosynthesis were also purified by using a heterologous *Escherichia coli* expression system, and polyclonal antibodies against the purified enzymes were prepared. Melanin biosynthesis is a developmentally regulated process specific to appressoria; conidia, germ tubes, penetration pegs, and infection hyphae are not melanized. The expression of melanin biosynthetic genes and enzymes during appressorium differentiation was investigated.

Cloning of Melanin Biosynthesis Genes

C. lagenarium synthesizes dihydroxynaphthalene (DHN) melanin via the polyketide pathway. We cloned three structural genes, *PKS1* (16,34), *SCD1* (19), and *THR1* (29), encoding for melanin biosynthesis enzymes and one regulatory gene *CMR1* (13).

PKS1

We cloned *PKS1* by complementation of an albino mutant with wild type cosmid library DNA. *PKS1* contained a deduced open reading frame of 2187 amino acids, consisting of three exons separated by two short introns (Fig. 1). The predicted PKS1 polypeptide was homologous with the type I polyketide synthase of *Aspergillus nidulans* (23) and 6-methylsalicylic acid synthase of *Penicillium patulum* (4). The data indicate that *PKS1* of *C. lagenarium* encodes for a large multifunctional polypeptide with type I polyketide synthase activity, containing beta-ketoacyl synthase, acyl/malonyl transferase, and two acyl carrier protein domains. Recently, we constructed a *PKS1* expression plasmid pTAPSG using a fungal expression vector pTAex3 (9), in which the introduced gene is expressed under the control of the *amyB* promoter in *A. oryzae*. From the culture medium of the transformed *A. oryzae*, we purified and identified 1,3,6,8-THN as a transformant-specific metabolite following acetylation treatment with acetic acid and pyridin. *PKS1* is, therefore, involved in the synthesis and cyclization of pentaketide. The final product of the reaction catalyzed by this enzyme is 1,3,6,8-THN.

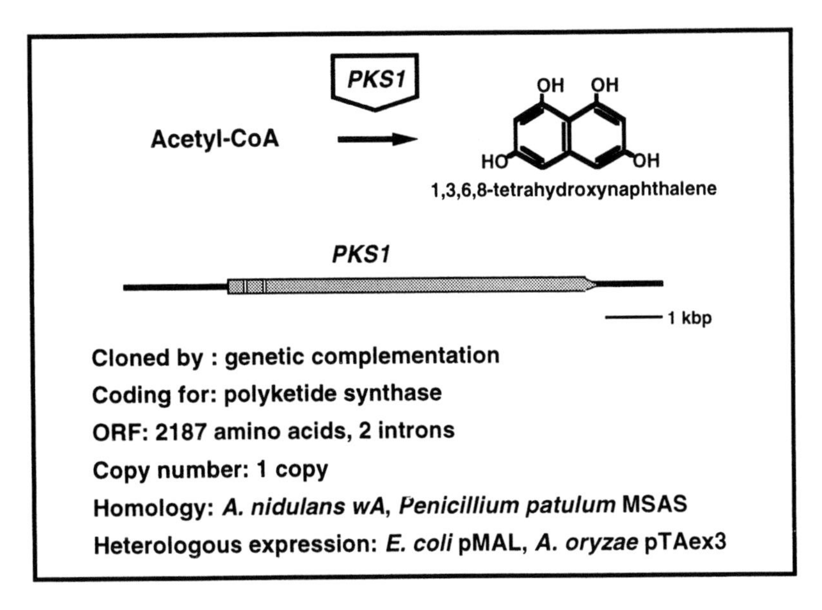

Fig. 1. Characterization of PKS1.

SCD1

SCD1 was cloned by heterologous hybridization using scytalone dehydratase cDNA of *M. grisea*. An Scd⁻ mutant which was obtained by gene disruption with a plasmid containing benomyl-resistant ß-tubulin as a selective marker could be complemented with the cloned *SCD1*. The N-terminal amino acid sequence of *SCD1* retained a sequence homology to that obtained from purified scytalone dehydratase of *Cochliobolus miyabeanus* (19,32). *SCD1* contained a deduced open reading frame of 188 amino acids consisting of three exons separated by two introns and encoding for scytalone dehydratase involved in the dehydration of scytalone to 1,3,8-THN (Fig. 2). This enzyme is also involved in the conversion of vermelone to 1,8-DHN. SCD1 had 62% amino acid sequence homology with the cloned scytalone dehydratase of *Magnaporthe grisea* (19,26).

THR1

THR1 was cloned by heterologous hybridization using *BRM2*, one of a cluster of genes involved in the conversion of 1,3,8-THN to vermelone in *Alternaria alternata* (14), as a probe. *THR1* contained a deduced open reading frame of 282 amino acids, consisting of five exons separated by four introns (Fig. 3). *THR1* encodes 1,3,8-THN reductase involved in the

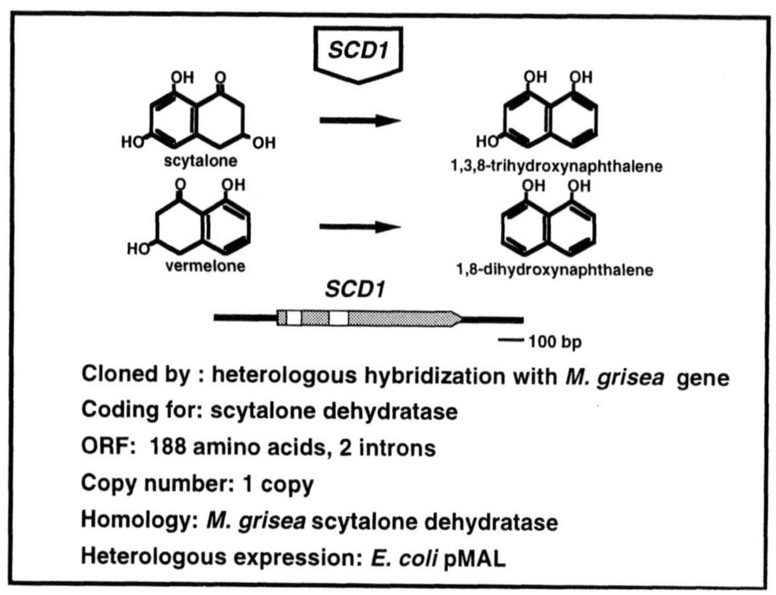

Fig. 2. Characterization of *SCD1*.

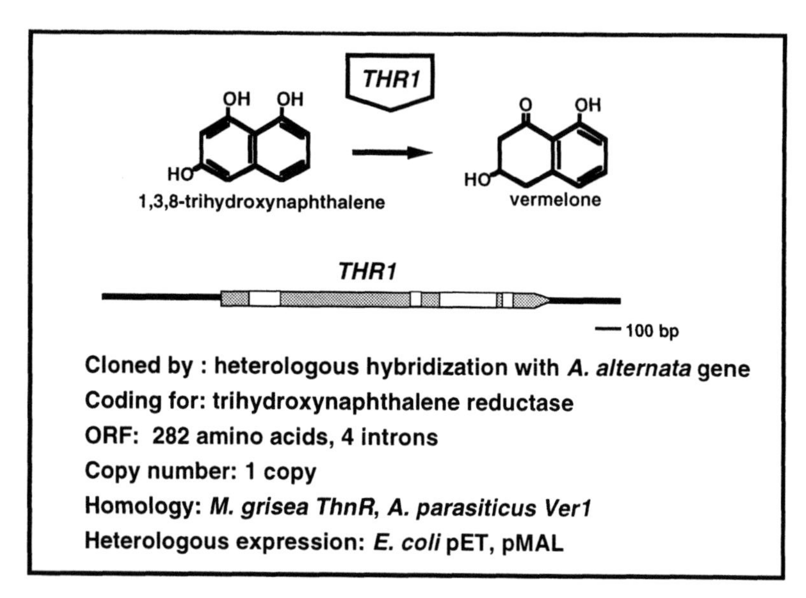

Fig. 3. Characterization of *THR1*.

reduction of 1,3,8-THN to vermelone. Gene disruption of the wild-type strain with a vector containing a piece of internal sequence of the *THR1* gene produced transformants with non-melanized appressoria with low infectivity. The deduced amino acid sequence of *THR1* showed 82% homology with the polyhydroxynapththalene reductase gene of *M. grisea* involved in converting 1,3,6,8-THN to scytalone and 1,3,8-THN to vermelone (38). Our recent genetic analysis indicated that *THR1* of *C. lagenarium* is also involved in the conversion of 1,3,6,8-THN to scytalone and that another type of reductase is involved in this conversion as well. (However, we have not cloned this presumed reductase.) The amino acid sequence of *THR1* also showed 56% homology with that of the *Ver1* gene of *A. parasiticus* associated with the conversion of versicolorin A to sterigmatocystin (31).

CMR1

We recently cloned a regulatory gene *CMR1* (GenBank accession number AB024516) involved in the expression of *SCD1* and *THR1* during mycelial melanin synthesis (13). *CMR1* was cloned by using *pig1* (GenBank accession number U38821) of *M. grisea* as a heterologous probe. *Pig1* is a gene controlling the amount of pigmentation in culture of *M. grisea*. We have previously isolated a conditional melanin-deficient mutant No. 8015 of *C. lagenarium* which showed little scytalone dehydratase activity during mycelial melanin synthesis and thus accumulated scytalone in the culture media, but showed normal melanization of appressoria indistinguishable from the wild-type strain (17,19). A cosmid containing *CMR1* complemented the mutant phenotype of No.8015, and gene-disrupted transformants with a vector containing an internal sequence of the *CMR1* gene showed a phenotype similar to mutant No. 8015. *CMR1* contained a deduced open reading frame of 958 amino acids consisting of five exons separated by four introns (Fig. 4).

The N-terminal region of the CMR1 protein consisted of a C_6 zinc binuclear cluster DNA binding motif which is most generally recognized among fungal transcription factors such as Gal4 of *Saccharomyces cerevisiae* and *aflR* of *A. nidulans*, *A. parasiticus*, and *A. flavus* (30,35) and a C_2H_2-type zinc finger, a DNA-binding motif of transcription factor reported first in TFIIIA of *Xenopus* (24) and later in *Saccharomyces* (3) (Fig. 5). Not many C_2H_2-type zinc finger transcription factors have been reported in fungi. A few examples are amdA (21) and amdX (27) of *A. nidulans* involved in expression of acetamidase, yeast ADR1 (10) involved in repression of alcohol dehydrogenase gene, yeast MIG1 (28), and *A. nidulans* CreA (7), negative regulators for carbon catabolite repression, *A. nidulans* BrlA (1), a transcriptional regulator for developmentally regulated

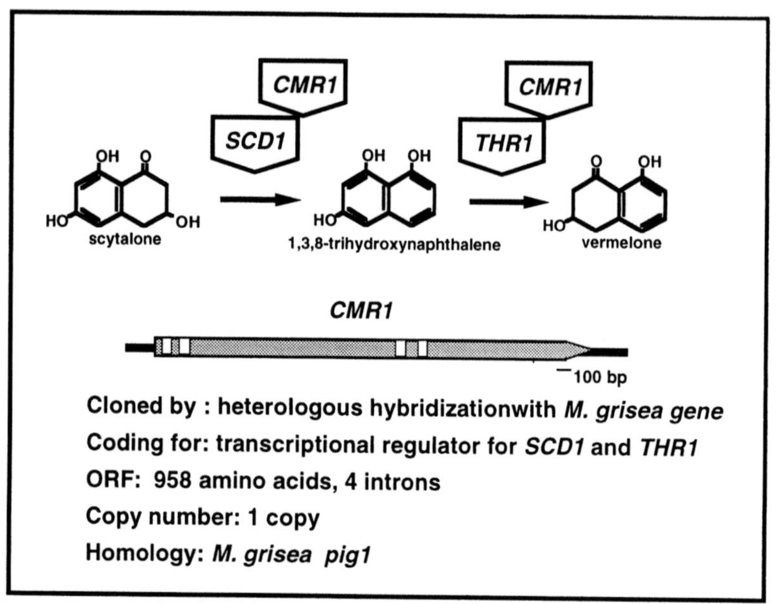

Fig. 4. Characterization of *CMR1.*

system (Fig. 5). A prominent feature of *CMR1* is the coexistence of both zinc binuclear and zinc finger motifs in a transcription factor; there is no precedent report on this. In CMR1 a domain rich in acidic amino acids such as aspartic acid and glutamic acid and also domains rich in glutamine or proline were recognized. Those sequences are generally considered as activation domains. The arrangement of the zinc finger DNA binding motif and those activation domains of *CMR1* was similar to a C2H2 type zinc finger transcription factor amdA of *A. nidulans* (Fig. 5).
The function of CMR1 expression on transcription of melanin biosynthesis genes in a *CMR1* disruptant was analyzed by RNA blotting. In the wild type, accumulation of transcripts of three melanin biosynthesis genes, polyketide synthase gene *PKS1*, scytalone dehydratase gene *SCD1* and trihydroxynaphthalene reductase gene *THR1* increased during mycelial melanization. However, in a *CMR1* disruptant, the accumulation of *SCD1* and *THR1* transcripts was quite low compared with the wild type. The level of accumulation of the *PKS1* transcripts was almost the same as the wild type. On the other hand, the level of accumulation of the three melanin biosynthesis genes during appressorial melanization was the same in the disruptant and the wild type. These results indicated that CMR1 is a positive transcriptional regulator for the expression of *SCD1* and *THR1* during mycelial melanin synthesis in *C. lagenarium.*

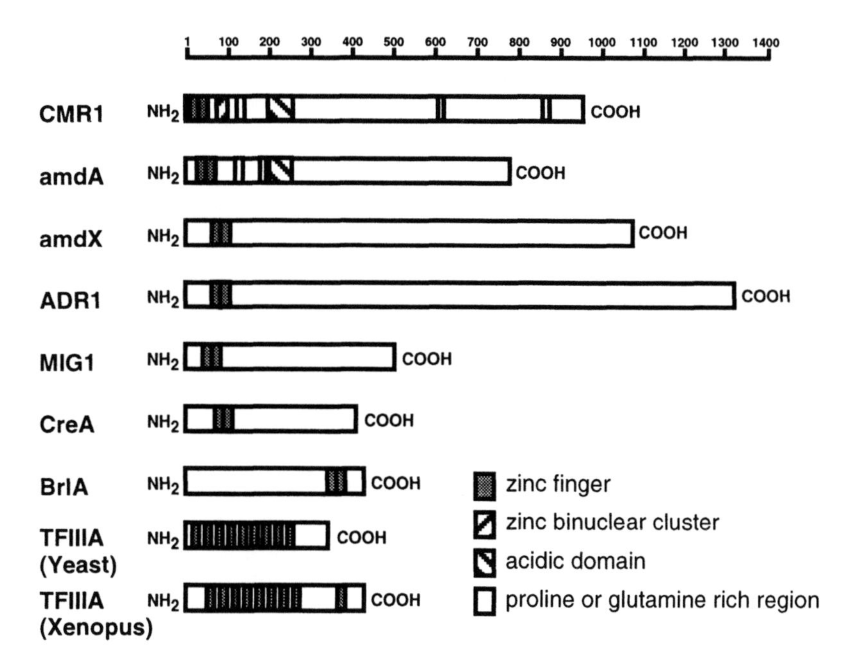

Fig. 5. Structure of fungal C_2H_2-type zinc finger transcriptional factors *CMR1* (13), amdA (21), amdX (27), ADR1 (10), MIG (28), CreA (7), BrlA (1), and TFIIIA (3,24).

Purification of Scytalone Dehydratase and 1,3,8-THN Reductase

We purified SCD1 and THR1 enzymes following heterologous expression of *SCD1* and *THR1* in *E. coli.* The SCD1 expression vector pMCSD1 was constructed with pMAL-C2. The maltose-binding protein-SCD1 fusion protein was purified, and the recombinant SCD1 was separated by cleavage with a proteinase Factor Xa. Following purification with a Mono-Q anion exchange column, a single protein band of 24 kDa was verified by SDS-PAGE (37). Crystal structural analysis of scytalone dehydratase showed that the enzyme is a trimer with a mass of 69 kDa (22). The association of SCD1 was also indicated from our recent study with a yeast two hybrid system. The most recent theoretical study of the mechanism of the dehydratase indicated the role of two active site tyrosine residues in the activity (39).

The purified recombinant SCD1 of *C. lagenarium* retained its activity. When the SCD1 protein was applied to a culture of a mutant defective in scytalone dehydratase activity, the mutant's ability to synthesize melanin was restored. In addition, in an *in vitro* reaction mixture using scytalone as a substrate, the purified enzyme converted scytalone to 1,3,8-THN. This *in vitro* reaction was inhibited by a novel melanin inhibiting anti-rice blast fungicide, carpropamid (20). From this experiment, the mode of action of the chemical is evidently the inhibition of scytalone dehydratase activity essential for melanin biosynthesis by appressoria (37). Salicylamide and 4-aminoquinazoline are also reported to inhibit scytalone dehydratase (11).

The THR1 expression vector pETTR1 was constructed using pET15b. The recombinant THR1 protein, which combined with six consecutive histidine residues at the N-terminal sequence, could be purified on a His-tag column. A single protein band of 31 kDa was verified by SDS-PAGE. Polyhydroxynaphthalene reductase of *M. grisea* was also purified with the weight of 29.9 kDa (38). Crystal structural analysis of the *M. grisea* reductase containing NADPH and tricyclazole, a reductase inhibitor, revealed structural details of the enzyme (2).

Regulation of *PKS1*, *SCD1*, and *THR1* During Appressorium Differentiation

Melanization in *C. lagenarium* is a developmentally regulated system confined to appressoria; conidia and germ tubes are not melanized. There is little information concerning the mechanism(s) that confines melanin to appressoria. To elucidate the regulatory mechanisms of melanin biosynthesis, we determined the temporal accumulation pattern of transcripts of *PKS1*, *SCD1*, and *THR1* as well as the translational products of SCD1 and THR1 in appressorium-differentiating and nondifferentiating germlings (33,36).

C. lagenarium wild-type strain 104-T produces melanized appressoria that display temperature-sensitive differentiation. Conidia germinated to form melanized appressoria at 24°C; however, at 32°C, germ tubes elongated without differentiating appressoria. Appressoria also failed to develop when conidia were incubated in 0.1% yeast extract solution (Fig. 6). At 24°C, 70% of the conidia incubated in water began to germinate within 2 to 4 hr, and 70% of the conidia produced appressoria by 6 hr. Visual melanization started at 6 hr, and mature melanized appressoria were observed at 12 hr (Fig. 7). We investigated the temporal transcriptional pattern of *PKS1*, *SCD1*, and *THR1*. In appressorium-differentiating germlings, no transcripts of the three melanin biosynthetic genes were detected at 0 hr. Transcripts of the three genes accumulated *de novo* 1 to 2 hr after the start of incubation, at an early developmental stage when germ tubes were not yet visible. The accumulation of transcripts of these genes increased until 4 hr, then began to decrease at 6 hr. By 18 hr, transcripts of the three genes were barely detectable (Fig. 7). In nondifferentiating germlings incubated at 32°C, the transcripts also accumulated *de novo*

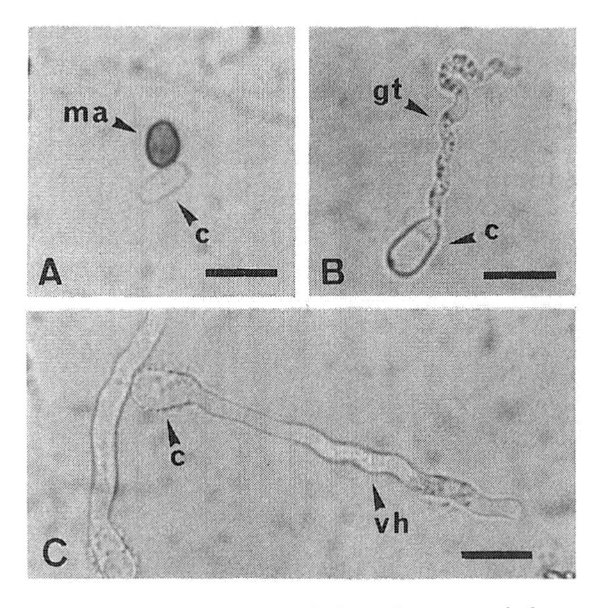

Fig. 6. Differentiation and nondifferentiation of appressoria by germlings of *C. lagenarium*. Conidia were incubated for 12 hr in water at 24°C (A), in water at 32°C (B), or in 0.1% yeast extract solution at 32°C (C).

similarly to differentiating conidia, even though these conidia did not produce any appressoria with melanin (Fig. 8). This result indicated the involvement of post-transcriptional regulation of melanin biosynthetic genes which could not function properly in nondifferentiating germlings. On the other hand, the transcripts of the three genes were weakly detected from conidia forming mycelia incubated in 0.1% yeast extract at 32°C. We found that conidia also formed mycelia when incubated in 0.1% tryptone solution at 32°C. We assume that the expression of the melanin biosynthetic genes was repressed in the vegetative hyphae produced in the presence of complex nutrients. As indicated in the previous section, if CMR1 acted as a positive regulator for the expression of *SCD1* and *THR1* during mycelial melanization, then CMR1 might inactivate a presumptive repressor during mycelial growth.

Western blot analysis was performed on appressorium-differentiating germlings at 24°C and nondifferentiating germlings at 32°C. In appressorium-differentiating germlings, neither the deduced 23-kDa scytalone dehydratase signal (SCD1) nor the 30-kDa 1,3,8-THN reductase signal (THR1) was detected by 4 hr after incubation. The deduced SCD1 and THR1 protein signals were detected from 6 hr after incubation, the time corresponding to the start of visual appressorial melanization. The intensity

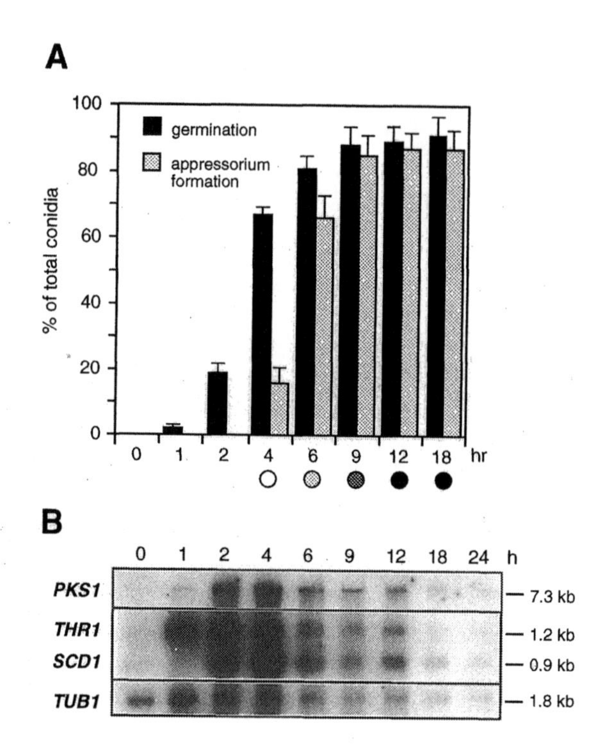

Fig. 7. Time-course of expression of the *PKS1*, *SCD1*, and *THR1* genes during appressorium formation by *C. lagenarium*. (A) Time-course of germination and appressorium formation by *C. lagenarium* conidia incubated in water at 24°C. The shading in the circles represents the degree of appressorial melanization. (B) RNA blot analysis showing time-course of *PKS1*, *SCD1*, *THR1*, and *TUB1* gene expression during appressorial differentiation.

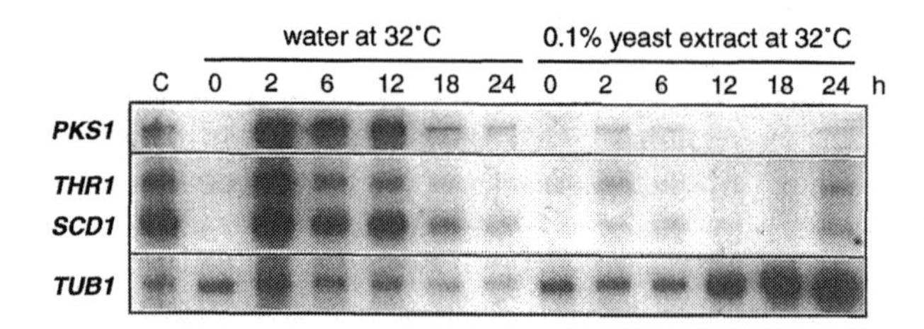

Fig. 8. RNA blot analysis showing time-course of expression of the *PKS1*, *SCD1*, *THR1*, and *TUB1* genes during development of nondifferentiating germlings. Conidia were incubated in water or 0.1% yeast extract solution at 32°C.

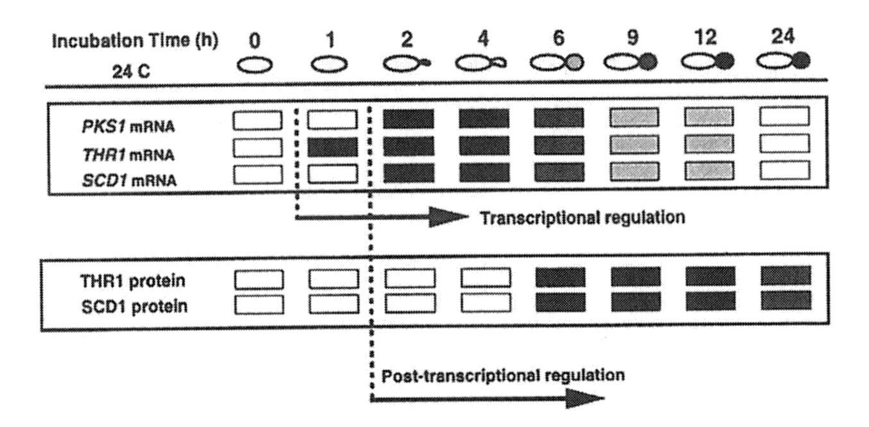

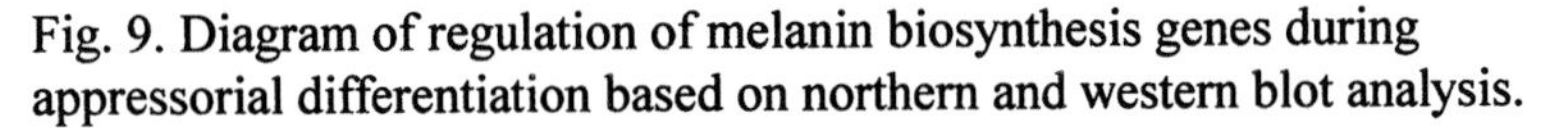

Fig. 9. Diagram of regulation of melanin biosynthesis genes during appressorial differentiation based on northern and western blot analysis.

of the band was constant from 6 to 24 hr. In nondifferentiating germlings incubated at 32°C, the deduced SCD1 and THR1 protein signals were not detected. However, the 38-kDa and 60-kDa bands which seem to be related to SCD1 and THR1, respectively, were detected from 2 to 24 hr after incubation. In appressorium-differentiating germlings incubated at 24°C, bands with a similar size were also detected from 2 to 4 hr after incubation, at the growth stage before appressorium differentiation which corresponds to the transcriptional periods of *SCD1* and *THR1* genes. The transcripts of melanin biosynthesis genes are considered to be translated before 2 hr after incubation because appressorium formation and melanization were not affected by the addition of cycloheximide, a protein synthesis inhibitor, later than 2 hr after the start of the incubation (18). These results suggest the possibility that post-translational regulation, as well as transcriptional regulation, is involved in melanin biosynthesis during appressorium formation (Fig. 9).

Conclusion

The work on regulation of melanin biosynthesis during appressorium development of *C. lagenarium* indicated the existence of an integrated regulation system which is coordinated with appressorium development. Transcriptional and post-transcriptional regulation was involved in melanin biosynthesis. A regulatory gene *CMR1* was identified as a positive transcriptional regulator for *SCD1* and *THR1* in mycelia. Still, there may be other positive or negative transcriptional regulators involved in gene expression. The complete cascade leading to the expression of melanin biosynthetic genes has yet to be elucidated. Elucidating the type of post-transcriptional regulation of the melanin biosynthesis genes which are coordinated with appressorium differentiation will also be fascinating. Cytological analysis of appressorial melanization will also be important in understanding the regulation of melanin biosynthesis. Immunoelectron microscopy using antibodies against melanin biosynthetic enzymes could determine the intracellular localization of these enzymes during cellular differentiation and could help to unravel their post-transcriptional regulation. Further research on the development of melanin biosynthesis should clarify the unique regulatory system related to the morphogenesis of this fungal pathogen.

Acknowledgements

This work was supported, in part, by grants from the Ministry of Education, Science and Culture, Japan (Nos. 09460031 and 10760031). We are grateful to Dr. James Sweigard, Central Research and Development, E.I. DuPont de Nemours and Co. for kindly providing *pig1* cDNA of *Magnaporthe grisea*. We are also grateful to Dr. Isao Fujii, Tokyo University for identification of melanin intermediates.

Literature Cited

1. Adams, T.H., Boylan, M.T., and Timberlake, W.E. 1988. *brlA* is necessary and sufficient to direct conidiophore development in *Aspergillus nidulans*. Cell 54:353-362.
2. Andersson, A., Jordan, D., Schneider, G., and Lindqvist, Y. 1997. Crystalline structure of the ternary complex of 1,3,8-trihydroxynaphthalene reductase from *Magnaporthe grisea* with NADPH and an active site inhibitor. Structure 4:1161-1170.
3. Archambault, J., Milne, C.A., Schappert, K.T., Baum, B., Friesen, J.D., and Segal, J. The deduced sequence of the transcription factor THIIIA from *Saccharomyces cerevisiae* reveals extensive divergence from *Xenopus* TFIIIA. J. Biol. Chem. 267:3282-3288.
4. Beck, J., Ripka, S., Siegner, A., Schiltz, E., and Schweizer, E. 1990. The multifunctional 6-methylsalicylic acid synthase gene of *Penicillium patulum*: Its gene structure relative to that of other polyketide synthases. Eur. J. Biochem. 192:487-498.
5. Bell, A.A. and Wheeler, M.H. 1986. Biosynthesis and function of fungal melanins. Annu. Rev. Phytopathol. 24:411-451.

6. DeJong, J.C., McCormack, B.J., Smirnoff, N., and Talbot, N.J. 1977. Glycerol generates turgor in rice blast. Nature 389:244-245.
7. Dowzer C.E.A. and Kelly, J.M. 1991. Analysis of the *creA* gene, a regulator of carbon catabolite repression in *Aspergillus nidulans*. Mol. Cell. Biol. 11:5701-5709.
8. Emmett, R.W. and Parbery, D.G. 1997. Appressoria. Annu. Rev. Phytopathol. 13:147-167.
9. Fujii, I., Ono, Y., Tada, H., Gomi, K., Ebizuka, Y., and Sankawa, U. 1996. Cloning of the polyketide synthase gene *atX* from *Aspergillus terreus* and its identification as the 6-methylsalicylic acid synthase gene by heterologous expression. Mol. Gen. Genet. 253:1-10.
10. Hartshorne, T.A., Blumberg, H., and Young, R.T. 1986. Sequence homology of the yeast regulatory protein ADR1 with *Xenopus* transcription factor TFIIIA. Nature 320:283-287.
11. Hodge, C.N. and Pierce, J. 1993. A diazine heterocycle replaces a six membered array in the active site of scytalone dehydratase. Bioorg. Med. Chem. Lett. 3:1605-1608.
12. Howard, R.J. and Ferrari, M.A. 1989. Role of melanin in appressorium function. Exp. Mycol. 13:403-418.
13. Kenmochi, Y. 1997. Construction of genomic library and cloning of *CMR1* gene involved in regulation of melanin biosynthesis of *Colletotrichum lagenarium*. B.S. thesis, Kyoto Prefectural University, Kyoto, Japan. (in Japanese)
14. Kimura N. and Tsuge, T. 1993. Gene cluster involved in melanin biosynthesis of the filamentous fungus *Alternaria alternata*. J. Bacteriol. 175:4427-4435.
15. Kubo, Y. and Furusawa, I. 1991. Melanin biosynthesis: Prerequisite for successful invasion of the plant host by appressoria of *Colletotrichum* and *Pyricularia*. Pages 205-218 in: The Fungal Spore and Disease Initiation in Plants and Animals. G.T. Cole and H.C. Hoch, eds. Plenum Press, New York.
16. Kubo, Y., Nakamura, H., Kobayashi, K., Okuno, T., and Furusawa, I. 1991. Cloning of a melanin biosynthetic gene essential for appressorial penetration of *Colletotrichum lagenarium*. Mol. Plant-Microbe Interact. 4:440-445.
17. Kubo, Y., Suzuki, K., Furusawa, I., and Yamamoto, M. 1983. Scytalone as a natural intermediate of melanin biosynthesis in appressoria of *Colletotrichum lagenarium*. Exp. Mycol. 7: 208-215.
18. Kubo, Y., Suzuki, K., Furusawa, I., and M. Yamamoto 1984. Regulation of melanin biosynthesis during appressorium formation in *Colletotrichum lagenarium*. Exp. Mycol. 8:364-369.
19. Kubo, Y., Takano, Y., Endo, N., Yasuda, N., Tajima, S., and Furusawa, I. 1996. Cloning and structural analysis of the melanin biosynthesis

gene *SCD1* encoding scytalone dehydratase in *Colletotrichum lagenarium*. Appl. Environ. Microbiol. 62:4340-4344.

20. Kurahashi, Y., Sakawa, S., Kinbara, T., Tanaka, K., and Kagabu, S. 1997. Biological activity of carpropamid (KTU3616)—a new fungicide for rice blast disease. J. Pestic. Sci. 22:108-112.
21. Lints, R., Davis, M.A., and Hynes, M.J. 1995. The positively acting *amdA* gene of *Aspergillus nidulans* encodes a protein with two C2H2 zinc-finger motifs. Mol. Microbiol. 15:965-975.
22. Lundqvist, T., Rice, J., Hodge, C.N., Basarab, G.S., Pierce, J., and Lundqvist, Y. 1994. Crystal structure of scytalone dehydratase—a disease determinant of the rice pathogen *Magnaporthe grisea*. Structure 2:937-944.
23. Mayorga, M.E and Timberlake, W.E. 1992. The developmentally regulated *Aspergillus nidulans wA* gene encodes a polypeptide homologous to polyketide and fatty acid synthases. Mol. Gen. Genet. 235: 205-212.
24. Miller, J., McLachlan, A.D., and Klug, K. 1985. Repetitive zinc binding domains in the protein transcription factor IIIA from *Xenopus* oocytes. EMBO J. 4:1609-1614.
25. Money, N.P. 1997. Mechanism linking cellular pigmentation and pathogenicity in rice blast disease. Fungal Genet. Biol. 22:151-152.
26. Motoyama, T., Imanishi, K., and Yamaguchi, I. 1998. cDNA cloning, expression and mutagenesis of scytalone dehydratase needed for pathogenicity of the rice blast fungus *Pyricularia oryzae.* Biosci. Biotech. Biochem. 62:564-566.
27. Murphy, R.L. Andrianopoulos, A., Davis, M.A., and Hynes, M.J. 1997. Identification of *amdX*, a new Cys-2-His-2 (C2H2) zinc-finger gene involved in the regulation of the *amdS* gene of *Aspergillus nidulans.* Mol. Microbiol. 23:591-602.
28. Nehlin, J.O. and Ronne, H. 1990. Yeast *MIG1* repressor is related to the mammalian early growth response and Wilms' tumor finger proteins. EMBO J. 9:2891-2898.
29. Perpetua, N.S., Kubo, Y., Takano, Y., and Furusawa, I. 1996. Cloning and characterization of a melanin biosynthetic *THR1* reductase gene essential for appressorial penetration of *Colletotrichum lagenarium.* Mol. Plant-Microbe Interact. 9:323-329.
30. Schjerling, P. and Holmberg, S. 1996. Comparative amino acid sequence analysis of the C6 zinc cluster family of transcriptional regulators. Nucleic Acids Res. 23: 4599-4607.
31. Skory, C.D., Chang, P.K., Cary, J., and Linz, J.E. 1992. Isolation and characterization of a gene from *Aspergillus parasiticus* associated with the conversion of versicolorin A to sterigmatocystin in aflatoxin biosynthesis. Appl. Environ. Microbiol. 58:3527-3537.

32. Tajima, S., Kubo, Y., Furusawa, I., and Shishiyama, J. 1989. Purification of a melanin biosynthetic enzyme converting scytalone to 1,3,8-trihydroxynaphthalene from *Cochliobolus miyabeanus*. Exp. Mycol. 13:69-76.
33. Takano, Y., Kubo, Y., Kuroda, I., and Furusawa, I. 1997. The temporal transcriptional pattern of three melanin biosynthesis genes *PKS1*, *SCD1* and *THR1* in appressorium-differentiating and nondifferentiating conidia of *Colletotrichum lagenarium*. Appl. Environ. Microbiol. 63:351-354.
34. Takano, Y., Kubo, Y., Shimizu, K., Mise, K., Okuno, T., and Furusawa, I. 1995. Structural analysis of *PKS1*, a polyketide synthase gene involved in melanin biosynthesis in *Colletotrichum lagenarium*. Mol. Gen. Genet. 249:162-167.
35. Todd, R.B. and Andrianopoulos, A. 1997. Evolution of a fungal regulatory gene family: the Zn(II)2Cys6 binuclear cluster DNA binding motif. Fungal Genet. Biol. 21:388-405.
36. Tsuji, G. 1997. Molecular biological analysis of melanin biosynthetic enzymes of *Colletotrichum lagenarium* using recombinant enzymes expressed in *Escherichia coli*. M.S. Thesis, Kyoto Prefectural University, Kyoto, Japan.
37. Tsuji, G., Takeda, T., Furusawa, I., Horino, O., and Kubo, Y. 1997. Carpropamid, an anti-rice blast fungicide, inhibits scytalone dehydratase activity and appressorial penetration in *Colletotrichum lagenarium*. Pestic. Biochem. Physiol. 57:211-219.
38. Vidal-Cros, A., Viviani, F., Labesse, G., Bocara, M., and Gaudry, M. 1994. Polyhydroxynaphthalene reductase involved in melanin biosynthesis in *Magnaporthe grisea*. Eur. J. Biochem. 219: 985-992.
39. Zheng, Y.J. and Bruice, T.C. 1998. Role of a critical water in scytalone dehydratase catalyzed reaction. Proc. Nat. Acad. Sci. USA 95:4158-4163.

Chapter 8

Colletotrichum as a Model System for Defining the Genetic Basis of Fungal Symbiotic Lifestyles

Rusty J. Rodriguez and Regina S. Redman

Fungal Symbiotic Lifestyles

Symbiosis was first defined as "the living together of unlike organisms" by DeBary in 1879 (17,32). DeBary recognized that symbiotic associations between different microorganisms or animals and microorganisms could either be commensalistic (neutral to both symbionts), mutualistic (beneficial to both symbionts), or parasitic (beneficial to one, antagonistic to the other). Since that time, it has become apparent that fungi establish symbioses with plants that encompass commensalistic, mutualistic, and parasitic associations. The genetic and biochemical bases of these different fungal symbiotic lifestyles are not known, but they all require fungi to infect and colonize host tissues. Many symbiotic fungi, including *Colletotrichum* species, have evolved specialized infection structures (appressoria) to infect plant hosts. Once inside the host, the type of symbiosis that ensues is dependent on genetic and biochemical factors contributed by the host and symbiont; therefore, fungal symbiotic lifestyles differ in the events that transpire after infection. Although there have been extensive studies on the interactions between fungi expressing different symbiotic lifestyles and their respective hosts, the genetic elements controlling the outcome of these interactions remain unknown.

In an attempt to define the genetic and biochemical bases of different fungal symbiotic lifestyles, we needed a fungal system that was conducive to laboratory culture, molecular genetics, and genetic segregation analysis and was expressive of different symbiotic lifestyles. Although these attributes may not all be found in a single fungal species, they do occur among species from the genus *Colletotrichum*.

Colletotrichum as a Model System

The genus *Colletotrichum* contains many filamentous fungal species which collectively infect virtually all major agricultural crops grown worldwide. Species from this genus vary tremendously in host range with some infecting only a single plant species while others cause disease on more than fifty hosts (24). One major misconception with host range designations is that they include only the plant species on which fungi cause disease and do not take into account the species that fungi are capable of colonizing. Although it is not feasible to determine the true host range of every fungus, it is important to discriminate between disease host range (plants that express disease symptoms after infection by fungi) and true host range (plants colonized by fungi regardless of disease expression). This discrimination is particularly important in *Colletotrichum* because ecological studies from several laboratories have shown that *Colletotrichum* species are commonly isolated as epiphytes and asymptomatic endophytes from a large variety of plant species.

The physical process of host infection and colonization has been extensively studied in several species of *Colletotrichum* revealing three mechanisms of pathogenesis: hemibiotrophy, necrotrophy, and latency. Although the infection process has been studied in great detail, little information has been obtained to explain the genetic basis of host colonization and the type of symbiotic lifestyle expressed. Collectively, the genus *Colletotrichum* contains species that are asymptomatic endophytes, primary pathogens, and latent pathogens. Although it is not yet known if the asymptomatic endophytes are acting as commensals, mutualists, or latent pathogens, *Colletotrichum* species are excellent candidates for studying the genetic basis of plant/fungal symbiotic lifestyles.

Colletotrichum species have several attributes which facilitate their use as experimental systems. Mycelial cells and conidia are easily cultured on defined media, susceptible to mutagenic treatments, and conidia from cultures grown on solid media are $\geq$97.7% mononucleate (12,13,56,81,84). In addition, several species are amenable to genetic segregation analysis, molecular biological manipulations, and *in vitro* pathogenicity analysis (described later).

PATHOGENICITY OF *COLLETOTRICHUM*

The physical process of host colonization and pathogenesis has been extensively studied in several species of *Colletotrichum* and is similar in many plant pathogens (28,47,60,61,62) (also see Chapter 5 by O'Connell et al. in this volume). After a spore adheres to host tissue, the spore germinates to form a germ tube which, when stimulated to penetrate, differentiates into an infection structure called an appressorium.

Prior to penetration of the first host cell, the appressorium becomes black as it produces the pigment melanin which strengthens the appressorial wall (48). Next, an infection peg is formed which physically penetrates through the plant cell wall. Penetration may involve the assistance of extracellular enzymes which can degrade components of plant cell walls (1,3,20,21,40). After successful penetration by the fungus into the first plant cell, a compatible or incompatible interaction will ensue.

In a compatible interaction, the fungus emerges from the first infected cell without activating the host defense system (5,46). After the fungus has begun ramifying through host tissue, the initially infected cells disintegrate, and several plant and fungal metabolites are released which initiate a defense response at that location resulting in a necrotic lesion (5). As the fungus grows through the host, a path of necrotic tissue is left behind which further activates the defense system. The growing tip of the fungus remains spatially and/or temporally ahead of the plant defense systems.

In an incompatible interaction, the fungus enters the first cell, activates the host defense response system which results in the collapse of the plant cells surrounding the infection site, thereby "walling off" the pathogen (2,5,22,75). Presumably, in the initial stages of an incompatible reaction, receptors on the plant cell surface recognize specific pathogen-generated molecules which initiates the activation of cellular transduction pathways leading to the rapid accumulation of specific gene transcripts within the plant (19,39,46). Thereafter, the host activates complex biochemical pathways inducing a resistant reaction usually accompanied by a hypersensitive response (HR) (7,15,19,23,25,29,30,31,36,43,45,51,54,88).

One of the major differences between compatible and incompatible interactions appears to be the timing of activation of the defense system. In fact, Kuć and Strobel (52) have implied that susceptible cultivars are capable of resisting pathogen attack by altering the timing and magnitude of the defense response. After extensive analyses of fungal pathogenesis, many correlations have been established between host resistance or susceptibility, and the expression of specific pathogen and host enzymes or metabolites. However, the processes by which fungi cause disease in plants remains undefined.

Classical Genetics

The number of species in the genus *Colletotrichum* is still in question (for a detailed discussion of *Colletotrichum* systematics, see Chapter 1 by Cannon et al.). Different taxonomic guides indicate that the numbers of species range between 20 to greater than 700 (4,78), with the most recent report suggesting 39 species (24). Regardless of the exact number of species, only a small percentage have discernible sexual cycles which are classified as either homothallic (86), heterothallic (86), or intermediate to homothallic and heterothallic (6). The sexual stage of *Colletotrichum* species is designated *Glomerella* and has been reported for *C. gloeosporioides* (4,14), *C. gossypii* (24), *C. graminicola* (85), *C. lindemuthianum* (44), *C. destructivum* (55), *C. falcatum* (59), *C. orbiculare* (57), *C. magna* (38), *C. acutatum* (see Chapter 10 by Correll), and *C. musae* (70). Although sexual cycles have been reported for these species, meiotic recombination has been confirmed in only a few (13,14,70,85).

MOLECULAR GENETICS

Work from several laboratories around the world has shown that *Colletotrichum* species are amenable to all aspects of recombinant DNA technology. Several species have been transformed with foreign DNA including *C. lindemuthianum, C. magna,* and *C. musae* (66); *C. graminicola* (63), *C. trifolii,* and *C. gloeosporioides* (18). These transformations have been done with a number of different selectable genes and promoters including the *amdS* gene of *Aspergillus nidulans* (35), the *hygB* gene of *E. coli* (41), and benomyl resistance genes (63).

Initially, transformation efficiencies using these selectable genes varied between 10 and 50 transformants per microgram (Txs/µg) of vector DNA. However, a combination of protocol modifications and the use of a new transformation vector have increased transformation efficiency of *Colletotrichum* species to 1000-5000 Txs/µg DNA (66,71). The protocol modifications increased protoplast viability from 0.5% to 5.0% and increased transformation efficiency from 10-50 Txs/µg DNA to 100-200 Txs/µg DNA (71). Use of the vector pHA 1.3 (64) increased transformation efficiency to 1000-5000 Txs/µg (66). Depending on the *Colletotrichum* species, pHA1.3 can transform by any of three mechanisms: 1) integration into the genome, 2) autonomous replication, or 3) concomitant integration and autonomous replication. In addition, a high efficiency cosmid for *Colletotrichum* species which transforms by all three mechanisms has been constructed (unpublished data).

Although high transformation efficiencies have been obtained for *Colletotrichum* species, transformation efficiencies vary between species, and the transformation protocol needs to be optimized for each species. For example, initial studies with *C. magna* produced transformation efficiencies of 0-5 Txs/µg DNA. However, we have recently made several modifications in the transformation protocol such that efficiencies of 1000 Txs/µg are routinely obtained for *C. magna.*

In working with these fungi it has become routine to perform recombinant DNA procedures including production of genomic libraries, isolation of mRNA, hybridization analyses, cloning and sequencing, and expression studies, and protein analyses. Several genes have been cloned characterized from *Colletotrichum* species including genes encoding cutinase (77), tubulin (63), melanin biosynthetic enzymes (49,50,79,80), orotate phosphoribosyl transferase (65), protein kinase (9), glyceraldehyde-3-phosphate dehydrogenase (82), pectin lyase (83), endopolygalacturonase (11), and two appressoria-specific genes (33,34). However, there have not been any reports on the identification or cloning of pathogenicity genes; therefore, the molecular basis of the disease process remains unknown.

We have developed a fungal DNA minipreparation protocol which is rapid and economical, allowing for 90 samples to be processed in 6-hr (66,69). This procedure typically yields between 200 and 500 µg of large molecular weight DNA (100-200 kb) from a 100 ml fungal culture and purifies the DNA away from long chain polyphosphates which often co-purify with fungal DNA. Polyphosphate molecules synthesized by many fungi inhibit restriction enzymes, DNA ligase, and DNA polymerases (69). The DNA obtained from this procedure is adequate for molecular manipulations.

Defining the Genetic Basis of Fungal Symbiotic Lifestyles

To investigate the molecular genetic basis of fungal symbiotic lifestyles in this genus, *C. magna* was chosen as the experimental system. This species has a heterothallic mating system (87), is easily mutated (26), transformed (66,71), and assayed for pathogenicity in growth chambers (26). *C. magna* has a wide host range among different genera from the plant family Cucurbitaceae which allows us to address questions concerning host colonization and pathogenesis in several plant species (37).

C. magna was mutagenized with ultraviolet light (UV) to determine if pathogenesis could be eliminated without interfering with the infection and/or colonization processes. To efficiently screen putative nonpathogenic mutants, we developed a bioassay for *C. magna* on cucurbits (26). The standard leaf inoculation assays previously used for evaluating *Colletotrichum* species were time- and labor-intensive and required considerable greenhouse and/or growth chamber space (8,16,76). We developed a simple, rapid, and reliable method which allowed us to screen several hundred survivors of UV-light mutagenesis and isolate a nonpathogenic mutant of *C. magna* (26,27). The bioassay is based on exposing roots and lower stems of seedlings to fungal conidia in small beakers. Using this bioassay, three nonpathogenic mutants were isolated following UV mutagenesis. One of the mutants, designated path-1, behaved similarly to the wild type in laboratory culture studies.

CHARACTERIZATION OF A NON-PATHOGENIC MUTANT

The path-1 mutant did not cause disease symptoms on any cucurbit species, but it retained the ability to infect and colonize host tissues. In addition to its nonpathogenic endophytic phenotype, we observed that plants colonized with path-1 were protected 100% and 85% against disease caused by virulent isolates of *C. magna* or *Fusarium oxysporum*, respectively (27). These pathogens have different mechanisms of pathogenesis which suggested that the protection was not based on the physical interference of the virulent fungi by path-1 (27). *In vitro* assays indicated that path-1 was not producing chemicals that inhibited the virulent fungi (27). Recently, we demonstrated that path-1-colonized watermelon plants were protected 90-100% against disease caused by *C. orbiculare* (67). Collectively, these data indicated that the path-1 mutation converted a virulent pathogen to a nonpathogenic endophytic mutualist.

The occurrence of path-1-induced plant protection is dependent on the ability of path-1 to infect and colonize the plants. A host-range study revealed that path-1 retained the wild-type host range but caused no disease on any of the plants tested (27). From these data, we conclude that in *C. magna*, the genetic basis of pathogenicity and host colonization involve at least some unique genetic elements. In addition, genetic segregation analysis indicated that the path-1 phenotype involved the mutation of a single genetic locus required for the expression of disease symptoms after the fungus has infected the plant (27). Therefore, the mutation of a single genetic locus appears to have changed the symbiotic lifestyle of *C. magna* from pathogen to mutualist.

The ability of path-1 to grow through host tissue without eliciting disease may be dependent on either suppression or avoidance of the host defense systems. Since the path-1 colonized plants are resistant to virulent wild-type fungi (*C. magna, C. orbiculare,* and *F. oxysporum*), path-1 is probably not suppressing the host defense systems. We have performed biochemical studies which indicate that path-1 avoids activation of host defenses and that path-1-colonized plants when challenged with virulent isolates, activate host defenses more quickly than control plants (67). These results were determined by monitoring the activity of peroxidase and the deposition of lignin in path-1-colonized and control plants.

To determine the stability of the nonpathogenic phenotypes, 50 single-spore colonies were derived from path-1 and analyzed for pathogenicity. None of the clonal isolates were capable of expressing disease symptoms and the phenotypes have been maintained in laboratory culture for over 4 years (unpublished data).

REMI Mutant Isolation

To determine the genetic complexity of mutations capable of changing the symbiotic lifestyle of *C. magna,* we generated nonpathogenic mutants by gene disruption using REMI (restriction enzyme mediated integration) transformation, thereby tagging pathogenicity genes with a transformation vector (53). By optimizing REMI transformation for *C. magna,* we now routinely obtain efficiencies of 1000-2000 Txs/µg DNA (Redman and Rodriguez, in preparation). To obtain REMI mutants, the pHA1.3 plasmid was linearized with *Hind*III and transformed into *C. magna* isolate L2.5 in the presence of *Hind*III as described previously (66). Under these conditions, pHA1.3 transforms only by integration into the fungal genome (66). This strategy allowed us to generate 14,400 hygromycin-resistant REMI Txs (data not shown). To screen these REMI Txs for nonpathogenic mutants, we modified the bioassay to rapidly screen thousands of REMI Txs as follows: hygromycin-resistant REMI Txs were inoculated onto solid MS medium (84) and incubated under conditions to induce sporulation (26). After the cultures conidiated, spores were streaked along the length of MS medium in a 13-mm test tube slant and incubated to induce conidiation. After the slant cultures had sporulated, 4 ml of a 0.07% agarose solution was added and the spores suspended by vortexing. Three watermelon seedlings (7- to 10-day-old) were placed in the spore suspensions with the root systems and lower stems submerged. The inoculated plants were incubated in a growth chamber (operating at 23°C, 95% humidity, and 12 hr of mixed fluorescent and incandescent lights daily) for 5 days and assessed for disease. Since 72 tubes were contained in a single test tube rack, each growth chamber could hold approximately 2000 assays/shelf. Using this new protocol, we screened 14,400 REMI transformants and have isolated 176 nonpathogenic REMI mutants (Redman and Rodriguez, in preparation).

REMI Mutant Characterization

All of the nonpathogenic REMI mutants were characterized for sporulation levels, adhesion (74), germination, appressoria formation, growth rates, elicitation of plant disease (26), colonization of plant tissue, and the ability to protect plants against wild-type challenge (data not shown). All of the mutants expressed wild-type levels of sporulation, adhesion, germination, appressoria formation, and growth rates. None of the mutants elicited disease on watermelon seedlings in our bioassay but all were capable of colonizing plants (26). However, the mutants varied in their ability to protect watermelon plants when challenged with lethal conidial concentrations of virulent fungi (*C. magna* and *C. orbiculare*). The mutants represented three nonpathogenic phenotypes: A, colonized and fully protected plants against virulent fungi; B, colonized and partially protected plants against virulent fungi; and C, colonized but did not protect plants against virulent fungi. Phenotype B may represent several different phenotypes, however, that will be assessed after segregation analyses are completed.

Hybridization analysis of 20 nonpathogenic REMI mutants representative of phenotypes A, B, and C revealed that 43% of the mutants resulted from single site integrations. In addition, pathogenicity and hybridization data indicated that several different fungal genomic sequences were responsible for the nonpathogenic phenotypes.

As mentioned, *C. magna* has a sexual stage which has been designated *Glomerella magna* (38). The mating system of *C. magna* has been arbitrarily designated as + and –, and the fungus is described as a heterothallic ascomycete (38). We have confirmed that the meiotic cycle of *C. magna* is heterothallic and demonstrated that genetic markers segregate as Mendelian traits (27). Recently, we modified culture conditions so that the sexual stage of *C. magna* is consistently generated in 10-14 days. The new conditions involve the use of autoclaved young corn leaves as a solid substrate (85) placed in 0.1X PDA medium. Mycelial plugs or spores of opposite mating types were inoculated on opposite ends of the corn leaves and the cultures incubated at 23° C with 12 hr of fluorescent light daily. This resulted in the production of 20 to 40 perithecia on each corn leaf with each perithecium containing up to 12 asci and each ascus containing up to eight ascospores. Viability of ascospores from individual perithecia varied from 20 to 80%.

Two nonpathogenic REMI mutants (R1 and R21) with single site integrations have been genetically analyzed. These REMI mutants had the same phenotype (A), but the transformation vector integrated into different genomic locations. Segregation analysis involved 90 ascospore progeny that equally represented six perithecia for each REMI mutant. In both R1 and R21, the nonpathogenic phenotype segregated in a 1:1 manner, and was linked 100% to hygromycin resistance conferred by the transformation vector (pHA1.3). These data suggest that the transformation vector integrated into the genome and modified the expression of one or more pathogenicity genes.

CLONING REMI DNA

The integrated vector and flanking genomic DNA sequences in REMI mutant R1 was re-isolated and cloned (Redman and Rodriguez, unpublished) by digesting genomic DNA from R1 with the restriction enzyme *XhoI* which does not have a recognition site in the pHA1.3 vector. This created a linear DNA fragment containing the original transformation vector flanked by genomic DNA representing the integration site and the putative pathogenicity gene(s). The digested genomic DNA was ligated, transformed into *E. coli* strain DH5a, and transformants selected for ampicillin resistance conferred by pHA1.3. The resulting product was found to be approximately 11 kb by hybridization analysis and designated pGMR1.

Once recovered, REMI plasmids containing disrupted pathogenicity genes are useful tools for performing single-step gene disruption experiments in wild-type isolates (72). Gene disruption by the REMI plasmid is dependent on the homologous integration of the genomic DNA flanking the pHA1.3 vector in the REMI plasmid into an active gene in the genome. Transformation with vectors containing interrupted genes have been used to induce gene disruptions in several fungal systems (34,42,58). In *Colletotrichum* species, homologous integration of transformation plasmids occurs up to 80% of the time when the plasmid contains 0.5-1.0 kb of DNA from the the same species (68). Gene disruptions in this genus have been accomplished by homologous integration of transformation vectors containing either disrupted or truncated genes (10,34). To verify that the fungal DNA in pGMR1 contained a pathogenicity gene, we transformed wild-type *C. magna* with pGMR1 in an attempt to induce a gene disruption. A total of 140 Txs were screened for the loss of pathogenicity and 47% were found to be nonpathogenic and express the same phenotype as the R1 mutant. These results indicate that gene disruption events can occur at a high frequency in *C. magna*. Hybridization analyses are currently underway to define the molecular basis of the gene disruption events in the non-pathogenic pGMR1 Txs.

Conclusions

Colletotrichum has become a valuable experimental system to study genetic, molecular, and biochemical aspects of plant-fungal symbiotic associations. Current technologies have allowed us to isolate a variety of gene disruption mutants which expressed nonpathogenic endophytic phenotypes. Two of thcsc mutants (REMI phenotypes A and C) have identified genes that are responsible for changing the symbiotic lifestyle of *C. magna* from pathogenic to mutualistic or commensalistic, respectively. Although these mutations do not indicate the evolutionary direction of these different symbiotic lifestyles (i.e., did pathogens evolve from commensal or mutualists), they do indicate that lifestyle conversions may occur with genetic simplicity (single-gene disruptions). The biochemical ramifications of the gene disruptions resulting in lifestyle conversions may be very significant. If the mutations are pleiotrophic, the impacts may not be easily observed and may not affect the ability of these fungi to perform as commensals or mutualists. The information presented in this chapter has begun to define the fine genetic balance that exists between different types of symbiosis. As more of the genes responsible for the nonpathogenic REMI mutants are cloned and analyzed the genetic complexity of these phenotypes will be defined. More importantly, this system will lead to an understanding of the genetic basis of commensalistic, mutualistic, and pathogenic lifestyles. Once the genetic basis of these symbiotic lifestyles is defined, the role of fungi in ecosystem structure and dynamics will be better understood.

Acknowledgements

We would like to thank Dr. Gail Brown and Dr. Alison Colwell for critical reviews of this manuscript. This research was supported in part by a BARD grant, a USDA grant, a joint NSF, DOE USDA grant (RJR as Co-PI), and the USGS.

Literature Cited

1. Anderson, A.J. 1978. Extracellular enzymes produced by *Colletotrichum lindemuthianum* and *Helminthosporium maydis* during growth on isolated bean and corn walls. Phytopathology 68:1585-1589.
2. Anderson, A.J. 1988. Elicitors, the hypersensitive response, and phytoalexins. Pages 103-110 in: Physiology and Biochemistry of Plant-Microbial Interactions. N.T. Keen, T. Kosuge, and L.L. Walling, eds. American Society of Plant Physiologists, Rockville, Maryland, USA.
3. Anderson, D.W. and Nicholson, R.L. 1996. Characterization of a laccase in the conidial mucilage of *Colletotrichum graminicola*. Mycologia 88:996-1002.
4. Arx, J.A. von. 1957. Die Arten der Gattung *Colletotrichum* Cda. Phytopath. Z. 29:413-468.
5. Bailey, J.A., O'Connell, R.J., Pring, R.J., and Nash, C. 1992. Infection strategies of *Colletotrichum* species. Pages 88-120 in: *Colletotrichum*: Biology, Pathology and Control. J.A. Bailey and M.J. Jeger, eds. CAB International, Wallingford, UK.
6. Beraha, L. and Garber, E.D. 1985. Relative heterothallism and production of hybrid perithecia by auxotrophic mutants of *Glomerella cingulata* from apple. Phytopath. Z. 112:32-39.
7. Bell, A.A. 1981. Biochemical mechanisms of disease resistance. Annu. Rev. Plant Physiol. 32:21-81.

8. Biles, C.L., Abeles, F.B., and Wilson, C.L. 1990. The role of ethylene in anthracnose of cucumber, *Cucumis sativus*, caused by *Colletotrichum lagenarium*. Phytopathology 80:732-736.
9. Buhr, T.L., Oved, S., Truesdell, G.M., Huanag, C., Yarden, O., and Dickman, M.B. 1996. A kinase-encoding gene from *Colletotrichum trifolii* complements a colonial growth mutant of *Neurospora crassa*. Mol. Gen. Genet. 112:787-792.
10. Bowen, J.K., Templeton, M.D., Sharrock, K.R., Crowhurst, R.N., and Rikkerink, E.H.A. 1995. Gene inactivation in the plant pathogen *Glomerella cingulata*: Three strategies for the disruption of the pectin lyase gene *pnlA*. Mol. Gen. Genet. 246:196-205.
11. Centis, S., Dumas, B., Fournier, J., Marolda, M., and Esquerre-Tugaye, M.T. 1996. Isolation and sequence analysis of *Clpg1*, a gene coding for an endopolygalacturonase of the phytopathogenic fungus *Colletotrichum lindemuthianum*. Gene 170:125-129.
12. Chilton, S.J.P. and Wheeler, H.E. 1948. Genetics of *Glomerella*. VI. Linkage. Am. J. Bot. 36:270-273.
13. Chilton, S.J.P. and Wheeler, H.E. 1949. Genetics of *Glomerella*. VII. mutations and segregation in plus cultures. Am. J. Bot. 36:717-721.
14. Cisar, C.R., Spiegel, F.W., TeBeest, D.O., and Trout, C. 1994. Evidence for mating between isolates of *Colletotrichum gloeosporioides* with different host specificities. Curr. Genet. 25:330-335.
15. Dangl, J.L., Dietrich, R.A., and Richberg, M.H. 1996. Death don't have no mercy: Cell death programs in plant-microbe interactions. Plant Cell 8:1793-1807.
16. Dean, R.A. and Kuć, J. 1986. Induced systemic protection in cucumber: Effects of inoculum density on symptom development caused by *Colletotrichum lagenarium* in previously infected and uninfected plants. Phytopathology 76:186-189.
17. DeBary, A. 1879. Die Erschenung Symbiose. Karl J. Trubner, Strassburg.
18. Dickman, M.B. 1988. Whole cell transformation of the alfalfa fungal pathogen *Colletotrichum trifolii*. Curr. Genet. 14:241-246.
19. Dixon, R.A., Harrison, M.J., and Lamb, C.J. (1994). Early events in the activation of plant defense responses. Annu. Rev. Phytopathol. 32:479-501.
20. English, P.D., Maglothin, A., Keegstra, K., and P. Albersheim. 1972. A cell wall-degrading endopolygalacturonase secreted by *Colletotrichum lindemuthianum*. Plant Physiol. 49:293-297.
21. English, P.D., Jurale, J.B., and P. Albersheim. 1971. Host-pathogen interactions II. Parameters affecting polysaccharide-degrading enzyme secretion by *Colletotrichum lindemuthianum* grown in culture. Plant Physiol. 47:1-6.

22. Erb, K., Gallegly, M.E., and Leach, J.G. 1973. Longevity of mycelium of *Colletotrichum lindemuthianum* in hypocotyl tissue of resistant and susceptible bean cultivars. Phytopathology 63:1334-1335.
23. Esquerre-Tugaye, M.T., Mazau, D., Brthe, J.P., Lafitte, C., and Touze, A. 1992. Mechanisms of resistance to *Colletotrichum* species. Pages 121-133 in: *Colletotrichum*: Biology, Pathology and Control. J.A. Bailey and M.J. Jeger, eds. CAB International, Wallingford, UK.
24. Farr, D.F., Bills, G.F., Chamuris, G.P., and Rossman, A.Y. 1989. Fungi on plants and plant products in the United States. APS Press, St. Paul, Minnesota, USA.
25. Fernandez, M.R. and Heath, M.C. 1985. Cytological responses induced by five phytopathogenic fungi in a nonhost plant, *Phaseolus vulgaris.* Can. J. Bot. 64:648-657.
26. Freeman, S. and Rodriguez, R.J. 1992. A rapid, reliable bioassay for pathogenicity of *Colletotrichum magna* on cucurbits and the isolation of non-pathogenic mutants. Plant Dis. 76:901-905.
27. Freeman, S. and Rodriguez, R.J. 1993. Genetic conversion of a fungal plant pathogen to a non-pathogenic endophytic mutualist. Science 260:75-78.
28. Green, J.R., Pain, N.A., Cannell, M.E., Jones, G.L., Leckie, C.P., McCready, S., Mendgen, K., Mitchell, A.J., Callow, J.A., and O'Connell, R.J. 1995. Analysis of differentiation and development of the specialized infection structures formed by biotrophic fungal plant pathogens using monoclonal antibodies. Can. J. Bot. 73 (suppl. 1): s408-s417.
29. Hammerschmidt, R. and Kuć, J. 1982. Lignification as a mechanism for induced systemic resistance in cucumber. Physiol. Plant Pathol. 20:61-71.
30. Hammerschmidt, R., Nuckles, E.M., and Kuć, J. 1982. Association of enhanced peroxidase activity with induced systemic resistance of cucumber to *Colletotrichum lagenarium*. Physiol. Plant Pathol. 20:73-82.
31. Hammond-Kosack, K.E. and Jones, J.D.G. 1996. Resistance gene-dependent plant defense responses. Plant Cell. 8:1773-1791.
32. Hertig, M., Taliaferro, W.H., and Schwartz, B. 1937. The Terms Symbiosis, Symbiont, and Symbiote. J. Parasitol. 23:235-239.
33. Hwang, C.S. and Kolattukudy, P. 1995. Isolation and characterization of genes expressed uniquely during appressorium formation by *Colletotrichum gloeosporioides* condia induced by the host surface wax. Mol. Gen. Genet. 247:282-294.
34. Hwang, C.S., Flaishman, M.A., and Kolattukudy, P. 1995. Cloning of a gene expressed during appressorium formation by *Colletotrichum gloeosporioides* and a marked decrease in virulence by disruption of this gene. Plant Cell 7:183-193

35. Hynes, M.J., Corrick, C.M., and King, L.A. 1983. Isolation of genomic clones containing the *amdS* gene of *Aspergillus nidulans* and their use in the analysis of structural and regulatory mutations. Mol. Cell. Biol. 3:1430-1439.
36. Jakobek, J.L. and Lindgren, P.B. 1993. Generalized induction of defense responses in bean is not correlated with the induction of the hypersensitive reaction. Plant Cell 5:49-56.
37. Jenkins, S.F. Jr. 1963. A host range study of *Glomerella magna.* Univ. Georgia Coastal Plain Expt. Sta. Mimeo N.S. 176. 8pp.
38. Jenkins Jr., S.F. and N.N. Winstead. 1964. *Glomerella magna,* cause of a new anthracnose of cucurbits. Phytopathology 54:452-454.
39. Kamoun, S., Young, M., Glascock, C.B., and Tyler, B.M. 1993. Extracellular protein elicitors from *Phythophthora*: Host-specificity and induction of resistance to bacterial and fungal phytopathogens. Mol. Plant-Microbe Interact. 6:15-25.
40. Karr, A.L. and Albersheim, P. 1970. Polysaccharide-degrading enzymes are unable to attack plant cell walls without prior action by a "wall-modifying enzyme." Plant Physiol. 46:69-80.
41. Kaster, K.R., Burgett, S.G., and Ingolia, T.D. 1984. Hygromycin B resistance as a dominant selectable marker in yeast. Curr. Gen. 8, 353-358.
42. Kelkar, H.S., Skloss, T.W., Haw, J.F., Keller, N.P., and Adams, T.H. 1997. *Aspergillus nidulans stcL* encodes a putative cytochrome P-450 monooxygenase required for bisfuran desaturation during aflatoxin/sterigmatocystin biosynthesis. J. Biol. Chem. 272:1589-1594.
43. Kessman, H., Staub, T., Hofmann, C., Maetzke, T., Herzog, J., Ward, E., Uknes, S., and Ryals, J. (1994). Induction of systemic acquired disease resistance in plants by chemicals. Annu. Rev. Phytopathol. 34:439-459.
44. Kimati, H. and Galli, F. 1970. *Glomerella cingulata* (Stonem.) Spauld. et Schrenk. f. sp. *phaseoli* N.F., ascogenic stage of the causal agent of anthracnose in the bean plant. Anais Da Escola Superior De Agricultura Luis De Queiroz. 27:411-437.
45. Klement, Z. (1982). Hypersensitivity. Pages 149-177 in: Phytopathogenic Prokaryotes. M.S. Mount and G.H. Lacy, eds. Academic Press, New York.

46. Knogge, W. 1996. Fungal infection of plants. Plant Cell 8:1711-1722.
47. Kolattukudy, P.E., Rogers, L.M., Li, D., Hwang, CS., and Flaishman, M.A. 1995. Surface signaling in pathogenesis. Proc. Natl. Acad. Sci. 92:4080-4087.
48. Kubo, Y., Suzuki, K., Furusawa, I., Ishida, N., and Yamamoto, M. 1982. Relation of appressorium pigmentation and penetration of nitrocellulose membranes by *Colletotrichum lagenarium*. Phytopathology 72:498-501.
49. Kubo, Y., Nakamura, H., Kobayashi, K., Okuno, T., and Furusawa, I. 1991. Cloning of a melanin biosynthetic gene essential for appressorial penetration of *Colletotrichum lagenarium*. Mol. Plant-Microbe Interact. 4:440-445.
50. Kubo, Y., Takano, Y., Endo, N., Yasuda, N., Tajima, S., and Furusawa, I. 1996. Cloning and structural analysis of the melanin biosynthesis gene *SCD1* encoding scytalone dehydratase in *Colletotrichum lagenarium*. Appl. Environ. Micro. 62:4340-4344.
51. Kuć, J. 1990. A case for self defense in plants against disease. Phytoparasitica 18:3-8.
52. Kuć, J. and Strobel, N.E. 1992. Induced resistance using pathogens and nonpathogens. Pages 295-303 in: Biological Control of Plant Diseases, E.S. Tjamos, ed. Plenum Press, New York.
53. Kuspa, A. and Loomis, W.F. 1992. Tagging developmental genes in *Dictostelium* by restriction enzyme-mediated integration of plasmid DNA. Proc. Natl. Acad. Sci. 89:8803-8807.
54. Legrand, M., Kauffmann S., Geoffroy, P., and Fritig, B. 1987. Biological function of pathogenesis-related proteins: Four tobacco pathogenesis-related proteins are chitinases. Proc. Natl. Acad. Sci. 84:6750.
55. Manadhar, J.B., Hartman, G.L., and Sinclair, J.B. 1986. *Colletotrichum destructivum*, the anamorph of *Glomerella glycines*. Phytopathology 76:282-285.
56. Markert, C.L. 1952. Radiation induced nutritional and morphological mutants of *Glomerella*. Genetics 37:339-352.
57. McLean, D.M. 1966. Sexual expression in *Colletotrichum orbiculare*. Plant Dis. Reptr. 50:871-873.
58. McNeil, J.B., Zhang, F., Taylor, B.V., Sinclair, D.A., Pearlman, R.E., Bognar, A.L. 1997. Cloning and molecular characterization of the *GCV1* gene encoding the glycine cleavage T-protein from *Saccharomyces cerevisiae*. Gene 186:13-20.
59. Mordue, J.E.M. 1967. *Glomerella tucumanensis*. C.M.I. descriptions of pathogenic fungi and bacteria, number 133. The Eastern Press Ltd., London.

60. Mercer, P.C., Wood, R.K.S., and Greenwood, A.D. 1975. Ultrastructure of the parasitism of *Phaseolus vulgaris* by *Colletotrichum lindemuthianum*. Physiol. Plant Pathol. 5:203-214.
61. O'Connell, R.J. and J.A. Bailey. 1986. Cellular interactions between *Phaseolus vulgaris* and the hemibiotrophic fungus *Colletotrichum lindemuthianum*. Pages 49-48 in: Biology and Molecular Biology of Plant-Pathogen Interactions. J. Bailey, ed. Springer Verlag, Heidelberg.
62. O'Connell, R.J., Bailey, J.A., and Richmond, D.V. 1985. Cytology and physiology of infection of *Phaseolus vulgaris* by *Colletotrichum lindemuthianum*. Physiol. Plant Pathol. 27:75-98.
63. Panaccione, D.G., McKiernan, M., and Hanau, R.M. 1988. *Colletotrichum graminicola* transformed with homologous and heterologous benomyl-resistance genes retains expected pathogenicity to corn. Mol. Plant-Microbe Interact. 3:113-120.
64. Powell, W.A. and Kistler, H.C. 1990. In vivo rearrangement of foreign DNA by *Fusarium oxysporum* produces linear self-replicating plasmids. J. Bacteriol. 172:3163-3171.
65. Rasmussen, J.B., Panaccione, D.G., Fang, G.C., and Hanau, R.M. 1992. The *PYR1* gene of the pathogenic fungus *Colletotrichum graminicola*: Selection by intraspecific complementation and sequence analysis. Mol. Gen. Genet. 235:74-80.
66. Redman, R.S. and Rodriguez, R.J. 1994. Factors which affect efficient transformation of *Colletotrichum* species. Exp. Mycol. 18:230-246.
67. Redman, R.S., Freeman, S., Clifton, D.R., Morrel, J., Brown, G.S., and Rodriguez, R.J. 1999. Biochemical analysis of plant protection afforded by a non-pathogenic endophytic mutant of *Colletotrichum magna*. Plant Physiol. 119:795-804.
68. Rikkerink, E.H.A., Solon, S., Crowhurst, R.,and Templeton, M.D. 1994. Integration of vectors by homologous recombination in the plant pathogen *Glomerella cingulata*. Curr. Genet. 25:202-208.
69. Rodriguez, R.J. 1992. Polyphosphate present in DNA preparations from fungal species of *Colletotrichum* inhibits restriction endonucleases and other enzymes. Anal. Biochem. 209:291-297.
70. Rodriguez, R.J. and Owen, J.L. 1992. Isolation of *Glomerella musae* [Teleomorph of *Colletotrichum musae* (Berk. & Curt.) Arx.] and segregation analysis of ascospore progeny. Mycol. Res. 16:291-301.
71. Rodriguez, R.J. and Redman, R. 1992. Molecular transformation and genome dynamics of *Colletotrichum* species. Pages 47-66 in: The Biology of *Colletotrichum* Species. J. Bailey and R. Shattock, eds. Cambridge Press, England.
72. Rothstein, R. 1991. Targeting, disruption, replacement, and allele rescue: integrative DNA transformation in yeast. Meth. Enzymol. 94:281-301.

73. Sambrook, J., Fritsch, E.F. and Maniatis, T. 1989. Molecular Cloning: A Laboratory Manual. Cold Spring Harbor Press, Cold Spring Harbor, NY.
74. Sela-Buurlage, M.B., Epstein, L., and Rodriguez, R.J. 1992. Adhesion of ungerminated *Colletotrichum musae* conidia. Physiol. Molec. Plant Pathol. 39:345-352.
75. Skipp, R.A. and Deverall, B.J. 1972. Relationships between fungal growth and host changes visible by light microscopy during infection of bean hypocotyls (*Phaseolus vulgaris*) susceptible and resistant to physiological races of *Colletotrichum lindemuthianum*. Physiol. Plant Pathol. 2:357-374.
76. Smith, B.J., Black, L.L., and Galletta, G.J. 1990. Resistance to *Colletotrichum fragariae* in strawberry affected by seedling age and inoculation method. Plant Dis. 74:1016-1021.
77. Soliday, C.L., Dickman, M.B., and Kolattukudy, P.E. 1989. Structure of the cutinase gene and detection of promoter activity in the 5'-flanking region by fungal transformation. J. Bacteriol. 171:942-1951.
78. Sutton, B.C. 1980. The Coelomycetes. Commonwealth Mycological Institute, Kew, England. pp. 524-537.
79. Takano, Y., Kubo, Y., Shimizu, K., Mise, K., Okuno, T., and Furusawa, I. 1995. Structural analysis of *PKS1*, a polyketide synthase gene involved in melanin biosynthesis in *Colletotrichum lagenarium*. Mol. Gen. Genet. 248:162-167.
80. Takano, Y., Kubo, Y., Kuroda, I., and Furusawa, I. 1997. Temporal transcriptional pattern of three melanin biosynthesis genes, *PKS1*, *SCD1*, and *THR1*, in appressorium-differentiating and nondifferentiating conidia of *Colletotrichum lagenarium*. Appl. Environ. Micro. 63:351-354.
81. TeBeest, D.O., Shilling, C.W., Hopkins, Riley, L., and Weidmann, G.J. 1989. The number of nuclei in spores of three species of *Colletotrichum*. Mycologia 81:147-149.
82. Templeton, M.D., Rikkerink, E.H.A., Solon, S.I., and Crowhurst, R.N. 1992. Cloning and molecular characterization of the glyceraldehyde-3-phosphate dehydrogenase-encoding gene and cDNA from the plant pathogenic fungus *Glomerella cingulata*. Gene 122:225-230.
83. Templeton, M.D., Sharrock, K.R., Bowen, J.K., Crowhurst, R.N., and Rikkerink, E.H.A. 1994. The pectin lyase-encoding gene (*pnl*) family from *Glomerella cingulata*: Characterization of *pnlA* and its expression in yeast. Gene 142:141-146.
84. Tu, J.C. 1985. An improved Mathur's medium for growth, sporulation and germination of spores of *Colletotrichum lindemuthianum*. Microbios 44:87-93.

85. Vaillancourt, L.J. and Hanau, R.M. 1991. A method for genetic analysis of *Glomerella graminicola* (*Colletotrichum graminicola*) from maize. Phytopathology 81:530-534.
86. Wheeler, H.E. 1954. Genetics and evolution of heterothallism in *Glomerella*. Phytopathology 44:342-345.
87. Winstead, N.N., Jenkins, S.F. Jr., Lucas, L.T., Campbell, G.J., and Bone, H.T. 1966. Influence of light on perithecial formation in *Glomerella magna*. Phytopathology 56:134-135.
88. Yamada, T., Hashimoto, H., Shiraishi, T., and Oku, H. (1989). Suppression of pisatin, phenylalanine ammonia-lyase mRNA, and chalcone synthase mRNA accumulation by a putative pathogenicity factor from the fungus *Mycosphaerella pinodes*. Mol. Plant-Microbe Interact. 2:256-261.

Chapter 9

Genetic Diversity and Host Specificity of *Colletotrichum* Species on Various Fruits

Stanley Freeman

Phytopathogenic fungi of the genus *Colletotrichum* and its teleomorph *Glomerella* cause major damage to crops in tropical, subtropical, and temperate regions worldwide. Cereals, grasses, legumes, ornamentals, vegetables, and fruit trees may be seriously affected by the pathogen. Although many cultivated fruit crops are infected by *Colletotrichum* species worldwide, the most significant economic losses are incurred when the fruiting stage is attacked. *Colletotrichum* species cause typical disease symptoms known as anthracnose, characterized by sunken necrotic tissue where masses of orange conidia are produced. Anthracnose diseases appear in both developing and mature plant tissues and affect developing fruit in the field as well as mature fruit during storage. Species of the pathogen appear predominantly on aboveground plant tissues; however, below-ground organs, such as roots and tubers, may also be affected.

Importance of Species Identification

Identifying the *Colletotrichum* species responsible for disease epidemics is vital for developing and implementing effective control strategies and for understanding the epidemiology of these diseases. Of equal importance is investigating whether the same or different pathogens are associated with diseases on different host tissues. Furthermore, breeding for resistance to a specific species depends on accurate identification of the causal agent.

Single hosts may be infected by multiple species of *Colletotrichum*, whereas multiple hosts may also be infected by single species of the pathogen.

Single Host Infected by Multiple Species

Several *Colletotrichum* species or genotypes can be associated with a single host. Some examples are listed:

1. Citrus can be affected by four different *Colletotrichum* diseases (30,49): (i) postbloom fruit drop caused by *C. acutatum*; (ii) key lime anthracnose caused by *C. acutatum*, (iii) postharvest decay of fruit caused by *C. gloeosporioides*, and (iv) shoot dieback and leaf spot caused by *C. gloeosporioides.*
2. Avocado and mango anthracnose caused by both *C. acutatum* and *C. gloeosporioides* affect fruit predominantly as postharvest diseases (21,34,35).
3. Strawberry may be infected by three *Colletotrichum* species, namely *C. fragariae*, *C. acutatum,* and *C. gloeosporioides* causing anthracnose of fruit and other plant parts (25).
4. Deciduous fruits such as apple, peach, pecan, and almond may be infected by either *C. acutatum* or *C. gloeosporioides* (1,3,40,42).
5. Other hosts affected by multiple *Colletotrichum* species include coffee, cucurbits, legumes, pepper, and tomato.

Multiple Hosts Infected by Single Species

Single species of *Colletotrichum* are known to infect multiple hosts. A number of examples are listed:

1. *C. gloeosporioides* (Penz.) Penz. & Sacc. in Penz. (teleomorph: *Glomerella cingulata* [Stoneman] Spauld. & H. Schrenk), which is considered a group species and forms the sexual stage in some instances, is found on a wide variety of fruits including apple, avocado, citrus, papaya, peach, pecan, mango, and strawberry (4,11,47,51).
2. *C. acutatum* J.H. Simmonds has been reported to infect a large number of fruit crops including almond, avocado, blueberry, mango, peach, pecan, citrus, grape, and strawberry (1,3,6,12,23).
3. *C. coccodes* has a multiple host range, including mainly vegetables such as pepper, potato, squash, and tomato (10).
4. Other species with multiple host ranges include *C. lindemuthianum, C. capsici, C. dematium, C. graminicola,* and *C. truncatum* (28).

Almond, Avocado, and Strawberry Anthracnose

This chapter will deal in particular with host-pathogen relationships and host specificity of *Colletotrichum* species on almond, avocado, and strawberry and demonstrate the use of classical and molecular tools for differentiating between species and genotypes. The three pathosystems chosen represent different lifestyles of *Colletotrichum*.

Species of the pathogen on almond, which is a temperate, deciduous tree, infect at initial fruiting stages, but do not appear to attack mature fruit or other plant organs (1,5,9,40). In contrast, avocado is a subtropical crop, and the avocado anthracnose pathogens cause postharvest decay from latent or quiescent infections originating in the field (4,36,37). In contrast to almond and avocado, strawberry is predominantly an annual crop, with *Colletotrichum* anthracnose responsible for serious damage on foliar and fruiting plant parts, as well as root necrosis (12,23,25,44). Damage to strawberry crops is not limited to a certain growth period or storage conditions, and anthracnose may appear throughout most of the season (15,31).

Although almond and avocado crops are cultivated in close proximity in Israel, where anthracnose of both crops, caused by *C. gloeosporioides*, occurs during the spring season, whether cross-infection resulted in the field was unclear. Likewise, strawberry, infected by *C. acutatum*, is cultivated in the same geographic regions and under the same climatic conditions as those of avocado and almond in Israel. Furthermore, a recent report has characterized *C. acutatum* as the pathogen responsible for almond anthracnose in California, in areas where strawberry and peach anthracnose also occur (1). The question raised, therefore, is whether the pathogens responsible for anthracnose in these locations are host-specific or whether the same pathogen is the causal agent of disease on all three crops.

Host Specificity

Differentiation between *Colletotrichum* species based on host range may not be reliable, since taxa such as *C. gloeosporioides*, *C. dematium*, *C. acutatum*, *C. graminicola* infect a broad range of host plants. Some taxa appear to be restricted to host families, genera, species, or even cultivars within those families, e.g. bean (43; Melotto et al., Chapter 21 of this volume), whereas others have more extensive host ranges. On the other hand, certain highly specific strains of *Colletotrichum* have been successfully developed as commercial mycoherbicides (e.g., *C. gloeosporioides* f. sp. *aeschynomene* for the control of northern jointvetch weed in rice and soybean [48]).

Cross-infection potential has been reported among different species of *Colletotrichum* and genotypes of *C. gloeosporioides*, on a variety of tropical, subtropical, and temperate fruits under artificial inoculation conditions. For example, isolates of *C. acutatum* and *C. gloeosporioides* from a variety of temperate fruits caused disease symptoms which were visually indistinguishable on detached peach fruits (3). *C. gloeosporioides* isolates from seven tropical fruit crops were also shown to cross-infect detached leaves and fruits, and infection appeared to be dependent on inoculum density (2). In studies in Israel, it was of interest to determine whether avocado anthracnose was caused by the same genotype of *C. gloeosporioides* as that responsible for almond anthracnose, since almond groves are situated alongside avocado plantations in some cultivation areas. Similarly, other crops, such as anemone, apple, mango, and strawberry, are known to be susceptible to anthracnose in Israel. To determine the potential of cross-infection, isolates from different crops were cross-inoculated on various hosts. Our study showed that *C. gloeosporioides* isolates from almond, apple, avocado, and mango, and *C. acutatum* isolates from anemone, apple and peach infected detached fruits of the various hosts, namely apple (two varieties), avocado, almond, mango, and nectarine (18). This study showed, therefore, the potential for cross-infection between two species, *C. gloeosporioides* (including representatives of distinct genotypes from almond, apple, avocado, and mango) and *C. acutatum* from apple and peach on a variety of hosts. These results should be considered in areas where *Colletotrichum* species cause anthracnose on a number of hosts.

Methods for Species Identification

Characterization of *Colletotrichum* species has relied on a number of criteria, including morphology, optimal growth temperature, vegetative compatibility, and fungicide sensitivity. In this section, we elaborate on the molecular approaches used to differentiate between isolates of *Colletotrichum* responsible for anthracnose on almond, avocado, and strawberry crops in Israel.

Traditional Methods

Discrimination of *Colletotrichum* species by traditional means has relied primarily on colony color, size and shape of conidia, optimal temperature, growth rate, presence or absence of setae, and existence of the teleomorph, *Glomerella* (20,49). However, because of environmental influences on the stability of morphological traits and the existence of intermediate forms that can be associated with storage, especially with frequent subculturing, these criteria are not always adequate for reliable differentiation between *Colletotrichum* species.

Vegetative Compatibility Grouping (VCG)

VCGs have been used to examine genetic relatedness in a number of plant pathogenic fungi (29,38) and for genotype characterization of *Colletotrichum* species including *C. dematium* from spinach (7); *C. orbiculare* from cucurbits (50); *C. gloeosporioides* from almond, anemone, and avocado (27); and *C. acutatum* from strawberry (12). A detailed description on using this method for determining population genetics is described in Chapter 4 by Katan.

MOLECULAR TECHNIQUES

Molecular methods have been utilized successfully for differentiating between species and genotypes of *Colletotrichum* from many hosts in general, and, in particular, from almond, avocado, and strawberry in Israel (12,14,18). Such methods exploit polymorphisms in nuclear DNA, A+T-rich DNA associated with mitochondrial DNA (mtDNA), arbitrarily primed PCR (ap-PCR), and ribosomal DNA (rDNA) to reliably discriminate between populations of *C. acutatum, C. gloeosporioides, C. graminicola, C. coccodes, C. fragariae, C. kahawae, C. lindemuthianum, C. magna, C. orbiculare,* and other species (6,8,16,41,46). In addition, molecular markers indicated that two distinct clonal populations appear to exist in Australian isolates of *C. gloeosporioides* on the tropical pasture legume, *Stylosanthes* spp. (22,32,33). Furthermore, the genetic complexity of *Colletotrichum* strains infecting various temperate and tropical fruits has been shown using these techniques (3,14,18,34).

A. Nuclear DNA polymorphisms. Repetitive DNA elements have proved useful for grouping various isolates of *Colletotrichum* species. By using a dispersed repeat sequence, two distinct clonal populations became evident among Australian isolates of *C. gloeosporioides* on a tropical pasture legume (32). Similarly, a repetitive nuclear DNA element (GcpR1), originally from the bean anthracnose pathogen *C. lindemuthianum* (39), was used successfully to differentiate between 10 species of *Colletotrichum* (16). GcpR1 was subsequently used to differentiate between *C. gloeosporioides* isolates from almond and avocado in Israel (14). The band patterns of the Israeli almond isolates were uniform and distinct from those of avocado isolates. Polymorphic fragments were observed within the avocado isolates, confirming results obtained by morphology and VCG that indicate heterogeneity in this population.

B. Analysis of A+T-rich DNA associated with mtDNA. Polymorphisms of A+T-rich DNA are generated by the restriction enzyme *Hae*III, which recognizes and cleaves the DNA sequence GGCC. A+T-rich DNA is cleaved infrequently by *Hae*III, whereas most of the nuclear DNA is digested to fragments of less than 2 kb in size (16). A+T-rich DNA was used successfully to characterize representative *Colletotrichum* isolates from 10 species (24). This approach subsequently helped to differentiate between *C. gloeosporioides* isolates from almond and avocado crops grown in Israel, supporting data obtained using GcpR1, which indicated that the almond population is clonal compared to a heterogeneous avocado population (14). A+T-rich DNA was also applied for grouping *C. acutatum* isolates from apple and peach, which were identical, whereas polymorphisms were observed when representative *C. gloeosporioides* isolates from almond, apple, avocado, and mango were compared (18). Additional studies based on mtDNA have been used in the past to discern populations of *Colletotrichum* species. Populations of *Colletotrichum* causing coffee berry disease were differentiated into *C. gloeosporioides* and *C. kahawae* using mtDNA (46). Similarly, species of *Colletotrichum* causing cucurbit anthracnose were analyzed using a mtDNA probe originating from *C. orbiculare* (8). Likewise, mtDNA was similar for isolates of *C. gloeosporioides* from mango fruits originating from both the eastern and western hemispheres (24).

C. Arbitrarily-primed polymerase chain reaction (ap-PCR). Ap-PCR or random amplified polymorphic DNA (RAPD) has been used extensively for identification and characterization of isolates in *Colletotrichum* (2,8,45,46). Ap-PCR has been applied routinely, using primers derived from mini-satellite or repeat sequences (CAGCAGCAGCAGCAG, TGTCTGTCTGTCTGTC, GACACGACACGACAC, GACAGACAGACAGACA) for characterizing *Colletotrichum* species from almond, avocado, strawberry, and other hosts. This method was used to accurately and reliably differentiate between *C. acutatum, C. gloeosporioides,* and *C. fragariae* from strawberry (17) and to identify isolates of *C. acutatum* that were responsible for a recent epidemic of strawberry anthracnose in Israel (12). Ap-PCR was used to determine the uniformity of Israeli isolates of *C. gloeosporioides* from almond, to differentiate them from the US *C. acutatum* population from almond, and to show heterogeneity among *C. gloeosporioides* isolates from avocado from Israel (Fig. 1). Varying band patterns were also observed for ap-PCR of numerous *C. acutatum* isolates from multiple hosts, including anemone, almond, apple, peach, pecan, and strawberry, indicating heterogeneity with this species, similar to that demonstrated for *C. gloeosporioides* (Fig. 2). This method was also useful for taxonomic identification of strawberry isolates from South Carolina as *C. acutatum* based on unique amplification products and for demonstrating heterogeneity within *C. gloeosporioides* in general, as illustrated by multiple band patterns (Fig. 3).

D. Ribosomal DNA analysis and sequence. rDNA genes appear as multiple copies in the genome. Due to lesser conservation of sequence, the non-transcribed and internal transcribed spacer (ITS) regions between the small and large nuclear rDNA subunits are suitable target sites for detection of recent evolutionary divergence within *Colletotrichum*. Sequence analysis has been used to discriminate between various species including *C. acutatum, C. fragariae, C. gloeosporioides, C. musae,* and *C. orbiculare* (26,41,45). In our work, limited restriction digest analyses of PCR-amplified rDNA failed to distinguish between isolates of *C. gloeosporioides* from various hosts including almond, apple, avocado, and strawberry, whereas the rDNA of *C. acutatum* was different (14).

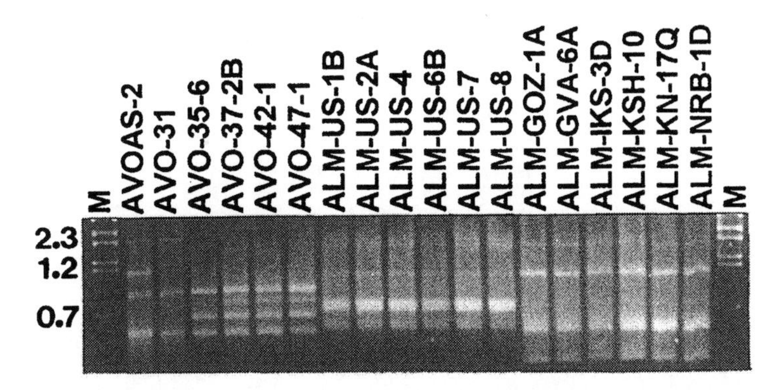

Fig. 1. Inter-species band patterns of ap-PCR amplified genomic DNA from isolates of *Colletotrichum* from avocado and almond using primer $(GACA)_4$ represented as follows: AVO-AS-2; AVO-31; AVO-35-6; AVO-37-2B; AVO-42-1; AVO-47-1 (*C. gloeosporioides* from avocado, Israel); ALM-US-1B; ALM-US-2A; ALM-US-4; ALM-US-6B; ALM-US-7; ALM-US-8 (*C. acutatum* from almond, California); and ALM-GOZ-1A; ALM-GVA-6A; ALM-IKS-3D; ALM-KSH-10; ALM-KN-17Q; ALM-NRB-1D (*C. gloeosporioides* from almond; Israel). Lanes M contain DNA size markers in kb.

Species-specific primers have been designed primarily according to dissimilarities in the sequence of the ITS regions of representative isolates of *Colletotrichum* from different species. These results demonstrate that species-specific primers are valuable for diagnostics and pathogen detection *in planta*, but not for studies on population diversity within a species. Recently, specific primers were designed to differentiate between *C. gloeosporioides* and *C. acutatum* from citrus (6). Furthermore, isolates of *C. acutatum* from a broad host range were reliably grouped and delineated from *C. gloeosporioides* isolates using the *C. acutatum*-specific primers (Fig. 4A)(13). Similarly, using *C. gloeosporioides*- and *C. fragariae*-specific primers, isolates from these two species were distinguished from those of *C. acutatum* (Fig. 4B) (13). This method was also able to distinguish *C. acutatum* isolates from almond, apple, peach, and strawberry, and *C. gloeosporioides* isolates from papaya and citrus (1).

Summary and Future Outlook

Because *Colletotrichum* is a broad-range pathogen with either multiple species on a single host or a single species on multiple hosts, the complexity related to host specificity should be determined for each host at every given

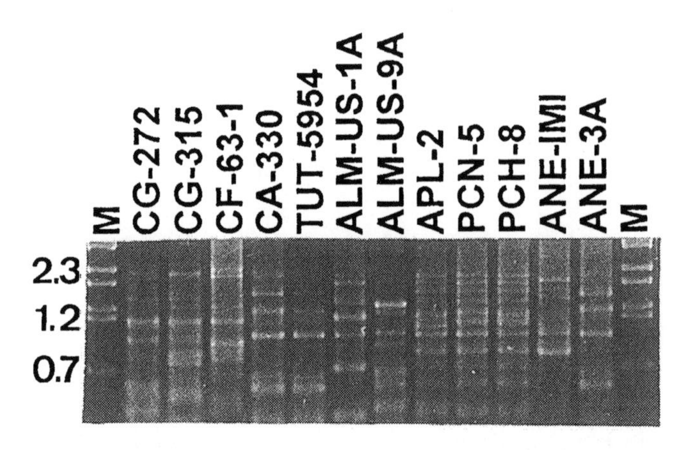

Fig. 2. Inter-species band patterns of ap-PCR amplified genomic DNA from isolates of *Colletotrichum* from various hosts using primer $(CAG)_5$ represented as follows: CG-272 and CG-315 (*C. gloeosporioides* from strawberry, Florida, and Nova Scotia); CF-63-1 (*C. fragariae* from strawberry, Mississippi); CA-330 (*C. acutatum* from strawberry, Tennessee); TUT-5954 (*C. acutatum* from strawberry, Israel); ALM-US-1A (*C. acutatum* from almond, California); ALM-US-9A (unidentified species from almond, California); APL-2 (*C. acutatum* from apple, S. Carolina); PCN-5 (*C. acutatum* from pecan, Alabama); PCH-13 (*C. gloeosporioides* from peach, S. Carolina); PCH-8 (*C. acutatum* from peach, S. Carolina); ANE-IMI (IMI-223120 *C. acutatum* from anemone); and ANE-3A (*C. acutatum* from anemone, Israel). Lanes M contain DNA size markers in kb.

location. This chapter has shown that certain populations of *Colletotrichum* appear to be host-specific, for example, *C. gloeosporioides* from almond in Israel. However, additional species may attack the same host, as recently demonstrated with *C. acutatum* causing almond anthracnose in California. Cross-infection potential has been reported between different species of *Colletotrichum* on a multitude of hosts, although these studies were mainly conducted on detached plant material, such as fruit and leaves. Artificial host inoculation is usually not reliable enough for determining host specificity, but indicates the potential for infection. This point has been emphasized in this chapter, showing that *C. gloeosporioides* from almond in Israel was able to infect various hosts under artificial inoculation conditions. However, isolations from naturally infected fruit from the field provided no evidence of cross-infection. Therefore, host specificity should primarily be determined by natural infections.

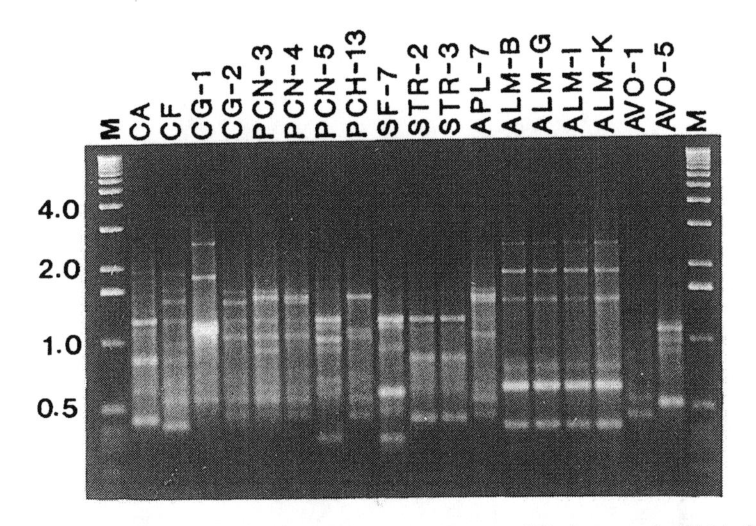

Fig. 3. Inter-species band patterns of ap-PCR amplified genomic DNA from isolates of *Colletotrichum* from various hosts using primer $(GACAC)_3$ represented as follows: CA (*C. acutatum* from strawberry, California); CF (*C. fragariae* from strawberry, Mississippi); CG-1 and CG-2 (*C. gloeosporioides* from strawberry, Florida); PCN-3 and PCN-4 (*C. gloeosporioides* from pecan, Louisiana); PCN-5 (*C. acutatum* from pecan, Alabama); PCH-13 (*C. gloeosporioides* from peach, S. Carolina); SF-7 (*C. acutatum* from peach, S. Carolina); STR-2 and STR-3 (*C. acutatum* from strawberry, S. Carolina); APL-7 (*C. gloeosporioides* from apple, North Carolina); ALM-B, ALM-G, ALM-I and ALM-K (*C. gloeosporioides* from almond, Israel); and AVO-1 and AVO-5 (*C. gloeosporioides* from avocado, Israel). Lanes M contain DNA size markers in kb.

Identification of species by traditional methods (morphology, colony color, conidial size, etc.) may not be accurate enough for identification of species and subspecies which is critical for disease management, breeding, and pathogen control. Vegetative compatibility and molecular approaches offer complementary means for characterization of taxa. Molecular approaches have been utilized successfully for identification and determination of genetic diversity of *Colletotrichum* populations. In a number of recent studies, species-specific primers have been used to differentiate between isolates of *C. acutatum* and *C. gloeosporioides* from various hosts. Simple and rapid DNA extraction techniques are necessary to use this approach more successfully for pathogen diagnostics *in vivo*, to facilitate multiple sampling and testing on a daily basis.

A sexual lifestyle is presumed to contribute towards the genetic diversity of a fungal population. Understanding and observing the sexual stage is important in determining whether a population is homogeneous or

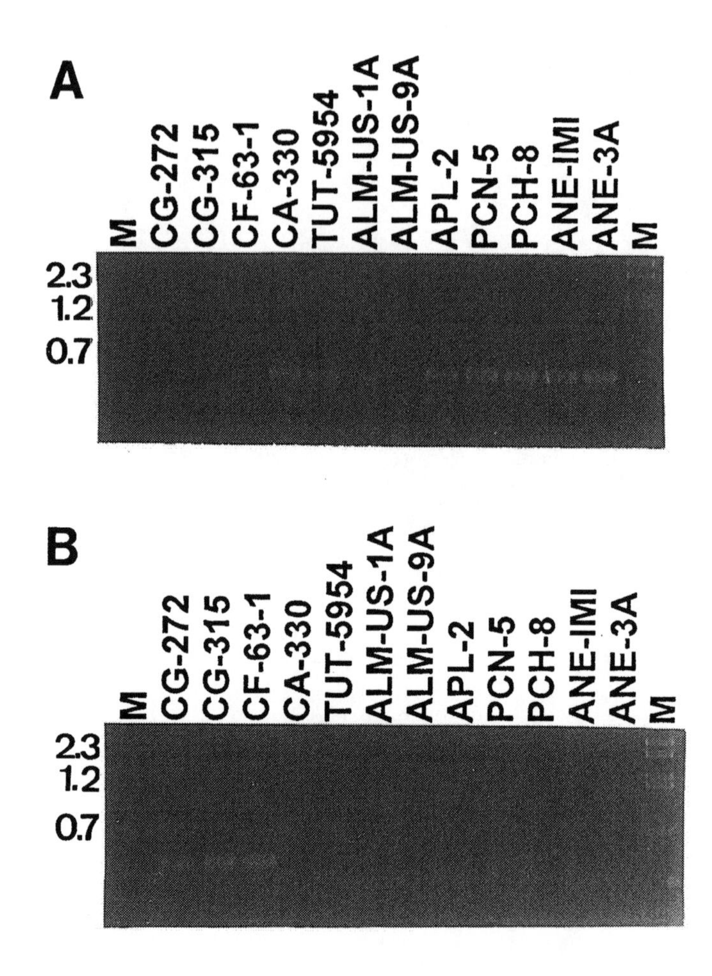

Fig. 4. Amplification of specific DNA fragments from isolates of *Colletotrichum* as specified in Figure 2. Target primers *Ca*Int2 and ITS4 are specific to *C. acutatum* (A), while primers CgInt and ITS4 are specific to *C. gloeosporioides* (B). Lanes M contain DNA size markers in kb.

heterogeneous and for developing a sexual genetic system. The use of classical morphology and molecular tools may contribute to the understanding of the complexity of certain populations, such as that of *C. gloeosporioides* from avocado, which produces the teleomorph in culture and may survive and proliferate via a similar lifestyle in nature. Similar to *C. gloeosporioides,* multiple genotypes of *C. acutatum* exist, and its isolates appear on multiple hosts. In addition, the teleomorph was recently reported

(19). Although clonal populations of *C. acutatum* have been observed (e.g., strawberries in Israel) it remains to be seen whether the presence of the sexual stage may be an important contribution to genetic diversity within *C. acutatum*, as hypothesized for *C. gloeosporioides*.

In general, molecular tools should not be used alone for classification of species. For meaningful identification and characterization, a large-scale study should be based on adequate sampling of a population as an element for reproducibility and should include morphological criteria.

Literature Cited

1. Adaskaveg, J.E. and Hartin, R.J. 1997. Characterization of *Colletotrichum acutatum* isolates causing anthracnose of almond and peach in California. Phytopathology 87:979-987.
2. Alahakoon, P.W., Brown, A.E., and Sreenivasaprasad, S. 1994. Cross-infection potential of genetic groups of *Colletotrichum gloeosporioides* on tropical fruits. Physiol. Mol. Plant Pathol. 44:93-103.
3. Bernstein, B., Zehr, E.I., Dean, R.A., and Shabi, E. 1995. Characteristics of *Colletotrichum* from peach, apple, pecan, and other hosts. Plant Dis. 79:478-482.
4. Binyamini, N. and Schiffmann-Nadel, M. 1972. Latent infection in avocado fruit due to *Colletotrichum gloeosporioides*. Phytopathology 62:592-594.
5. Brizi, U. 1896. Eine neue krankheit (anthracnosis) des mandelbaums. Z. Pflanzenkr. Pflanzenshutz. 6:65-72.
6. Brown, A.E., Sreenivasaprasad, S., and Timmer, L.W. 1996. Molecular characterization of slow-growing orange and key lime anthracnose strains of *Colletotrichum* from citrus as *C. acutatum*. Phytopathology 86:523-527.
7. Correll, J.C., Guerber, J.C., and Morelock, T.E. 1993. Vegetative compatibility and virulence of the spinach anthracnose pathogen, *Colletotrichum dematium*. Plant Dis. 77:688-691.
8. Correll, J.C., Rhoads, D.D., and Guerber, J.C. 1993. Examination of mitochondrial DNA restriction fragment length polymorphisms, DNA fingerprints, and randomly amplified polymorphic DNA of *Colletotrichum orbiculare*. Phytopathology 83:1199-1204.
9. Czarnecki, H. 1916. A *Gloeosporium* disease of the almond probably new to America. Phytopathology 6:310.
10. Dillard, H.R. 1992. *Colletotrichum coccodes*: The pathogen and its hosts. Pages 225-236 in: *Colletotrichum*: Biology, Pathology and Control. J.A. Bailey and M.J. Jeger, eds. CAB International, Wallingford, UK.
11. Fitzell, R.D. 1987. Epidemiology of anthracnose disease of avocados. S. Afr. Avocado Grow. Assoc. Yrbk. 10:113-116.

12. Freeman, S. and Katan, T. 1997. Identification of *Colletotrichum* species responsible for anthracnose and root necrosis of strawberry in Israel. Phytopathology 87:516-521.
13. Freeman, S., Katan, T., and Shabi, E. 1998. Characterization of *Colletotrichum* species responsible for anthracnose diseases of various fruits. Plant Dis. 82:596-605.
14. Freeman, S., Katan, T., and Shabi, E. 1996. Characterization of *Colletotrichum gloeosporioides* isolates from avocado and almond fruits with molecular and pathogenicity tests. Appl. Environ. Microbiol. 62:1014-1020.
15. Freeman, S., Nizani, Y., Dotan., S., Even, S., and Sando, T. 1997. Control of *Colletotrichum acutatum* in strawberry under laboratory, greenhouse and field conditions. Plant Dis. 81:749-752.
16. Freeman, S., Pham, M., and Rodriguez, R.J. 1993. Molecular genotyping of *Colletotrichum* species based on arbitrarily primed PCR, A+T-rich DNA, and nuclear DNA analyses. Exp. Mycol. 17:309-322.
17. Freeman, S. and Rodriguez, R.J. 1995. Differentiation of *Colletotrichum* species responsible for anthracnose of strawberry by arbitrarily primed PCR. Mycol. Res. 99:501-504.
18. Freeman, S. and Shabi, E. 1996. Cross-infection of subtropical and temperate fruits by *Colletotrichum* species from various hosts. Physiol. Mol. Plant Pathol. 49:395-404.
19. Guerber, J.C. and Correll, J.C. 1997. The first report of the teleomorph of *Colletotrichum acutatum* in the United States. (Abstr.). Plant Dis. 81:1334.
20. Gunnell, P.S and Gubler, W.D. 1992. Taxonomy and morphology of *Colletotrichum* species pathogenic to strawberry. Mycologia 84:157-165.
21. Hartill, W.F.T. 1992. Postharvest rots of avocado in New Zealand and their control. Pages 1157-1162 in: Brighton Crop Protection Conference, Brighton, UK.
22. He, C.Z., Masel, A.M., Irwin, J.A.G., Kelemu, S., and Manners, J.M. 1995. Distribution and relationship of chromosome-specific dispensable DNA sequences in diverse isolates of *Colletotrichum gloeosporioides*. Mycol. Res. 99:1325-1333.
23. Henz, G.P., Boiteux, L.S., and Lopes C.A. 1992. Outbreak of strawberry anthracnose caused by *Colletotrichum acutatum* in central Brazil. Plant Dis. 76:212.
24. Hodson, A., Mills, P.R, and Brown, A.E. 1993. Ribosomal and mitochondrial DNA polymorphisms in *Colletotrichum gloeosporioides* isolated from tropical fruits. Mycol. Res. 97:329-335.
25. Howard, C.M., Maas, J.L., Chandler, C.K., and Albregts, E.E. 1992. Anthracnose of strawberry caused by the *Colletotrichum* complex in Florida. Plant Dis. 76:976-981.

26. Johnston, P.R. and Jones, D. 1997. Relationships among *Colletotrichum* isolates from fruit-rots assessed using rDNA sequences. Mycologia 89:420-430.
27. Katan, T. and Shabi, E. 1996. Vegetative compatibility among isolates of *Colletotrichum gloeosporioides* from almond in Israel. Eur. J. Plant Pathol. 102:597-600.
28. Lenné, J. M. 1992. *Colletotrichum* diseases of legumes. Pages 134-166 in: *Colletotrichum*: Biology, Pathology and Control. J.A. Bailey and M.J. Jeger, eds. CAB International, Wallingford, UK.
29. Leslie, J.F. 1993. Fungal vegetative compatibility. Annu. Rev. Phytopathol. 31:127-150.
30. Liyanage, H.D., McMillan, R.T. Jr., and Kistler, H.C. 1992. Two genetically distinct populations of *Colletotrichum gloeosporioides* from citrus. Phytopathology 82:1371-1376.
31. Maas, J.L. 1984. Anthracnose fruit rots (black spot). Pages 57-60 in: Compendium of Strawberry Diseases. J.L. Maas, ed. APS Press, St. Paul, Minnesota.
32. Manners, J.M., Masel. A., Braithwaite, K.S., and Irwin, J.A.G. 1992. Molecular analysis of *Colletotrichum gloeosporioides* pathogenic on the tropical legume *Stylosanthes*. Pages 250-269 in: *Colletotrichum*: Biology, Pathology and Control. J.A. Bailey and M.J. Jeger, eds. CAB International, Wallingford, UK.
33. Masel, A.M., He, C., Poplawski, A.M., Irwin, J.A.G., and Manners, J.M. 1996. Molecular evidence for chromosome transfer between biotypes of *Colletotrichum gloeosporioides.* Mol. Plant Microbe Interact. 9:339-348.
34. Mills, P.R., Hodson, A., and Brown, A.E. 1992. Molecular differentiation of *Colletotrichum gloeosporioides* isolates infecting tropical crops. Pages 269-288 in: *Colletotrichum*: Biology, Pathology and Control. J.A. Bailey and M.J. Jeger, eds. CAB International, Wallingford, UK.
35. Prior, C., Elango, F., and Whitewell, A. 1992. Chemical control of *Colletotrichum* infection in mangoes. Pages 326-336 in: *Colletotrichum*: Biology, Pathology and Control. J.A. Bailey, and M.J. Jeger, eds. CAB Interantional, Wallingford, UK.
36. Prusky, D. 1996. Pathogen quiescence in postharvest diseases. Ann. Rev. Phytopathol. 34:413-434.
37. Prusky, D. and Keen, N.T. 1993. Involvement of preformed antifungal compounds in the resistance of subtropical fruits to fungal decay. Plant Dis. 77:114-119.
38. Puhalla, J.E. 1985. Classification of strains of *Fusarium oxysporum* on the basis of vegetative compatibility. Can. J. Bot. 63:179-183.
39. Rodriguez, R.J. and Yoder, O.C. 1991. A family of conserved repetitive DNA elements from the fungal plant pathogen *Glomerella cingulata* (*Colletotrichum lindemuthianum*). Exp. Mycol. 15:232-242.

40. Shabi, E. and Katan, T. 1983. Occurrence and control of anthracnose of almond in Israel. Plant Dis. 67:1364-1366.
41. Sherriff, C., Whelan, M.J., Arnold, G.M., Lafay, J.-F., Brygoo, Y., and Bailey, J.A. 1994. Ribosomal DNA sequence analysis reveals new species groupings in the genus *Colletotrichum*. Exp. Mycol. 18:121-138.
42. Shi, Y., Correll, J.C., Guerber, J.C., and Rom, C.R. 1996. Frequency of *Colletotrichum* species causing bitter rot of apple in the southeastern United States. Plant Dis. 80:692-696.
43. Sicard, D., Michalakis, Y., Dron, M., and Neema, C. 1997. Genetic diversity and pathogenic variation of *Colletotrichum lindemuthianum* in the three centers of diversity of its host, *Phaseolus vulgaris*. Phytopathology 87:807-813.
44. Smith, B.J. and Black, L.L. 1990. Morphological, cultural, and pathogenic variation among *Colletotrichum* species isolated from strawberry. Plant Dis. 74:69-76.
45. Sreenivasaprasad, S., Brown, A.E., and Mills, P.R. 1992. DNA sequence variation and interrelationships among *Colletotrichum* species causing strawberry anthracnose. Physiol. Mol. Plant Pathol. 41:265-281.
46. Sreenivasaprasad, S., Brown, A.E., and Mills, P.R. 1993. Coffee berry disease pathogen in Africa: Genetic structure and relationship to the group species *Colletotrichum gloeosporioides*. Mycol. Res. 97:995-1000.
47. Sutton, B.C. 1992. The genus *Glomerella* and its anamorph *Colletotrichum*. Pages 1-26 in: *Colletotrichum*: Biology, Pathology and Control. J.A. Bailey and M.J. Jeger, eds. CAB International, Wallingford, UK.
48. Templeton, G.E. 1992. Use of *Colletotrichum* strains as mycoherbicides. Pages 358-380 in: *Colletotrichum*: Biology, Pathology and Control. J.A. Bailey and M.J. Jeger, eds. CAB International, Wallingford, UK.
49. von Arx, J.A., 1957. Die Arten der Gattung *Colletotrichum* Cda. Phytopathol. Z. 29:413-468.
50. Wasilwa, L.A., Correll, J.C., Morelock, T.E., and McNew, R.E. 1993. Reexamination of races of the cucurbit anthracnose pathogen *Colletotrichum orbiculare*. Phytopathology 83:1190-1198.
51. Zulfiqar, M., Brlansky, R.H. and Timmer, L.W. 1996. Infection of flower and vegetative tissues of citrus by *Colletotrichum acutatum* and *C. gloeosporioides*. Mycologia 88:121-128.

Chapter 10

Inter- and Intra-Species Variation in *Colletotrichum* and Mechanisms which Affect Population Structure

James C. Correll, John C. Guerber, Lusike A. Wasilwa,
Jennifer F. Sherrill, and Teddy E. Morelock

Fungi are ideal organisms to examine macroevolutionary events, such as speciation, and microevolutionary events, such as population diversity and host adaptation (6,9,16). Recent efforts by a wide range of scientists have begun to dissect such events in the genus *Colletotrichum* (2,3,5,22,26,27, 28,30,38,39,41,48,51,52,54,59,67,70,71,72). However, much remains to be learned about inter- and intra-species variation in this economically important genus. Often, species delineations within the genus remain unresolved and in need of substantial clarification (see Cannon et al., Chapter 1 of this volume). In addition, our understanding of the mechanisms involved in vertical (sexual) and horizontal (asexual) genetic exchange within species lags even further behind. Thus, it is hoped that future efforts will continue to help resolve species delineations and address genotype variation within a species, as well as within and between populations. Moreover, developing a more complete understanding of the genetic mechanisms which affect gene flow within a species, and perhaps between species, is critical. A more thorough understanding of genotype diversity within *Colletotrichum* will greatly facilitate our understanding of the biology of this cosmopolitan plant pathogen and thereby aid in our efforts to control diseases caused by this genus.

Phenotypic variation exists below the genus level in fungi such that a continuum of variation may exist down to the individual (Fig. 1). A species, therefore, represents a means by which we demarcate groups of fungi along this continuum. Traditionally, species demarcations were based on morphological criteria, and designations were assigned based on a defined set of rules with the assumption being that the morphological criteria were a reflection of the phylogenetic, or evolutionary, relationships between species. Molecular based techniques indicate that phylogenetic relationships of organisms do not necessarily support conventional systematics. Consequently, individuals which belong to a "morphospecies" sometimes are only distantly related evolutionarily. More current species concepts involve both morphological and molecular criteria. Perhaps of equal

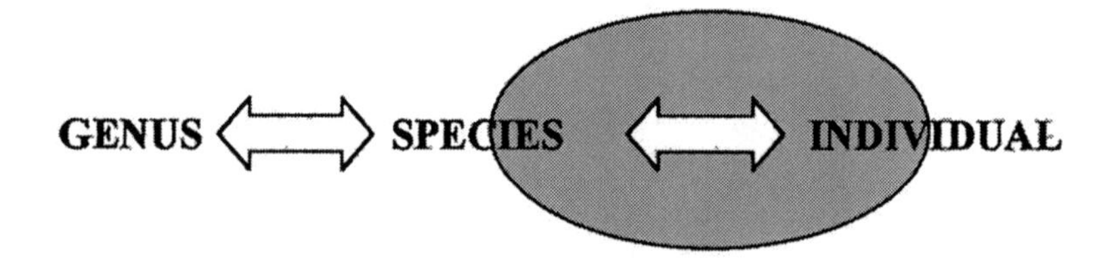

Fig. 1. Uncertainties in species and subspecies delineations remain in *Colletotrichum* and genetically defined populations are often unclear.

importance in circumscribing species in *Colletotrichum* is the overall biology and ecology of the organisms in question. Without such an approach, we may find that our current molecular-based methodologies have some of the same inherent problems as morphologically based assumptions had 50 years ago. Our lack of a clear understanding of molecular-based clocks, mutation rates, mobile genetic elements, and the potential for horizontal gene transfer may represent significant short-comings in our current understanding of phylogenetic relationships based strictly on molecular data.

The *C. orbiculare* "Species Complex" and Anthracnose of Cucurbits

Von Arx (82,83) originally recognized 13 species in the genus *Colletotrichum*, based on conidial shape and size, host range, and type of appressorium produced. Within *C. gloeosporioides*, nine specialized forms were recognized. Sutton (74,75) later elevated each of these nine forms to a distinct species, four of which included *C. orbiculare* (Berk. & Mont.) von Arx (=*C. lagenarium* (Pass.) Ell. & Halst.) from cucurbits, *C. trifolii* Bain. & Essary from alfalfa (*Medicago sativa*), *C. lindemuthianum* (Sacc. & Magnus) Briosi & Cavara from common bean (*Phaseolus vulgaris* L.), and *C. malvarum* (Braun & Casp.) Southw. from prickly sida *(Sida spinosa)*.

The production of a septum in germinating conidia of *Colletotrichum* species following mitosis also has been considered a useful taxonomic criterion (57,58). *C. orbiculare*, *C. trifolii*, *C. lindemuthianum*, and *C. malvarum* generally have similar shaped conidia (67) and, consequently, conidial morphology cannot be used to distinguish these species. However, the absence of a septum in germinated conidia is characteristic for *C. orbiculare*, *C. trifolii*, *C. lindemuthianum*, and *C. malvarum* (2,58,67). Appressoria were also found of similar shapes and dimension (67).

Molecular technologies based on sequence analysis of DNA have been very useful in examining the relationships of plant pathogenic fungi. Based upon molecular and morphological studies, *C. orbiculare* from cucumber, *C. trifolii* from alfalfa, *C. malvarum* from prickly sida, and *C. lindemuthianum* from bean are closely related (2,58,59,60,67). Based on spore morphology, appressorium development, and sequence similarities of the rDNA, Sherriff et al. (1994) proposed that *C. orbiculare*, *C. trifolii*, *C. lindemuthianum*, and *C. malvarum* should be considered a single species and indicated that *C. orbiculare* should be retained as the species designation. They also suggested that, within the broader species concept of *C. orbiculare*, isolates which showed host specialization should be classified as distinct formae speciales of *C. orbiculare*.

Uncertainties about the taxa involved in anthracnose of cucurbits have created confusion in the literature. Anthracnose of cucurbits, typically caused by the fungal pathogen *Colletotrichum orbiculare* (=*C. lagenarium*), is an important disease of cucurbits worldwide (31,69,78). It is most destructive on cucumber (*Cucumis sativus*), watermelon (*Citrullus lunatus*) and cantaloupe (*Cucumis melo*). Anthracnose also occurs on other cucurbits such as gourds (*Lagenaria siceraria*), pumpkin (*Cucurbita* spp.), squash (*Cucurbita* spp.), gherkin (*Cucumis anguria* L.) and chayote (*Sechium edule*) (31,80).

Three taxa, *Colletotrichum orbiculare*, *Glomerella cingulata* var. *orbiculare*, (the previously reported putative teleomorph of *C. orbiculare*), and *Glomerella magna* have been reported to cause cucurbit anthracnose, (29,43,44,45,46,73). However, both *G. cingulata* var. *orbiculare* and *C. magna* have only been observed to cause cucurbit anthracnose under laboratory or greenhouse conditions (29,45,46,73). In contrast, *C. orbiculare* is more commonly recognized as the cucurbit anthracnose pathogen (69).

C. orbiculare also infects cocklebur (*Xanthium* spp.) (56). Isolates of *C. orbiculare* obtained from cocklebur caused leaf spotting on some cultivars of watermelon (*Citrullus lanatus* var. *lanatus*, Candy Red, and Warpaint) and honeydew melon (*Cucumis melo*) as well as a fruit rot of cucurbits (84). Cocklebur isolates also were pathogenic on safflower (*Carthamus tinctorius*) and celery (*Apium graveolens* var. *dulce*).

Previous reports have examined a geographically diverse collection of isolates from cucurbits for virulence, mtDNA RFLPs, RAPDs, and vegetative compatibility (22,86). Pathogenic isolates belonged to a single mtDNA haplotype. All isolates examined belonged to one of three vegetative compatibility groups (VCGs) identified, and there was a strong correspondence between VCG, virulence, and host origin. All pathogenic isolates of *C. orbiculare* recovered from cucumber had a race 1 phenotype and belonged to two VCGs (VCG 1001 and 1003). All pathogenic isolates of *C. orbiculare* recovered from watermelon had a race 2 phenotype and belonged to VCG 1002. Examination of a limited number of isolates showed that *C. orbiculare* from cocklebur, *C. magna* from cucurbits, and *G. cingulata* from cucurbits were avirulent or weakly virulent on cucurbit foliage and had mtDNA RFLPs haplotypes distinct from pathogenic isolates of *C. orbiculare.*

The objectives of this study were to examine representative isolates of the *Colletotrichum orbiculare* "complex", namely *C. lindemuthianum, C. trifolii*, and *C. malvarum,* as well as an expanded collection of geographically diverse isolates recovered from foliage and fruit of numerous cucurbit hosts and cocklebur, for their host specificity on cucurbits in greenhouse inoculation assays, for vegetative compatibility diversity, and mtDNA RFLPs.

ISOLATES

All cucurbit isolates used in this study were either recovered from symptomatic tissue by the authors, received from other researchers, obtained from the American Type Culture Collection (ATCC, Rockville, MD), or obtained from the *Colletotrichum* Collection, University of Arkansas (Table 1). Single-spore isolates were stored on desiccated filter paper at 4°C. The isolates of *C. orbiculare* selected were representative of the genetic diversity previously identified within a worldwide collection of *C. orbiculare* (22,86). Representative isolates of *C. lindemuthianum,* recovered from common bean (*Phaseolus vulgaris*), *C. trifolii* from alfalfa (*Medicago sativa* L.), and *C. malvarum* from prickly sida (*Sida spinosa*) were included in the study (Table 1).

Isolates of *Colletotrichum orbiculare, Glomerella cingulata*, and *C. magna* were recovered from acorn squash (*Cucurbita maxima*), cantaloupe [*Cucumis melo* (Reticulatus group)], cucumber (*Cucumis sativus*), cocklebur (*Xanthium spinosum*), cucuzzi gourd [*Lagenaria siceraria* (Mol.) Standl.], gooseberry gourd (*Cucumis myriocarpus*), melon [*Cucumis melo* (Inodorus watermelon [*Citrullus lanatus* (Thunb.) Matsum and Nakai]. The isolates were collected from throughout the USA, or from Africa, Australia, China, France, Japan, and Taiwan between 1990-1998.

VEGETATIVE COMPATIBILITY TESTS

Isolates were characterized for vegetative compatibility using nitrate nonutilizing (*nit*) mutants as previously described (18,63,86; Katan, Chapter 4, this volume). Vegetative compatibility tests were performed by pairing *nit* mutants in all possible combinations. Vegetative compatibility tests were conducted at least twice. Pairing among isolates where a distinct heterokaryon developed as a result of complementation was considered

Table 1. Summary of host specificity, phenotypes, and genotypes of *Colletotrichum orbiculare*, allied species, and taxa recovered from cucurbits.

		Origin				Disease Reaction[a]				
Species	Isolate	Host	Geographic	mtDNA RFLP haplotype[b]	VCG[c]	Marketer	H19	Black Diamond	Charleston Grey	Fruit[d]
C. orbiculare f.sp. *cucurbitacearum*	JC1	cucumber	AR	A	1001 (race 1)	+++	+++	+++	+	++
	MH2	cucumber	NC	A	1003 (race 1)	+++	+++	+++	+	++
	CP6	watermelon	TX	A	1002 (race 2)	+++	+	+++	+++	++
	CR2	watermelon	FL	A	1002 (race 2B)	+++	+	+++	+	++
	JX10	melon	Taiwan	A	1004 (race 2B)	+++	+	+++	+	++
C. trifolii	14-2-63	alfalfa	MD	A	CT-1	-	-	-	-	+
C. malvarum	3-7-11	Sida	AR	A	CM-1	-	-	-	-	+
C.orbiculare	LW1	cockleburr	Australia	A1	1051	-	-	-	-	+
C. lindemutheanum	14-2-37	bean	Canada	A2	CL-1	-	-	-	-	+
C. magna	AK2	squash	SC	B	C1	+/-	-	+/-	-	+++
Glomerella cingulata	HD3	honeydew	OK	C	GC2	+/-	-	+/-	-	++
C. sp.	JD7	watermelon	OK	D	-	+/-	-	+/-	-	++

[a]Relative disease reactions on cotyledons of cucumber (*C. sativus*, Marketer and H19) and watermelon (*C. lanatus*, Black Diamond and Charleston Grey) cultivars (86).
[b]All isolates within a mtDNA RFLP haplotype had the same restriction fragment patterns when probed with two *C. orbiculare* mtDNA clones (4U40 and 2U18) (22).
[c]VCG = vegetative compatibility group.
[d]Cucumber fruit.

vegetatively compatible, and the pairs were placed into the same VCG. VCG identity was used in selecting isolates for examining additional characteristics.

Mitochondrial DNA RFLP Analysis

The mtDNA RFLP haplotypes were determined as previously described (22). Two large non-overlapping mtDNA clones, 4U40 (13.7 kb) and 2U18 (10.1 kb), of *C. orbiculare* (65% of mitochondrial genome) were combined in equimolar concentrations and used for hybridization (22).

Foliar Pathogenicity and Virulence Tests

Representative isolates of *C. orbiculare*, *C. trifolii*, *C. lindemuthianum*, and *C. malvarum* were evaluated for their ability to cause disease (pathogenicity) and for differences in aggressiveness (virulence) using a cucurbit cotyledon inoculation assay (86). Four differential cultivars, cucumber cultivars Marketer (Harris Moran Co.) and Arkansas Little Leaf [H19 (Peto Seed)] and watermelon cultivars Black Diamond (Northrup King Co.) and Charleston Gray (Northrup King Co.) that have been shown to distinguish races 1, 2, and 2B when cotyledon inoculations were used (85,86,87). Evidence of disease symptoms and severity ratings were obtained by visually assessing the area of the cotyledons showing symptoms of infection (chlorosis and necrosis) as described by Wasilwa et al. (86). Disease reactions on the differential cultivars were classified after 8 days as resistant (disease severity ratings 0.0-2.5) or susceptible (disease severity ratings 5.0) based on the mean disease ratings.

Fruit Pathogenicity and Virulence tests

Representative isolates of *C. orbiculare*, *C. malvarum*, *C. trifolii*, and *C. lindemuthianum* also were selected based on their vegetative compatibility group, mtDNA RFLPs, race, host and geographic origin and used in a fruit inoculation assay. Fruit of the cucumber cultivars, Marketer and H19, and the watermelon cultivars, Black Diamond and Charleston Grey, were used. In addition, mature fruit from a "slicing type" cucumber cultivar purchased from a grocery store were used. Seven days after inoculation, lesion diameters were recorded by measuring two perpendicular axes of the surface lesion. A cross section through the center of the lesion was made, and lesion depth and width of the symptomatic tissue were measured.

Results and Discussion

MtDNA RFLPs

Cucurbit isolates of the four known VCGs of *C. orbiculare* (VCGs 1001, 1002, 1003, and 1004), of races 1, 2, and 2B, from geographically diverse origins had a common mtDNA RFLP haplotype (haplotype A) when examined with several restriction enzymes and probed with two mtDNA clones of *C. orbiculare* [4U40 (13.7 kb) and 2U18 (10.1 kb)] (Table 1, Figs. 2 and 3). Also, representative isolates of *C. orbiculare* from cocklebur from Australia had a similar mtDNA haplotype to haplotype A, with one polymorphism detected with the enzyme *Pvu*II. This polymorphism was previously shown to be due to the occurrence of one additional restriction site in the mitochondrial genome among isolates from cocklebur (22).

Isolates of *C. trifolii*, *C. lindemuthianum*, and *C. malvarum* had an identical mtDNA haplotype to haplotype A of *C. orbiculare* when examined with two different restriction enzymes (*Eco*RI and *Pvu*II). The isolates of *C. orbiculare*, *C. trifolii*, and *C. malvarum* also had an identical mtDNA haplotype (haplotype A) when examined with a third restriction enzyme (*Hae*III). However, a single mtDNA RFLP was detected among the five isolates of *C. lindemuthianum* with the enzyme *Hae*III (Fig. 2).

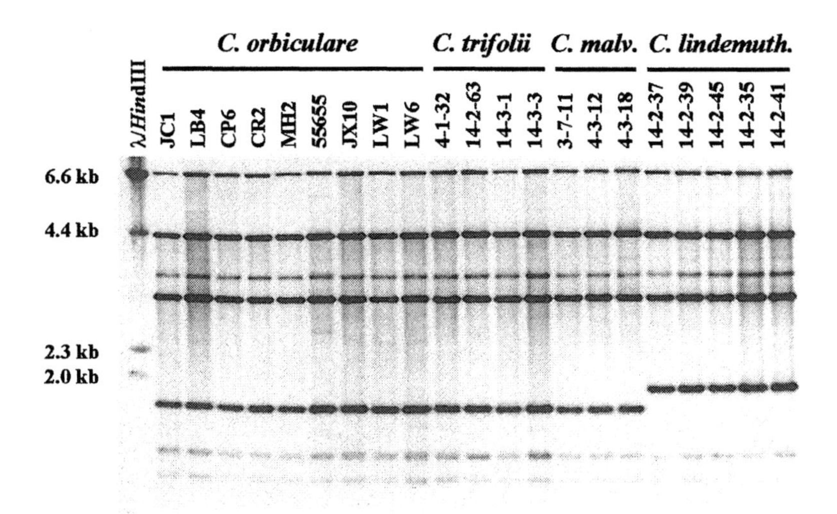

Fig. 2. MtDNA RFLPs of *Colletotrichum orbiculare* and allied species. Total DNA was restricted with *Hae*III and probed with two mtDNA clones (4u40 and 2u18) of *C. orbiculare* (22).

Foliar Pathogenicity and Virulence Tests

Only isolates of *C. orbiculare* which originated from cucurbit hosts produced distinct disease reactions on the cotyledons of the four differential cultivars under standardized conditions (85,86,87). Isolates of *C. trifolii*, *C. lindemuthianum*, *C. malvarum*, and *C. orbiculare* from cocklebur were considered either weakly virulent or avirulent on the susceptible cucurbit differentials Marketer and Black Diamond (Table 1). The representative isolates of *C. orbiculare* which originated from cucurbit hosts were significantly more virulent than the other taxa, including *C. orbiculare* from cocklebur, on all four differentials and could be divided into three distinct races (Table 2). Thus, the data indicate that isolates of *C. orbiculare* from cucurbits show distinct host specificity on cucurbit foliage relative to other taxa in the *C. orbiculare* species complex.

C. orbiculare, *Glomerella cingulata* var. *orbiculare* (the putative teleomorph of *C. orbiculare*), and *C. magna* have been reported to cause anthracnose of cucurbits. However, in this study only isolates of *C. orbiculare* from cucurbits were pathogenic on the susceptible cucurbit differentials Marketer and Black Diamond and considered foliage pathogens of cucurbits (Table 1). Among the three distinct races (race 1, 2, and 2B) of *C. orbiculare*, there was a correspondence between vegetative compatibility group (VCG) and virulence whereby all pathogenic isolates of

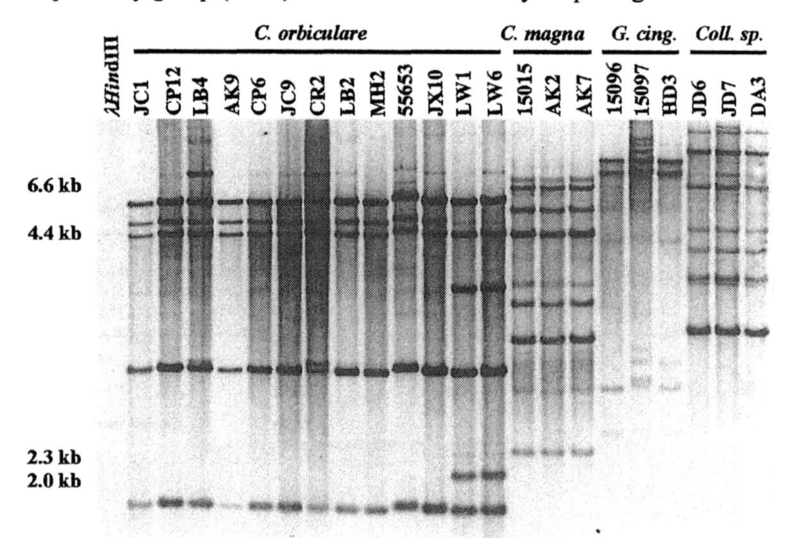

Fig. 3. MtDNA RFLPs of anthracnose isolates recovered from cucurbit foliage and fruit. LW1 and LW6 are isolates from cockleburr. Total DNA was restricted with *Pvu*II and probed with two mtDNA clones (4u40 and 2u18) of *C. orbiculare* (22).

Table 2. Summary of disease reactions of a worldwide collection of isolates of *Colletotrichum orbiculare f.sp. cucurbitacearum* from diverse hosts in four vegetative compatibility groups (VCGs) on cucumber and watermelon differential cultivars.

mtDNA RFLP haplotype[b]	VCG[c]	Race	Host[d]	Disease reactions[a]			
				Cucumber		Watermelon	
				Marketer	H19	Black Diamond	Charleston Gray
A	1001	1	cucumber cantaloupe pumpkin	S	S	S	R
A	1002	2	watermelon cantaloupe	S	R	S	S
A	1002	2B	watermelon	S	R	S	R
A	1002	2B	cucuzzi gourd	S	R	S	R
A	1003	1	cucumber	S	S	S	R
A	1004	2B	melon	S	R	S	R

[a]Disease rating scale of 0 - 7 on cotyledons (see Wasilwa et al., 1993); R = resistant (disease rating ≤ 2.5); S= susceptible (≥ 5.0).

[b]All isolates within a mtDNA RFLP haplotype had the same restriction fragment patterns when probed with two *C. orbiculare* mtDNA clones (22).

[c] VCG = vegetative compatibility group.

[d] Cantaloupe (*Cucumis melo.*), cucumber (*C. sativus*), melon (*C. melo*), cucuzzi gourd (*Lagenaria siceria*), pumpkin (*Cucurbita* spp.), and watermelon (*Citrullus lanatus*).

C. orbiculare belonged to one of four VCGs (1001, 1002, 1003, and 1004). Race 1 consisted of cucumber, cantaloupe, and pumpkin isolates from the USA, Australia, France, and Japan and belonged to VCG 1001 and VCG 1003; race 2 consisted of watermelon isolates from the USA and China and belonged to VCG 1002; and race 2B consisted of watermelon and cucuzzi gourd isolates from the USA and Taiwan and also belonged to VCG 1002.

The limited VCG diversity in the anthracnose population of *C. orbiculare* along with the widespread distribution of VCG 1001 throughout the U. S. and the worldwide distribution of VCGs 1002 and 1003 indicate that the cucurbit anthracnose pathogen may be reproducing predominately asexually. However, more isolates from cucurbits need to be examined.

Fruit Pathogenicity and Virulence Tests

Many fungi are capable of causing fruit rots of succulent fruit including a number of *Colletotrichum* spp. Distinct host specificity was not observed in the fruit inoculation assay. All four taxa, *C. orbiculare*, *C. trifolii*, *C. lindemuthianum*, and *C. malvarum*,were able to colonize wounded cucumber fruit (Table 1). However, the majority of the isolates of *C. orbiculare* from cucurbits had significantly larger lesion diameters and lesion volumes than isolates of *C. orbiculare* from cocklebur, *C. trifolii*, *C. lindemuthianum*, and *C. malvarum*.

Isolates of *C. magna*, *G. cingulata* and an unidentified *Colletotrichum* sp. were not pathogenic on cucurbit foliage and could be distinguished from *C. orbiculare* based on mtDNA RFLPs (Fig. 3). Two ATCC isolates of *G. cingulata* var. *orbiculare*, reported as the teleomorph of *C. orbiculare*, and two isolates recovered from honeydew fruit in Oklahoma belonged to mtDNA RFLP haplotype C, and were vegetatively incompatible with all isolates of *C. orbiculare.* Ten isolates of *C. magna*, from various cucurbit hosts (watermelon, cucumber, acorn squash, and spaghetti squash) from the USA and China belonged to mtDNA RFLP haplotype B, and were vegetatively incompatible with all isolates of *C. orbiculare*. Four isolates of an unknown *Colletotrichum* sp. recovered from watermelon fruit in Oklahoma belonged to a fourth mtDNA RFLP haplotype (haplotype D).

Although *C. orbiculare* can cause fruit rots of cucurbits, several taxa including *G. cingulata* var. *orbiculare*, *C. magna*, and an unidentified *Colletotrichum* species apparently are more aggressive fruit rot pathogens. Further examination may reveal a number of as yet unidentified *Colletotrichum* species occurring on fruit of cucurbits. In this study, only minor differences in virulence were observed among isolates representing the three races (races 1, 2, and 2B) of *C. orbiculare* (VCGs 1001, 1002, 1003, and 1004) on fruit of both cucumber and watermelon differential cultivars. However, results indicated that race 1 (VCG 1001) and race 2B (VCG 1002), were more aggressive than race 2 on cucumber fruit, and that race 2 was more aggressive than races 1 and 2B (VCGs 1002 and 1004, respectively) on watermelon cultivar Charleston Gray.

Data from mtDNA RFLPs provide supporting evidence that the taxa studied have a common ancestry and likely represent a single phylogenetic species. However, only isolates of *C. orbiculare* from cucurbits were pathogenic on cucurbit foliage. Because certain isolates within the broader species concept of *C. orbiculare* also clearly show virulence to cucurbits, it is proposed that isolates virulent on cucurbit foliage be given a *forma specialis* designation, *C. orbiculare* f.sp. *cucurbitacearum*.

Glomerella cingulata: Mechanisms which Affect Population Structure

Sexual and vegetative incompatibility are two major barriers to gene flow in fungi, but are not well understood in *Colletotrichum*. In the ascomycete fungi, sexual compatibility is typically controlled by a single mating locus with two alleles (15,32,33,55). Isolates can be heterothallic whereby monoconidial isolates are self-sterile, and successful mating occurs between isolates with different alleles at the *mat* locus. In homothallic systems, isolates are self-fertile and monoconidial or single ascospore cultures can produce fertile perithecia. In most heterothallic ascomycete fungi, mating is controlled by one *mat* locus with two alleles (55). The different *mat* alleles are dissimilar in gene sequence and, consequently, the alleles are referred to as idiomorphs (55).

Research on the genetics of sexuality in *C. gloeosporioides* (teleomorph: *Glomerella cingulata*) began with Edgerton (24,77). Edgerton isolated two strains of *G. cingulata* which could reproduce sexually and designated these strains "plus" and "minus." In mating studies, the plus and minus characteristics were verified to be produced within a single ascus. Wheeler (88,89), however, reported that mating was controlled by many loci, with each locus controlling a specific step in the sexual process. Wheeler and McGahen (90) described the genes *A* and *B*, which controlled switching between perithecial and conidial stages and perithecial distribution, respectively. These early studies provided a foundation for *G. cingulata* as a model system for the study of mating in the ascomycetes (8). However, this early research ceased, perhaps because of the genetic complexity of the mating system in *G. cingulata*. The lack of good genetic and molecular markers was a significant weakness in some of these earlier studies. More recently, Cisar and TeBeest (12) reported that mating in *G. cingulata* was controlled by a single locus with multiple alleles. Isolates from different hosts had the ability to mate and produce viable progeny on both agar plates and host tissue in artificial inoculations (13,14). Analysis of crosses between isolates of *C. musae* (teleomorph: *Glomerella musae*) indicated that this species was heterothallic and had a minimum of four mating types

(66). No further studies were performed to determine the genetic control of sexual compatibility in *C. musae*.

Little is known about the relative importance of sexual reproduction to the survival of *C. gloeosporioides* in nature, the effect on population structure, or the potential for genetic exchange between geographically diverse isolates from a wide range of hosts.

Vegetative incompatibility, the inability of two fungal individuals to fuse and form a heterokaryon, represents an asexual barrier to genetic exchange in ascomycete fungi (50). The ability to fuse and form a heterokaryon is controlled by vegetative incompatibility (*vic*) loci (also called heterokaryon incompatibility, or *het*, loci) and is heterogenic, meaning genetic differences at the *vic* loci inhibit heterokaryon formation between two strains (25). Vegetative incompatibility does not interfere with sexual compatibility in most fungi, but, in a few cases (*e.g., Neurospora crassa*), the mating alleles also function as *vic* genes in the vegetative phase (65).

Vegetative incompatibility subdivides a population into vegetative compatibility groups (VCGs) and has been used to characterize genetic diversity in many fungal populations (50). However, the actual genetic control of vegetative compatibility has only been examined in a few fungal species. Vegetative compatibility has been used in a number of studies examining population diversity in species of *Colletotrichum* (1,4,7,17,20,26,49,52,81,86; Freeman, Chapter 9 of this volume), but the genetic basis of vegetative incompatibility in *Colletotrichum* is largely unknown. Vaillaincourt and Hanau (81) studied the segregation ratios of VCGs in crosses of *Glomerella graminicola* (teleomorph of *C. graminicola*). Based on crosses of two wild-type isolates, they estimated that at least five unlinked *vic* loci were segregating. No comparable studies have been published for *C. gloeosporioides.*

Several techniques can be employed to determine if two isolates are vegetatively compatible (50). The preferred method for many fungi has been the use of complementation of auxotrophic or pigmentation mutants (64) because the heterokaryon can be easily distinguished from the parental isolates (50). In many fungi, vegetative compatibility has been studied using nitrate non-utilizing (*nit*) mutants (18,23,63). When two phenotypically distinct *nit* mutants of vegetatively compatible isolates are paired on minimal medium, a heterokaryotic zone of wild-type growth appears because of the complementation of the mutations. Sulfate non-utilizing (*sul*) mutants, which can be generated by culturing an isolate on a medium amended with selenate (19,40,42) can also be used to assess vegetative compatibility in *Colletotrichum*.

Several *Colletotrichum* species can cause bitter rot of apple, an economically important disease in the southeastern United States (68,76). Bitter rot is caused by *C. gloeosporioides* (heterothallic isolates), *G. cingulata* (homothallic isolates of *C. gloeosporioides*), and *C. acutatum*

(68). The three taxa can be distinguished based on mtDNA RFLPs (Fig. 4) (17). Populations of the three bitter rot taxa have been characterized for VCG diversity, which has served as a useful genetic framework for genetic studies of these taxa. Isolates of *C. gloeosporioides* from avocado, lime, lychee, mallow, mango, pecan, and *Stylosanthes* were examined for their ability to mate with four apple reference isolates, all of which were capable of mating with each other. In addition, to test the hypothesis that *C. gloeosporioides* contained multiple *vic* loci and that vegetative incompatibility was heterogenic in nature, a series of backcrosses was conducted with the apple reference isolates to produce isogenic fungal strains which contained different alleles at a single *vic* locus. A second goal was to determine if sexual mating type could segregate independently of the *vic* loci. Similar studies were conducted with *C. acutatum*.

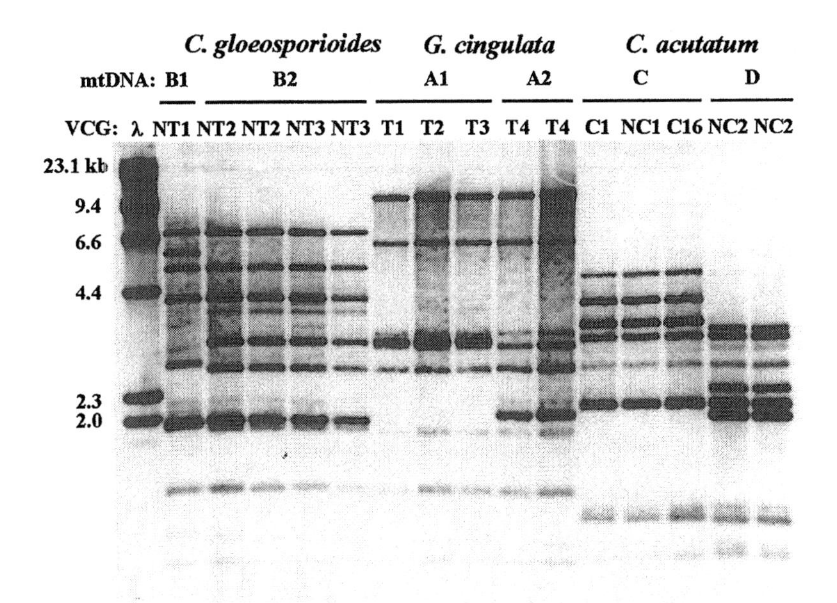

Fig. 4. MtDNA RFLPs of *Colletotrichum gloeosporioides* (self-sterile), *Glomerella cingulata* (self-fertile), and *C. acutatum* from apple. MtDNA haplotypes and vegetative compatibility group designations are indicated. Total DNA was restricted with *Pal*I and probed with two mtDNA clones (4u40 and 2u18) of *C. orbiculare* (22).

ISOLATES

Four self-sterile reference isolates of *C. gloeosporioides* recovered from apple (*Malus domestica* Borkh.) (68) were used in the mating studies. Two of the apple reference isolates (A6 and FC226) were collected in Arkansas, and two (NC329 and NC131) in North Carolina. The four apple isolates each represented a unique VCG (NT1, NT2, NT3, and NT4) and contained one of four different alleles at the *mat* locus (37) (Tables 3 and 4). The isolates belonged to mtDNA haplotypes B1 (A6), B2 (FC226 and NC329), and B3 (NC131) (Fig. 3)(17). Isolates of *C. gloeosporioides* from pecan (*Carya illinoensis*) from Georgia and Louisiana and isolates from avocado (*Persea americana*), lychee (*Litchi chinensis*), mango (*Mangifera indica*), and Persian lime (*Citrus aurantifolia*) from Florida also were examined. One mallow (*Malva pusilla*) isolate (815) originated from Canada, and one isolate (388) from *Stylosanthes hamata* was obtained from the American Type Culture Collection (Table 3). The mtDNA RFLP haplotype was determined for all isolates (17) (Fig. 5). All wild-type cultures and mutants were stored desiccated on Whatman filter paper at 4°C as previously described (21).

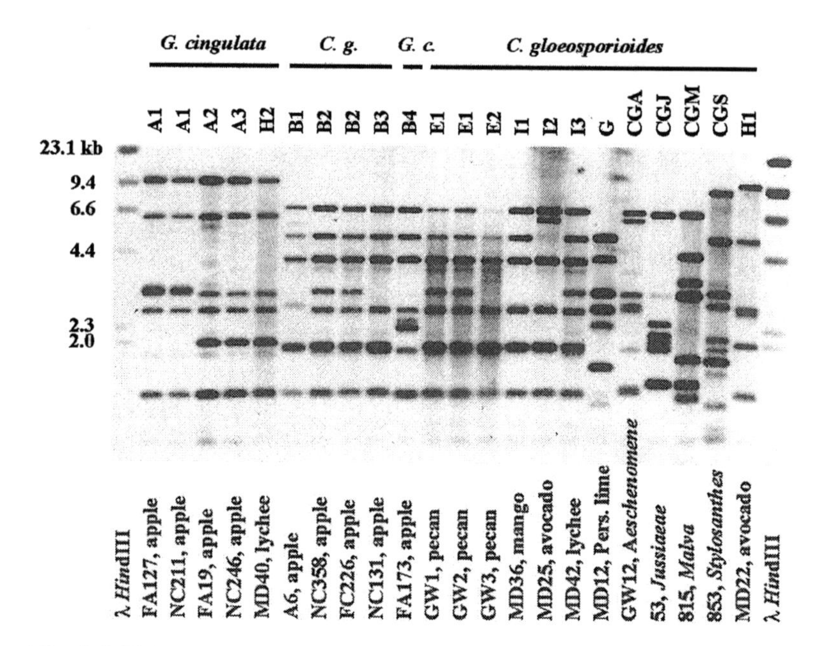

Fig. 5. MtDNA RFLPs of *Colletotrichum gloeosporioides* (self-sterile), *Glomerella cingulata* (self-fertile). MtDNA haplotype designations are indicated at top and isolate and host origin at the bottom. Total DNA was restricted with *Hae*I and probed with two mtDNA clones (4u40 and 2u18) of *C. orbiculare* (22).

Table 3. Isolates of *Colletotrichum gloeosporioides* used in the mating studies.

Isolate	Origin [a] Host	Origin [a] Geographical	mtDNA haplotype [b]
A6	Apple (*Malvus domestica*)	AR	B1
FC226	Apple	AR	B2
NC131	Apple	NC	B3
NC329	Apple	NC	B2
LC3	Pecan (*Carya illinoensis*)	AL	B3
LC5A	Pecan	AL	B2
LC6A	Pecan	AL	B5
LC10B	Pecan	AL	B5
NUT1	Pecan	AL	B3
NUT4	Pecan	AL	B3
HO6	Pecan	GA	B2
HO15	Pecan	GA	B2
HO30	Pecan	GA	B2
HO31	Pecan	GA	B3
GW1	Pecan	LA	B2
GW4	Pecan	LA	B2
HO2	Pecan	LA	B2
HO18	Pecan	LA	B3
MD16	Avocado (*Persea americana*)	FL	H1
MD22	Avocado	FL	H1
MD23	Avocado	FL	I1
MD25	Avocado	FL	I2
MD26	Avocado	FL	G
MD27	Avocado	FL	I2
MD41	Lychee (*Litchi chinensis*)	FL	I1
MD42	Lychee	FL	I3
MD35	Mango (*Mangifera indica*)	FL	I1
MD38	Mango	FL	I1
MD1	P. Lime (*Citrus aurantifolia*)	FL	g
MD5	P. Lime	FL	G
815	Mallow (*Malva pusilla*)	Canada	Cgm
388	*Stylosanthes hamata*	ATCC	Cgs

[a] The host and geographical origin of the isolate.

[b] Mitochondrial DNA RFLP haplotypes were determined by probing total DNA with two mtDNA clones from *C. orbiculare* (22).

Table 4. Results of apple by apple and apple by non-apple isolate crosses.

Host		Apple Reference Isolates			
		A6	NC329	FC226	NC131
Apple	A6	-/-[a]	+/+	+/+	+/+
	NC329		-/-	+/+	+/+
	FC226			-/-	+/+
	NC131				-/-
Pecan	HO15	+/+	+/+	+/-	+/-
	HO31	+/+	+/+	+/+	+/+
	LC5A	+/+	+/+	+/+	+/+
	NUT1	+/+	+/+	+/-	+/+
	NUT4	+/+	+/+	+/-	+/+
	GW1	-/-	+/+	+/-	+/+
	HO2	+/+	+/+	+/-	-/-
	HO18	+/+	+/+	+/+	-/-
	HO30	-/-	+/+	+/+	-/+
	LC3	-/-	+/+	+/+	+/+
	HO6	-/-	+/+	-/-	+/-
	GW4	-/-	-/+	-/-	-/-
	LC10B	-/-	-/-	+/-	-/-
	LC6A	-/-	-/-	-/-	-/-
Avocado	MD23	+/+	+/+	+/-	+/+
	MD16	+/-	+/-	-/-	+/-
	MD22	+/+	-/-	-/+	-/+
	MD27	+/+	+/+	-/-	-/+
	MD25	+/+	-/+	-/-	-/-
	MD26	-/-	-/-	-/-	-/-
Lychee	MD41	+/+	+/-	-/-	-/+
	MD42	+/+	-/-	-/-	-/+
Mallow	815	-/-	-/-	-/-	-/-
Mango	MD35	+/+	+/+	+/+	+/-
	MD38	-/-	+/-	+/-	-/+
P.lime	MD1	-/-	-/-	-/-	-/-
	MD5	-/-	-/-	-/-	-/-
Stylosanthes	388	-/-	-/-	-/-	-/-

[a] The first symbol (+/) represents the apple (wildtype) by non-apple (*nit* mutant) crosses. The second symbol (/+) represents the reciprocal apple (*nit* mutant) by non-apple (wildtype) crosses. "+" = fertile perithecia, "-" = no fertile perithecia.

Mating Conditions

Sexual matings were performed on Czapek-Dox minimal salts medium (CD; 0.01 g $FeSO_4$, 2 g $NaNO_3$, 1 g K_2HPO_4, 0.5 g $MgSO_4$, 0.5 g KCl, and16 g Difco granulated agar per liter of H_2O). Three sterile 58 x 1 mm^2 birch toothpicks (Diamond Brand, Inc., Minneapolis, MN) were placed on the agar surface of each plate to act as a substrate for perithecia formation. Two toothpicks were on either side of a 9-cm-diameter plate and one diagonally in the middle to form an 'N'. The isolates to be crossed were removed from storage and grown on Difco potato dextrose agar at room temperature. Mycelial plugs of 2- to 7-day-old isolates were transferred to opposite sides of the CD plate. The edges of the plates were wrapped with a 5-cm wide strip of plastic food wrap. Cultures were incubated in a growth chamber at 20°C (±1°C) under 24 hr light (four 61-cm long 40 W cool light fluorescent bulbs on the sides and two 30-cm long 40 W cool light fluorescent bulbs approximately 30 cm above the cultures). Plates were checked for the presence of perithecia after 3, 4, and 5 weeks, and squash mounts of
perithecial-shaped structures were examined at 400X magnification for the presence of asci and ascospores. The plastic wrap seal was removed from all mating plates after 3 weeks, prior to checking the plates for mature perithecia; after which the plates remained unsealed. Five mature perithecia were selected using a dissecting stereoscope at 25X, crushed with a dissecting needle in a sterile microcentrifuge tube containing 150 ml sterile water, and spread onto a water agar plate. After 24 hrs, single germinating ascospores were transferred to a minimal medium.

Mating Studies

Fourteen pecan and 14 isolates from several other tropical fruit were crossed with each of the four apple reference isolates (Table 4). The wild type of each of the four apple reference isolates was crossed with a *nit* mutant of the 14 pecan or 14 tropical isolates on CD medium as described. The reciprocal cross, whereby a *nit* mutant of the four apple reference isolates was crossed with a wild-type non-apple isolate, was also performed. Single ascospore progeny from one to five perithecia were collected from successful crosses. The phenotype of the F_1 progeny was determined on minimal medium (18).

In almost all cases, *nit* mutant and wild-type progeny were recovered from a given perithecium. However, to confirm recombination in sexual crosses, vegetative compatibility tests were performed as previously described. In wild-type (apple) by *nit* mutant (non-apple) crosses, the F_1 progeny with a *nit* phenotype were paired with a *sul* mutant of the original apple parental isolate and with the complementary *nit* mutant of the non-apple parental isolate (when available) on minimal media. After 7, 10, and

Table 5. The estimated number of vegetative incompatibility (*vic*) loci segregating in progressive backcrosses.

Cross [a]	Recurrent parent	Number of progeny vegetatively compatible with the recurrent parent	Total progeny	Estimated number of *vic* loci segregating [b]
Cross: A6/3b X NC329[c]	---	14	63[c]	3
First backcross: A6/3b X P2F7	A6/3b	41	110	2
Second backcross: A6/3b X BC30A19	A6/3b	19*	52	1
	A6/3b	13*	22	1
	A6/3b	19*	40	1
	A6/3b	16*	35	1
		67*	**149**	**1**
Cross: NC329 X A6/3b	---	9	63	3
First backcross: NC329/2c X P2C8	NC329/2	51	135	2
Second backcross: NC329/2 X BC35C3	NC329/2	13*	25	1
	NC329/2	5*	10	1
	NC329/2	10*	25	1
	NC329/2	11*	22	1
		39*	**82**	**1**

[a] The first isolate listed was the recurrent parent.

[b] The number of *vic* loci segregating in a cross was estimated based on the formula $1/X = 1/2^n$, where n is the number of loci segregating and 1/X is the proportion of vegetatively compatible progeny.

[c] A total of 189 progeny were recovered from this cross, 81 of which were *nit* mutants. Of the 81 *nit* mutants, 63 were examined in vegetative compatibility tests with the *sul* mutant of each parent.

*Ratio of vegetatively compatible:vegetatively incompatible progeny does not deviate significantly from a 1:1 ratio ($p<0.05$) based on a chi square analysis.

14 days, isolates were examined for heterokaryon formation. Any F_1 *nit* progeny which were vegetatively compatible with the original wild-type apple isolate were considered to be definitive evidence of recombination. In the non-apple (wild type) by apple (*nit* mutant) crosses, any F_1 *nit* progeny which were vegetatively compatible with the non-apple parent were considered to be definitive evidence of recombination.

Backcross Scheme and Generation of Isogenic Fungal Strains

To determine if vegetative compatibility in *C. gloeosporioides* was heterogenic at a single *vic* locus, isogenic fungal strains which contained different alleles at a single *vic* locus were generated. To accomplish this, a series of crosses and backcrosses were performed with two apple reference isolates. The proportion of ascospore progeny that were vegetatively compatible with the recurrent parent was then determined to estimate the number of *vic* loci which were segregating. The estimate of the number of loci segregating was based on the formula $1/X = 1/2^n$, where n is the number of loci segregating and 1/X is the proportion of the progeny vegetatively compatible with one of the parents. This model assumes that there are only two alleles for a given *vic* locus.

The original cross involved a *nit* mutant and a wild-type isolate. Subsequent backcrosses involved the original *nit* mutant parent and *nit* mutant progeny from the previous cross. Thus, all progeny from the *nit* mutant by *nit* mutant cross would be *nit* mutants and could then be tested for vegetative compatibility with the recurrent parent by pairing them with a *sul* mutant of the recurrent parent. The first cross was between isolate A6/3b (*nit* mutant) and NC329 (wild type) (Table 5). It had previously been shown that alleles at a minimum of three *vic* loci were segregating in crosses involving these two strains (35). Randomly selected ascospore progeny were recovered from one successful cross (A6/3b X P2F7). All F_1 progeny from this cross were then paired with the *sul* mutant of A6 to determine if they were vegetatively compatible with the recurrent parent. Ten progeny which were vegetatively incompatible with A6 were then backcrossed to A6/3b, the recurrent parent. Randomly selected ascospore progeny were recovered from one successful cross (A6/3b X BC30A19) and examined for vegetative compatibility with A6. Twenty of the progeny from cross A6/3b X BC30A19, ten which were vegetatively compatible and ten which were vegetatively incompatible with A6/3b, were then crossed with A6/3b to determine if the mating type was segregating independently of the *vic* loci. A parallel study was conducted with the same cross involving A6/3b (a *nit* mutant) and NC329 (wild type), but NC329/2c (the *nit* mutant of NC329) was used as the recurrent parent (Table 5).

Results and Discussion

MATING STUDIES

Perithecia were only observed on the center toothpick where certain isolates came in contact with one another. No perithecia were observed on the outer toothpicks for any isolates of *C. gloeosporioides* or *C. acutatum*, indicating that none of the isolates were capable of producing perithecia without being in intimate contact with another isolate.

Of all of the crosses observed, the apple-by-apple isolate crosses appeared to be the most fertile; these crosses generally produced more perithecia and had more ascospores per perithecium than did apple-by-non-apple isolate crosses. Crush mounts of the perithecia from apple-by-apple isolate crosses showed that almost all of the asci were filled with six to eight mature ascospores. Crush mounts of perithecia from apple-by-non-apple isolate crosses, however, showed that approximately one-fourth to one-half of the asci were empty, and asci with ascospores contained four to six ascospores. The fertility of the apple-by-pecan isolate and the apple-by-tropical isolate crosses were comparable in fertility, but a higher frequency of pecan isolates mated with the apple isolates.

The pecan isolates could be grouped based on the number of apple reference isolates with which an isolate was sexually compatible. For the apple (wild type) by pecan (*nit* mutant) isolate crosses, of the 14 pecan isolates tested, five pecan isolates (HO15, HO31, LC5A, NUT1, and NUT4) mated with all four apple reference isolates. Four pecan isolates (GW1, HO2, HO18, and LC3) mated with three of the apple reference isolates, two pecan isolates (HO6 and HO30) mated with two of the apple reference isolates, one isolate (LC10B) mated with one apple reference isolate, and two pecan isolates (GW4 and LC6A) did not mate with any of the apple reference isolates.

The *nit* mutation segregated in a 1:1 wild type:*nit* ratio in 31 of the 37 (83.8%) successful apple (wild type) by pecan (*nit* mutant) crosses and in 27 of the 31 (87.1%) successful pecan (wild type) by apple (*nit* mutant) crosses. Recombination was confirmed in the apple (wild type) by pecan (*nit* mutant) crosses if any F_1 *nit* progeny isolates were vegetatively compatible with the apple parental isolate *sul* mutant. Based on these criteria, recombination was confirmed in 17 of 37 (47.2%) of the crosses. For one atypical cross (FC226 X LC10B/3), the progeny segregated in a 1:1 wild type:*nit* ratio; however, all of the *nit* progeny were vegetatively compatible with LC10B (the *nit* parent). Although the possibility remains that contaminating conidia from either parent were recovered when removing the perithecia, this occurrence would have been unusual with two different perithecia and that the wild type:*nit* ratio would remain 1:1. Thus, the possibility that some type of induced selfing may have occurred in this fungus cannot be ruled out.

Recombination was confirmed in the pecan (wild type) by apple (*nit* mutant) crosses if any F_1 *nit* progeny isolates were vegetatively compatible with the pecan parental isolate complementary *nit* mutant. Recombination was confirmed based on these criteria in only 4 of 31 (12.9%) of the crosses.

Geographically and genetically distinct isolates of *C. gloeosporioides* were able to mate with each other under laboratory conditions. Isolates from avocado, mango, pecan, and lychee were able to mate with isolates from apple. None of the isolates examined in this study had the ability to self; all appeared to be strictly heterothallic. This heterothallism was in contrast to a subset of bitter rot isolates from apple which apparently were strictly homothallic (68).

The data are consistent with the hypothesis that the control of mating in *C. gloeosporioides* is under the genetic control of one *mat* locus with multiple alleles. However, the possibility that mating in *C. gloeosporioides* is under the genetic control of some other yet undescribed model cannot be ruled out (16,61). In *Neurospora crassa* and *Podospora anserina*, loci other than the *mat* locus are required for the development of perithecia (15). The same could be true for *C. gloeosporioides*. Furthermore, the genetic mechanisms controlling mating in other species of *Colletotrichum* may be different (Vaillancourt et al., Chapter 3 of this volume). Recombination was confirmed for 23 of the 107 successful crosses (21.5%).

Although recombination was not confirmed for all of the crosses, it does not necessarily indicate that recombination did not occur. The genetic markers used (the *nit* mutation and VCG) may not have detected recombination in all instances. In general, fewer of the F_1 progeny from the crosses were vegetatively compatible with the non-apple parent. The VCG tester used for the non-apple parents were *nit* mutants, whereas the VCG tester used for the apple parental isolates were *sul* mutants. When pairing F_1 progeny from apple-by-apple crosses, the *sul* tester is more robust and more F_1 progeny will show complementation with the *sul* tester than with the *nit* tester from the same isolate (37). More F_1 progeny from the apple-by-non-apple isolate crosses may in actuality be vegetatively compatible with the non-apple parent than is being detected by the *nit* mutant tester. Although a clear explanation of this discrepancy cannot be provided, it is plausible that a given *nit* mutation was perhaps modified during meiosis, thereby making it unable to complement the second *nit* mutation.

Generation of Isogenic Fungal Strains

In the first cross (A6/3b X NC329) a total of 189 F_1 progeny was recovered from five perithecia. Of these 189 progeny, the *nit*:wild type phenotype segregated 81:108, respectively. In vegetative compatibility tests using the complementary *nit* mutant (A6/4) as the VCG tester, none of the 81 *nit* progeny were vegetatively compatible with A6/3b. However, 14 out of 63 of the *nit* mutant progeny tested were vegetatively compatible with A6/3b using the *sul* mutant (A6/sul1) as the VCG tester (Table 5). Thus, between 1/8 (*i.e.* $1/2^3$) and 1/4 (*i.e.* $1/2^2$) of the progeny were vegetatively compatible with A6/3b which indicated that two to three *vic* loci were segregating in the cross. Ten of the F_1 *nit* mutant progeny which were vegetatively incompatible with A6/3b were then backcrossed with A6/3b. Five of the 10 isolates were sexually compatible with isolate A6/3b. Progeny from one of the backcrosses (A6/3b X P2F7) were recovered and further examined (Table 5).

In the first backcross (A6/3b X P2F7) a total of 110 BC_1F_1 progeny was recovered from five perithecia, all of which were *nit* mutants. In the vegetative compatibility tests using the recurrent parent *sul* mutant (A6/sul1) as the VCG tester, 41 of the progeny were vegetatively compatible with isolate A6/3b. Thus, between 1/4 (*i.e.* $1/2^2$) and 1/ 2 (*i.e.* $1/2^1$) of the progeny were vegetatively compatible with A6/3b, which indicated that one to two *vic* loci were segregating in the cross. Ten of the F_1 *nit* mutant progeny which were vegetatively incompatible with A6/3b were then crossed with A6/3b in a second backcross. Five of the ten isolates were sexually compatible with isolate A6/3b.

In the second backcross (A6/3b X BC30A19), a total of 149 BC_2F_1 progeny was recovered from a total of 16 perithecia from four replications of the same cross. All of the progeny were *nit* mutants. In vegetative compatibility tests using the recurrent parent *sul* mutant (A6/sul1) as the VCG tester, 67 of the 149 progeny were vegetatively compatible with isolate A6/3b, indicating that one *vic* locus was segregating in the cross (Table 5). Ascospores recovered from perithecia from all four replications did not significantly differ from a 1:1 segregation for vegetative compatibility with the recurrent parent in a chi-square test ($p<0.05$). A parallel study was performed using a *nit* mutant of NC329 (NC329/2c) as the recurrent parent in the backcrosses and similar results were obtained (Table 5).

To examine the segregation of mating type in the progeny after the second backcross, 10 of the F_1 *nit* mutant progeny which were vegetatively incompatible with A6/3b and 10 which were vegetatively compatible with A6/3b were mated with A6/3b. Of the 10 vegetatively incompatible progeny, five mated with A6/3b, two with BC30A19, while three did not mate with either parent (Table 6). Of the 10 vegetatively compatible progeny, seven mated with A6/3b and three mated with BC30A19. In the crosses in which the progeny did not mate with either of the parents, perithecium-like structures which did not contain asci or ascospores were observed to have formed with one parent.

Ten of the F_1 *nit* mutant progeny which were vegetatively incompatible with NC329/2c and 10 which were vegetatively compatible with NC329/2c were then crossed with NC329/2c to determine their mating types. Of the ten vegetatively incompatible progeny, five mated with NC329/2c and one with BC30A19, whereas four did not mate with either parent (Table 6). Of the 10 vegetatively compatible progeny, two mated with NC329/2c and seven with BC30A19, while one did not mate with either parent. In the crosses in which the progeny did not mate with either of the parents, perithecia-like structures which did not contain asci or ascospores were observed to have formed with one parent. As the backcrosses progressed, a general decrease in fertility was observed. Crosses from the first (A6/3b X P2F7) and second level (A6/3b X BC30A19) backcrosses produced fewer perithecia and had fewer ascospores per perithecium than did the original cross (A6/3b X NC329).

The genetics of vegetative incompatibility were examined in *C. gloeosporioides* by examining the proportion of F_1 progeny which were vegetatively compatible with parental isolates. In initial apple-by-apple

Table 6. The segregation of mating (*mat*) alleles in two isolates after a second backcross with the recurrent parent.

		Vegetative compatibility with recurrent parent [a]		*mat* [b]	
Recurrent parent	Number tested [c]	+	-	*mat1*	*mat2*
A6/3b (*mat1*)	10	10		3	7
	10		10	2	5
NC329/2c (*mat2*)	10	10		2	7
	10		10	5	1

[a] The "+" column represents isolates vegetatively compatible with the recurrent parent, while the "-" column represents isolates vegetatively incompatible with the recurrent parent which were tested for mating type.

[b] The allele at the *mat* locus of the isolates in each VCG.

[c] The number of F_1 progeny from the second backcrosses (A6/3b X BC30A19 and NC329/2c X BC35C3) tested for sexual compatibility with the recurrent parent and the other parent.

isolate crosses, three to eight *vic* loci were estimated to be segregating, depending upon the isolates used in a cross (37). A cross between two isolates (A6/3b X NC329) in which three loci appeared to be segregating was chosen for the backcross study. In a parallel study with each of two different recurrent parents, approximately one-fourth of the progeny was vegetatively compatible with the recurrent parent after the first backcross, indicating that two *vic* loci were segregating. After the second backcross, approximately one-half of the progeny was vegetatively compatible with each of the two different recurrent parents indicating that one *vic* locus was segregating in the cross. The data indicate that vegetative compatibility segregated 1:1 among isogenic strains of *C. gloeosporioides* which differed at a single *vic* locus confirming the heterogenic nature of vegetative incompatibility in this fungus.

The data from this study indicated that the genetic control of vegetative compatibility in *C. gloeosporioides* is consistent with that observed in other plant pathogenic ascomycete fungi. Furthermore, the number of 3-7 *vic* loci estimated in field isolates of *C. gloeosporioides* is consistent with estimates from other fungi (50).

Although sexual crosses between isolates of *C. gloeosporioides* from apple indicate that there is the potential for eight (i.*e.,* 2^3) to 128 (i.e., 2^7) VCGs in field populations, the population structure of field isolates examined indicate that one or few VCGs predominate in a given orchard (17; Correll, unpublished). These data are consistent with the population structure of several other *Colletotrichum* species examined including *C. orbiculare* from cucurbits (86), *C. dematium* from spinach (20), and *C. gloeosporioides* from almond (49). Thus, vegetative incompatibility in *C. gloeosporioides* from apple appears to be a significant isolating barrier which can effectively subdivide populations, whereby isolates within a VCG typically represent clonally related individuals (34). However, with *C. gloeosporioides*, strong circumstantial evidence suggests that cryptic asexual genetic exchange can occur between certain biotypes which are vegetatively incompatible (53,62). How individual *vic* loci can affect horizontal gene transfer needs further examination in *Colletotrichum*.

Although there may be some genetic linkage between certain *vic* loci and the *mat* locus in *C. gloeosporioides*, the *mat* locus examined among the isolates in this study apparently cannot function as a *vic* locus as it can in *Neurospora crassa*. However, additional crosses are needed before this can be confirmed.

The significance of sexual recombination and the effect on population structure in species of *Colletotrichum* remains largely unresolved. The presence of a relatively high number of VCGs in populations of *C. gloeosporioides* from coffee (4) and *C. acutatum* from apple (17,37) may indicate that sexual reproduction is operative in these populations. However, data from this study indicate that sexual fertility was substantially reduced after successive backcrosses. With each successive backcross, the number of perithecia along with the number of ascospores per perithecium generally decreased. Furthermore, nine progeny out of 40 in the second level backcrosses did not mate with one parent and only produced perithecial-like structures with the other. This reduction in sexual fertility and the formation of empty perithecia could be due to inbreeding depression among closely related isolates within a population.

Studies are also underway to examine the genetics of sexual and vegetative compatibility in *C. acutatum* (35,36). Isolates of *C. acutatum* from apple are self-sterile but can produce a teleomorph similar to *G. cingulata* when mated (Fig. 6). A considerable amount of mtDNA and nuDNA RFLP diversity also has been identified within selected isolates of this species (Figs. 7 and 8). Much remains to be resolved regarding the genetics of sexual and vegetative compatibility in *C. acutatum* and the effects of these mechanisms on population structure. However, initial studies indicate that genetics of sexual and vegetative compatibility in *C. gloeosporioides* and *C. acutatum* are quite similar.

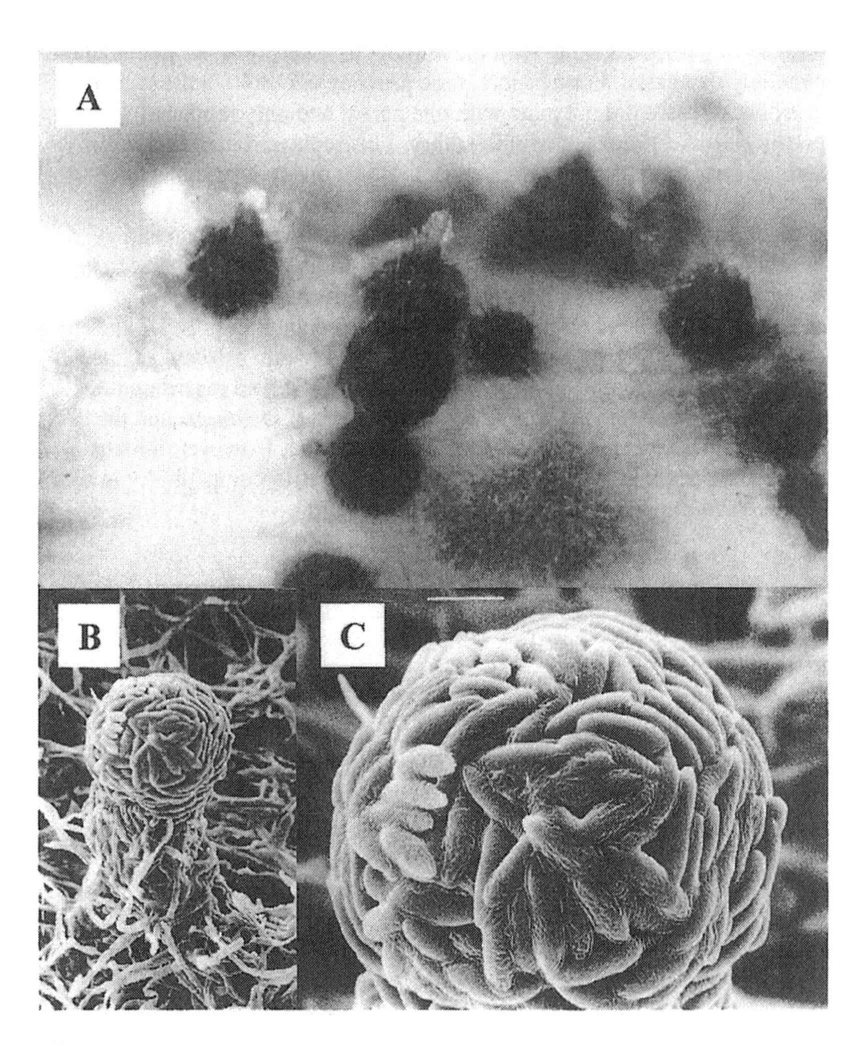

Fig. 6. Sexual reproduction in *C acutatum*. A. Mature perithecia from a mating of two isolates of *C. acutatum* (A5 x A7) on toothpick surface (60X). B-C. SEM of mature perithecial neck exuding ascospores (700 and 1500X).

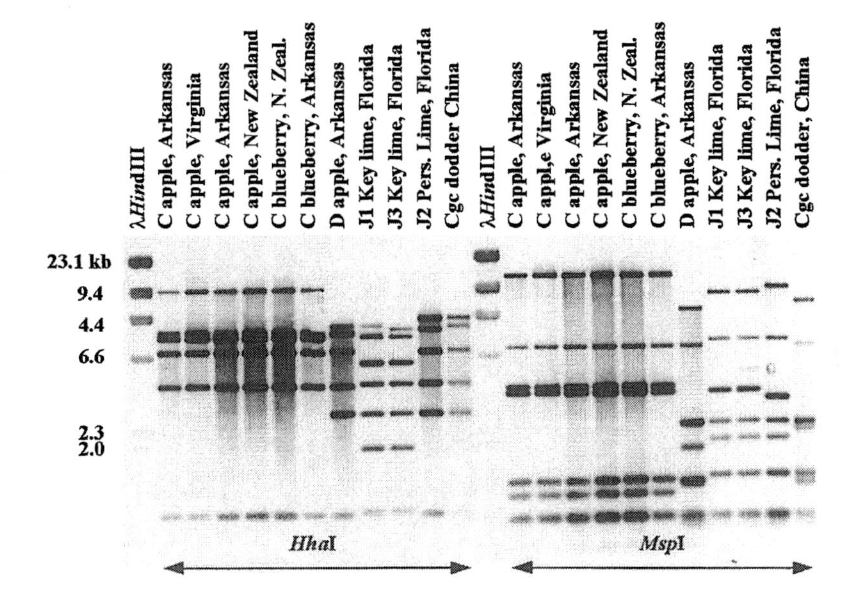

Fig. 7. Apple and non-apple isolates of *C. acutatum*. Total DNA was restricted with *Hha*I and *Msp*I and probed with two mtDNA clones (4u40 and 2u18) of *C. orbiculare* (22).

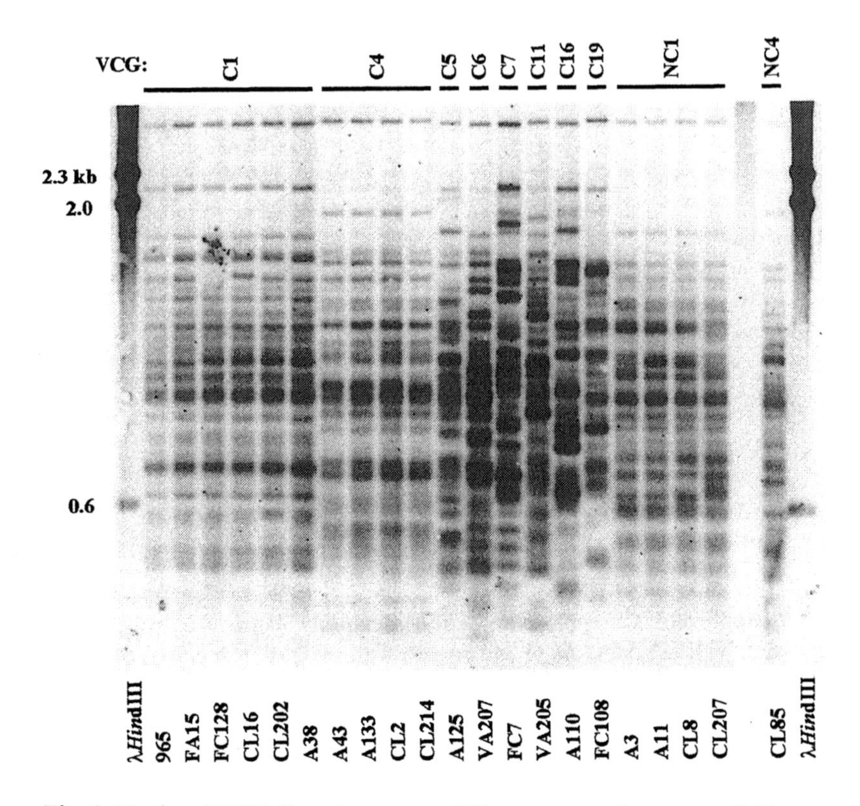

Fig. 8. Nuclear RFLP diversity among different vegetative compatibility groups of *C. acutatum.* Total DNA was restricted with *Hinf*I and probed with a repetitive DNA element (Ctrep1) recovered from *C. trifolii.* (10).

Literature Cited

1. Agostini, J.P. and Timmer, L.W. 1994. Population dynamics and survival of strains of *Colletotrichum gloeosporioides* on citrus in Florida. Phytopathology 84:420-425.
2. Bailey, J.A., Nash, C., Morgan, L.W., O'Connell, R.J., and TeBeest, D.O. 1996. Molecular taxonomy of *Colletotrichum* species causing anthracnose of Malvaceae. Phytopathology 86:1076-1083.
3. Bailey, J.A., Sherriff, C., and O'Connell, R.J. 1995. Identification of specific and intraspecific diversity in *Colletotrichum*. Pages 197-211 in: Disease analysis through genetics and biotechnology—interdisciplinary bridges to improved sorghum and millet crops. J.F. Leslie and R.A. Fredericksen, eds. Iowa State University Press, Ames, Iowa.
4. Beynon, S.M., Coddington, A., Lewis, B.G., and Varzea, V. 1995. Genetic variation in the coffee berry disease pathogen, *Colletotrichum kahawae*. Physiol. Mol. Plant Pathol. 46:457-470.
5. Bonde, M.R., Peterson, G.L., and Maas, J.L. 1991. Isozyme comparisons for identification of *Colletotrichum* species pathogenic to strawberry. Phytopathology 81:1523-1528.
6. Brasier, C.M. 1987. The dynamics of fungal speciation. Pages 231-260 in: Evolutionary Biology of the Fungi. A.D.M. Rayner, C.M. Brasier, and D. Moore, eds. Cambridge University Press, Cambridge, UK.
7. Brooker, N.L., Leslie, J.F., and Dickman, M.B. 1991. Nitrate non-utilizing mutants of *Colletotrichum* and their uses in studies of vegetative compatibility and genetic relatedness. Phytopathology 81:672-677.
8. Bryson, R.J., Caten, C.E., Hollomon, D.W., and Bailey, J.A. 1992. Sexuality and genetics of *Colletotrichum*. Pages 308-325 in: *Colletotrichum*: Biology, Pathology and Control. J.A. Bailey and M.J. Jeger, eds. CAB International, Wallingford, UK.
9. Carlile, M.J. and Watkinson, S.C. 1994. The Fungi. London, Academic Press. pp. 203-221.
10. Cawthon, D., Dyson, W., Correll, J.C., and Rhoads, D.D. 1994. RAPD isolation of fingerprint probes for analysis of *Colletotrichum* species. Phytopathology 84:1124.
11. Chacko, R.J., Weidemann, G.J., TeBeest, D.O., and Correll, J.C. 1994. The use of vegetative compatibility and heterokaryosis to determine potential asexual gene exchange in *Colletotrichum gloeosporioides*. Biol. Control 4:382-389.
12. Cisar, C.R. and TeBeest, D.O. 1996. Mating system of *Glomerella cingulata*. Phytopathology 86:S10. (Abstr.)
13. Cisar, C.R., Thornton, A.B., and TeBeest D.O. 1996. Isolates of *Colletotrichum gloeosporioides* (teleomorph: *Glomerella cingulata*) with different host specificities mate on northern jointvetch. Biol. Control 7:75-83.

14. Cisar, C.R., Spiegel, F.W., TeBeest, D.O., and Trout, C. 1994. Evidence for mating between isolates of *Colletotrichum gloeosporioides* with different host specificities. Curr. Genet. 25:330-335.
15. Coppin, C., Debuchy, R., Arnaise, S., and Picard, M. 1997. Mating types and sexual development in filamentous ascomycetes. Micro. and Mol. Biol. Rev. 61:411-428.
16. Correll, J.C. and Gordon, T.R. 1998. Population structure of ascomycetes and deuteromycetes: What we have learned from case studies. Pages 225-250 in: Structure and Dynamics of Fungal Populations. Ed. J. J. Worrall. Kluwer Academic Press, Dordrecht, The Netherlands.
17. Correll, J.C., Guerber, J.C., and Rhodes, D.D. 1994. Genetic and molecular diversity of populations of *Glomerella cingulata, Colletotrichum gloeosporioides,* and *Colletotrichum acutatum* on apple fruit. Proc. Fifth Inter. Mycol. Conf. Vancouver B. C., Canada. (Abstr.)
18. Correll, J.C., Klittich, C.J.R., and Leslie, J.F. 1987. Nitrate nonutilizing mutants of *Fusarium oxysporum* and their use in vegetative compatibility. Phytopathology 77:1640-1645.
19. Correll, J.C. and Leslie, J.F. 1987. Recovery of spontaneous selenate resistant mutants from *Fusarium oxysporum* and *Fusarium moniliforme*. Phytopathology 77:S1710. (Abstr.)
20. Correll, J.C., Morelock, T.E., and Guerber, J.C. 1993. Vegetative compatibility and virulence of spinach anthracnose pathogen, *Colletotrichum dematium*. Plant Dis. 77:688-691.
21. Correll, J.C., Puhalla, J.E., and Schneider, R.W. 1986. Identification of *Fusarium oxysporum* f.sp. *apii* on the basis of colony size, virulence, and vegetative compatibility. Phytopathology 76:396-400.
22. Correll, J.C., Rhoads, D.D., and Guerber, J.C. 1993. Examination of mitochondrial DNA restriction fragment length polymorphisms, DNA fingerprints, and randomly amplified polymorphic DNA of *Colletotrichum orbiculare*. Phytopathology 83:1199-1204.
23. Cove, D.J. 1976. Chlorate toxicity in *Aspergillus nidulans*. The selection and characterization of chlorate resistant mutants. Heredity 36:191-203.
24. Edgerton, C.W. 1914. Plus and minus strains in the genus *Glomerella*. Amer. J. Bot. 1:224-254.
25. Esser, K. and Blaich, R. 1994. Heterogenic incompatibility in fungi. Pages 211-229 in: The Mycota. J.G.H. Wessels and F. Meinhardt, eds. Springer-Verlag, Berlin.
26. Freeman, S. and Katan, T. 1997. Identification of *Colletotrichum* species responsible for anthracnose and root necrosis of strawberry in Israel. Phytopathology 87:516-521.
27. Freeman, S., Katan, T., and Shabi, E. 1996. Characterization of *Colletotrichum gloeosporioides* isolates from avocado and almond

fruits with molecular and pathogenicity tests. App. Environ. Microbiol. 62:1014-1020.

28. Freeman, S., Pham, M., and Rodriguez, R.J. 1993. Molecular genotyping of *Colletotrichum* species based on arbitrarily primed PCR, A + T-Rich DNA, and nuclear DNA analysis. Exp. Mycol. 17:309-322.
29. Freeman, S. and Rodriguez, R.J. 1992. A rapid, reliable bioassay for the pathogenicity of *Colletotrichum magna* on cucurbits and its use in screening for nonpathogenic mutants. Plant Dis. 76:901-905.
30. Freeman, S. and Rodriguez, R.J. 1995. Differentiation of *Colletotrichum* species responsible for anthracnose of strawberry by arbitrarily primed PCR. Mycol. Res. 99:501-504.
31. Gardner, M.W. 1918. Anthracnose of cucurbits. U.S. Dept. Agric. Bull. 727:1-68.
32. Glass, N.L. and Kuldau, G.A. 1992. Mating type and vegetative incompatibility in filamentous ascomycetes. Annu. Rev. Phytopathol. 30:201-224.
33. Glass, N.L. and Nelson, M.A. 1994. Growth, differentiation, and sexuality. Pages -306 in: The Mycota. J.G.H. Wessels and F. Meinhardt, eds. Springer-Verlag, Berlin.
34. Gordon, T.R. and Martyn, R.D. 1997. The evolutionary biology of *Fusarium oxysporum*. Annu. Rev. Phytopathol. 35:111-128.
35. Guerber, J.C. and Correll, J.C. 1997. The first report of the teleomorph of *Colletotrichum acutatum* in the United States. Plant Dis. 81:1334.
36. Guerber, J.C. and Correll, J.C. 1998. Segregation of genetic markers in *Colletotrichum acutatum*. Phytopathology 88:S34. (Abstr.)
37. Guerber, J., Sherrill, J.F., and Correll, J.C. 1997. Genetic analysis of sexual and vegetative compatibility in *Colletotrichum gloeosporioides*. (Abstr.) Phytopathology 87:S36.
38. Gunnell, P.S. and Gubler, W.D. 1992. Taxonomy and morphology of *Colletotrichum* species pathogenic to strawberry. Mycologia 84:157-165.
39. Guthrie, P.A.I., Magill, C.W., Frederiksen, R.A., and Odvody, G.N. 1992. Random amplified polymorphic DNA markers: A system for identifying and differentiating isolates of *Colletotrichum graminicola*. Phytopathology 82:832-835.
40. Harp, T.L. and Correll, J.C. 1998. Recovery and characterization of spontaneous, selenate-resistant mutants of *Magnaporthe grisea*, the rice blast pathogen. Mycologia 90:954-963.
41. Hodson, A., Mills, P.R., and Brown, A.E. 1992. Ribosomal and mitochondrial DNA polymorphisms in *Colletotrichum gloeosporioides* isolated from tropical fruits. Mycol. Res. 97:329-335.
42. Jacobson, D.J. and Gordon, T.R. 1988. Vegetative compatibility and self-incompatibility within *Fusarium oxysporum* f.sp. *melonis*. Phytopathology 78:669-672.

43. Jenkins, S.F. 1960. Studies on the morphology, taxonomy, physiology, and control of a new species of *Colletotrichum* affecting cucurbits. M.S. thesis. North Carolina State College, North Carolina. 68 pp.
44. Jenkins, J.F. 1962. Genetic, taxonomic, and physiological studies of *Glomerella* species on cucurbits. Ph.D. thesis. North Carolina State College. Raleigh, North Carolina. 45 pp.
45. Jenkins, S.F. and Winstead, N.N. 1961. Observations on the sexual stage of *Colletotrichum orbiculare*. Science 133:581-582.
46. Jenkins, S.F. Jr. and Winstead, N.N. 1964. *Glomerella magna*, cause of a new anthracnose of cucurbits. Phytopathology 54:452-454.
47. Jenkins, S.F. Jr., Winstead, N.N., and McCombs, C.L. 1964. Pathogenic comparisons of three new and four previously described races of *Glomerella cingulata* var. *orbiculare*. Pl. Dis. Reptr. 48:619-622.
48. Johnston, P.R. and Jones, D. 1997. Relationship among *Colletotrichum* isolates from fruit-rots assessed using rDNA sequences. Mycologia 89:420-430.
49. Katan, T. and Shabi, E. 1996. Vegetative compatibility among isolates of *Colletotrichum gloeosporioides* from almond in Israel. Eur. J. Plant Pathol. 102:597-600.
50. Leslie, J.F. 1993. Fungal vegetative compatibility. Annu. Rev. Phytopathol. 31:127-150.
51. Liyanage H.D., McMillan, R.T. Jr., and Kistler, H.C. (1992) Two genetically distinct populations of *Colletotrichum gloeosporioides* from citrus. Phytopathology 82:1371-1376.
52. Manners, J.M., Masel, A., Braithwaite, K.S., and Irwin, J.A.G. 1992. Molecular analysis of *Colletotrichum gloeosporioides* pathogenic on the tropical pasture legume *Stylosanthes*. Pages 250-268 in: *Colletotrichum*: Biology, Pathology, and Control. J.A. Bailey and M.J. Jager, eds. CAB International, Wallingford, UK.
53. Masel, A.M., He, C., Poplawski, A.M., Irwin, J.A.G., and Manners, J.M. 1996. Molecular evidence for chromosome transfer between biotypes of *Colletotrichum gloeosporioides*. Mol. Plant-Microbe Interact. 9:339-348.
54. Mills, P.R , Hodson, A. and Brown, A.E. 1992. Molecular differentiation of *Colletotrichum gloeosporioides* isolates infecting tropical fruits. Pages 269-288 in: *Colletotrichum*: Biology, Pathology, and Control. J.A. Bailey and M.J. Jager, eds. CAB International, Wallingford, UK.
55. Nelson, M.A. 1996. Mating systems in ascomycetes: a romp in the sac. Trends Genet. 12:69-74.
56. Nikandrow, A., Weidemann, G.J., and Auld, B.A. 1990. Incidence and pathogenicity of *Colletotrichum orbiculare* and *Phomopsis* sp. on *Xanthium* spp.). Plant Dis. 74:796-799.

57. O'Connell, R.J., Bailey, J.A., and Richmond, D.V. 1985. Cytology and physiology of infection of *Phaseolus vulgaris* by *Colletotrichum lindemuthianum*. Physiol. Pl. Pathol. 27:75-98.
58. O'Connell, R.J., Nash, C., and Bailey, J.A. 1992. Lectin cytochemistry: A new approach to understanding cell differentiation, pathogenesis and taxonomy in *Colletotrichum*. Pages 67-87 in: *Colletotrichum*: Biology, Pathology and Control. J.A. Bailey and M.J. Jeger, eds. CAB International, Wallingford, UK.
59. O'Neill, N.R., van Berkum, P., Lin, J., Kuo, J., Ude, G.N, Kenworthy, W., and Saunders, J.A. 1997. Application of amplified restriction fragment polymorphism for genetic characterization of *Colletotrichum* pathogens of alfalfa. Phytopathology 87:745-750.
60. Pain, N.A., O'Connell, R.J., Bailey, J.A., and Green, J.R. 1992. Monoclonal antibodies which show restricted binding of four *Colletotrichum* species: *C. lindemuthianum*, *C. malvarum*, *C. orbiculare*, and *C. trifolii*. Physiol. Mol. Pl. Pathol. 40:111-126.
61. Perkins, D.D. (1987) Mating type switching in filamentous ascomyectes. Genetics 115: 215-216.
62. Poplawski, A.M., He, C., Irwin, J.A.G., and Manners, J.M. 1997. Transfer of an autonomously replicating vector between vegetatively incompatible biotypes of *Colletotrichum gloeosporioides*. Curr. Genet. 32:66-72.
63. Puhalla, J.E. 1985. Classification of strains of *Fusarium oxysporum* on the basis of vegetative compatibility. Can. J. Bot. 63:179-183.
64. Puhalla, J.E. and Hummel, M. 1983. Vegetative compatibility groups within *Verticillium dahliae*. Phytopathology 73:1305-1308.
65. Puhalla, J.E. and Mayfield, J.E. 1974. The mechanism of heterokaryotic growth in *Verticillium dahliae*. Genetics 76:411-422.
66. Rodriguez, R.J. and Owen, J.L. 1997. Isolation of *Glomerella musae* [teleomorph of *Colletotrichum musae* (Berk. & Curt.) Arx.] and segregation analysis of ascospore progeny. Exp. Mycol. 16:291-301.
67. Sherriff, C., Whelan, M.J., Arnold, G.M., Lafay, J., Brygoo, Y., and Bailey, J.A. 1994. Ribosomal DNA sequence analysis reveals new species groupings in the genus *Colletotrichum*. Exp. Mycol. 18:121-138.
68. Shi, Y., Correll, J.C., and Guerber, J.C. 1996. Frequency of *Colletotrichum* species causing bitter rot of apple in the southeastern United States. Plant. Dis. 80:692-696.
69. Sitterly, W.R. and Keinath, A.P. 1996. Anthracnose. Pages 24-25 in: Compendium of Cucurbits Diseases. APS Press, St. Paul, Minnesota.
70. Sreenivasaprasad, S., Brown, A.E., Mills, P.R. 1992. DNA sequence variation and interrelationships among *Colletotrichum* species causing strawberry anthracnose. Physiol. Molec. Plant Pathol. 41:265-281.
71. Sreenivasaprasad, S., Brown, A.E., Mills, P.R. 1993. Coffee berry disease pathogen in Africa: Genetic structure and relationship to the

group species *Colletotrichum gloeosporioides*. Mycol. Res. 97:995-1000.

72. Sreenivasaprasad, S., Mills, P.R., and Brown, A.E., 1994. Nucleotide sequence of the rDNA spacer 1 enables identification of isolates of *Colletotrichum* as *C. acutatum*. Mycol. Res. 98:186-188.
73. Stevens, F.L. 1931. The ascigerous stage of *Colletotrichum lagenarium* induced by ultra-violet irradiation. Mycologia 23:134-139.
74. Sutton, B.C. 1980. The coelomycetes—Fungi imperfecti with pycnidia acervuli and stromata. Commonwealth Mycological Institute, Kew, Surrey, UK.
75. Sutton, B.C. 1992. The genus *Glomerella* and its anamorph *Colletotrichum*. Pages 523-537 in: *Colletotrichum*: Biology, pathology and control. J.A. Bailey and M.J. Jeger, eds. CAB International, Wallingford, UK.
76. Sutton, T.B. 1990. Bitter Rot. Pages 34-35 in: The Compendium of Apple and Pear Diseases. A.L. Jones and H.S. Aldwinckle, eds. APS Press, St Paul, Minnesota.
77. TeBeest, D.O., Correll, J.C., and Weidemann, G.J. 1997. Speciation and population biology in *Colletotrichum*. Pages 157-168 in: The Mycota, Vol. 5. K. Esser and P.A. Lemke, eds. Springer-Verlag, Berlin.
78. Thompson, D.C. and Jenkins, S.F. 1985. Influence of cultivar resistance, initial disease, environment, and fungicide concentration and timing on anthracnose development and yield loss in pickling cucumbers. Phytopathology 75:1422-1427.
79. Tuite, J. 1969. Plant Pathological Methods. Burgess Publ. Co., Minneapolis, Minnesota. 30 pp.
80. Ullasa, B.A. and Amin, K.S. 1986. Epidemiology of bottlegourd anthracnose, estimation of yield loss and fungicidal control. Trop. Pest Manag. 32:277-282.
81. Vaillancourt, L.J. and Hanau, R.M. 1994. Nitrate-nonutilizing mutants used to study heterokaryosis and vegetative compatibility in *Glomerella graminicola* (*Colletotrichum graminicola*). Exp. Mycol. 18:311-319.
82. von Arx, J.A. 1957. Die Arten der Gattung *Colletotrichum* Cda. Phytopathol. Z. 29:413-468.
83. von Arx, J. A. 1970. A revision of the fungi classified as *Gloeosporium*. Biblio. Mycol. 24:1-203.
84. Walker, J. Nikandrov, A., and Millar, G.D. 1991. Species of *Colletotrichum* on *Xanthium* (Asteraceae) with comments on some taxonomic and nomenclatural problems in *Colletotrichum*. Mycol. Res. 95:1175-1193.
85. Wasilwa, L.A. 1997. Characterization of *Colletotrichum orbiculare* and allied species using mtDNA RFLPs, vegetative compatibility, and virulence on cucurbits, and screening cucumber and watermelon for

anthracnose resistance. Ph.D. thesis. University of Arkansas, Fayetteville, Arkansas.

86. Wasilwa, L.A., Correll, J.C., Morelock, T.E., and McNew, R. 1993. Re-examination of races of the cucurbit anthracnose pathogen, *Colletotrichum orbiculare*. Phytopathology 83:1190-1198.
87. Wasilwa, L.A., Correll, J.C., and Morelock, T.E. 1996. Further characterization of *Colletotrichum orbiculare* for vegetative compatibility and virulence. Phytopathology 86:S63. (Abstr.)
88. Wheeler, H.E. 1950. Genetics of *Glomerella*. VIII. A genetic basis for the occurrence of minus mutants. Am. J. Bot. 37:304-312.
89. Wheeler, H.E. 1954. Genetics and evolution of heterothallism in *Glomerella*. Phytopathology 41:342-345.
90. Wheeler, H.E. and McGahen, J.W. 1952. Genetics of *Glomerella*. X. Genes affecting sexual reproduction. Am. J. Bot. 39:110-119.

Chapter 11

Gene Transfer and Expression in *Colletotrichum gloeosporioides* Causing Anthracnose on *Stylosanthes*

John M. Manners, Sally-Anne Stephenson, Chaozu He, and Don J. Maclean

Isolates of *Colletotrichum gloeosporioides* (Penz.) Penz. and Sacc. cause anthracnose disease on a wide range of hosts, including legumes and fruits, and are of particular economic importance in the tropics. Anthracnose is the most important disease of the tropical pasture legumes in the genus *Stylosanthes.* These legumes originate from South America and are grown extensively for cattle production in northern Australia and southeast Asia. Anthracnose diseases of *Stylosanthes* spp. are currently controlled by the deployment of new resistant varieties (5). However, the pathogen is genetically very variable and shows considerable pathogenic specialization (7,8). The interaction between *C. gloeosporioides* and *Stylosanthes* spp. in Australia has been studied in detail at the molecular level (22). The introduction of new biotypes and the generation of new races within biotypes by mutation may have both contributed to the development of new virulent isolates in Australia (22). Recently, studies of the interaction between *C. gloeosporioides* and *Stylosanthes* spp. have focused on molecular events associated with infection (39) and novel mechanisms that generate chromosome variation and transfer genes between normally incompatible fungal genotypes (15,26,35). Prior to describing new research findings from these recent studies and discussing their implications for fungal pathogenesis, we will provide a brief overview of previous research on the molecular characterization of *C. gloeosporioides* infecting *Stylosanthes.*

Anthracnose Disease on *Stylosanthes* spp. in Australia

In Australia, two *C. gloeosporioides*-incited anthracnose diseases, designated types A and B, were recognized by the different symptoms produced on the same host (*S. guianensis*) (17). These diseases are caused by two fungal biotypes, A and B. A biotype can be defined as a subgroup within a species having a like genetic make-up (21). Biotype A isolates have a wide host range, infecting species of *Aeschynomene, Desmodium, Indigofera,* and *Psoralia* in addition to *Stylosanthes* (47). The biotype A fungus produces limited lesions with a light-colored center and a dark margin. In contrast, biotype B is specific to *S. guianensis*, producing blight symptoms on leaves and stems. Pathogenic specialization has been reported within both biotype A and biotype B (7,8,18).

A Comparison of the Genomes of Biotypes A and B

So far, all highly virulent isolates of both biotypes have been anamorphic (asexual) precluding any studies of sexual compatibility (28). The possibility that heterokaryosis and parasexual recombination may occur between biotypes was first investigated using complementation studies with nitrate non-utilizing mutants (26,35). Although isolates within each biotype were vegetatively compatible, detecting the formation of heterokaryons between the two biotypes was not possible using standard techniques. On the basis of these results, Masel et al. (26) proposed that biotypes A and B form distinct vegetative compatiblity groups (VCGs). These conclusions were consistent with the results of several analyses that have compared the genomes of biotypes A and B.

Considerable differences in the molecular karyotypes of biotypes A and B were observed when the chromosomes of representative isolates were compared by pulsed-field electrophoresis (24). Sherrif et al. (38) demonstrated that the ribosomal DNA (rDNA) ITS1 sequences of representative isolates of biotypes A and B differed, and these biotypes did not cluster together when compared to several other *C. gloeosporioides* isolates. RFLP analysis using dispersed repeat and low copy sequences isolated from *C. gloeosporioides* and heterologous rDNA and DNA fingerprinting probe has demonstrated that the two biotypes are genetically distinct (2,3,4,14). For example, the genome of biotype B contains a widely dispersed retrotransposon that is completely absent in biotype A (14). In contrast, variation between isolates within the two biotypes has been shown to be either undetectable or very slight using neutral molecular markers. These observations suggest that two distinct clonally derived populations of *C. gloeosporioides* infecting *Stylosanthes* spp. exist in Australia. The distinction between biotypes A and B is sufficiently large to rule out the possibility that the two biotypes are of recent common descent. The two biotypes were probably derived by separate introductions from South America, most likely on imported seed (22,23).

New Methods for Comparing *Colletotrichum* Isolates

Genetic diversity of isolates of *C. gloeosporioides* from *Stylosanthes* has been studied intensively using a range of molecular techniques. However, direct comparisons between these isolates and those of other hosts are difficult because standardized techniques are not commonly used in the analysis of genetic diversity in *Colletotrichum,* except in the studies of rDNA sequences. Most laboratories conducting RAPD and RFLP analysis usually employ different primers, probes, enzymes, and gel running conditions. The value of a standard technique used by all laboratories is evident in the study of the rice blast fungus where the MGR repeat probe is universally used in diversity studies (20). However, DNA fragment analysis has little value for deducing phylogenies, and a technique that lends itself to both fragment analysis and DNA sequence analysis would be preferable.

We propose that *Colletotrichum* researchers test a system of DNA fragment analysis based on the amplification of a selected number of specific gene regions. The genes chosen for this analysis should be single copy, ubiquitous, and contain conserved and variable regions for amplification. Degenerate oligonucleotide primers specific to conserved regions could be used to amplify fragments in all *Colletotrichum* isolates using the PCR. Fragments could then either be examined directly or following digestion with restriction enzymes. The amplified fragments may also be sequenced depending on the amount of information required. The amplified region could span an intron to provide greater variation but could also span coding regions so that broader phylogenies could be assessed if necessary.

An example of a gene that may meet these criteria in *Colletotrichum* is glutamine synthetase (39). The glutamine synthetase gene has been used for evolutionary studies in a range of organisms (42). In biotype B this gene is single copy and contains introns and highly conserved regions for primer design (39). Degenerate primers to conserved sequences capable of amplifying the major intron of this gene have already been successfully used for diversity studies in *Colletotrichum* (P. Weeds, S. Chakraborty and J. Manners, unpublished).

Information on fragment length from several genes could be obtained simultaneously in a multiplex PCR. Such a system, particularly one based on DNA sequence information, would provide a universal database of information on genetic diversity within

Colletotrichum and permit a broad view of evolutionary relationships and population dynamics. It would also greatly assist in the ongoing debate of relationships within complex and diverse group species such as *C. gloeosporioides.*

Evidence for Chromosome Transfer Between Biotypes

The large genetic differences observed between biotypes A and B using RFLP markers and the high level of monomorphy detected amongst isolates within each biotype suggests that genetic recombination between biotypes is either rare or non-existent in the field in Australia (22). More recently, a supernumerary 2-Mb chromosome present in a biotype B isolate, termed Bx, was described (26). Molecular markers show that apart from the 2-Mb supernumerary chromosome, isolate Bx has a characteristic biotype B genetic background. The 2-Mb chromosome was either absent from other biotype B isolates or was represented by a homologous supernumerary 1.2-Mb chromosome (26). The 1.2-Mb and 2-Mb chromosomes appeared to be relatively recent additions to the biotype B genome because they lacked at least two dispersed repeat sequences that were monomorphic in the biotype B population and present on all other resolved chromosomes (25,26). Presumably the 1.2- and 2-Mb chromosomes were added to the genome after these repeat sequences proliferated (25,26). Using chromosome-specific markers, the 2-Mb chromosome in Bx was shown to be similar, if not identical to a 2-Mb chromosome present in biotype A isolates. No biotype A-like markers from other chromosomes were found in the Bx isolate, and the investigators (26) suggested that a form of "horizontal" transfer of the 2-Mb chromosome had recently occurred from biotype A to biotype B (26). The potential for limited vegetative genetic recombination between the biotypes was experimentally demonstrated by Popwalski *et al* (35) who showed that an autonomously replicating vector could transfer from biotype A isolates to biotype B in vegetative axenic culture. Subsequently, genetic recombination has been tested by vegetatively co-culturing transformants of biotype A with the hygromycin-resistance gene integrated into various chromosomes with biotype B carrying a phleomycin-resistance gene marker (15). In these experiments monoconidial cultures resistant to both antibiotics were obtained only when the biotype A transformant used as a parent contained the hygromycin-resistance gene in the 2-Mb chromosome (Fig. 1). In addition, the recombinant progeny appeared to contain only the 2-Mb chromosome in a biotype B genome. The transfer of the 2-Mb chromosome was, therefore, highly specific (15).

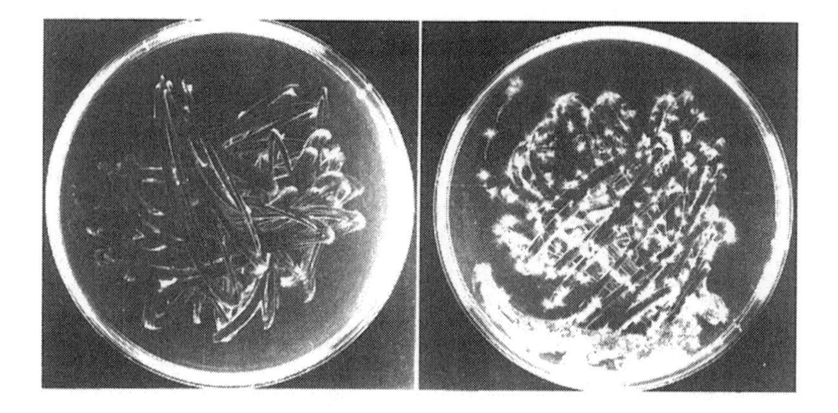

Fig. 1. The isolation of recombinant colonies carrying the 2-Mb chromosome of biotype A in a biotype B genetic background (15). A biotype A transformant carrying the hygromycin-resistance gene (*hph*) in either the 2-Mb chromosome or in other chromosomes was co-cultured with a phleomycin-resistant biotype B transformant. After collection, 5 x 10^8 conidia were streaked across a plate containing both antibiotics. The figure shows many double antibiotic-resistant colonies developing from the conidia where biotype A with the *hph* gene in the 2-Mb chromosome was a parent (right panel), but no colonies developed when the *hph* gene was located on other chromosomes in the biotype A parent (left panel).

The results of He et al. (15) are important for two reasons. First, they suggest a possible mechanism that may explain the origin of supernumerary chromosomes in fungal isolates (11). Second, they suggest that the standard techniques used for investigating vegetative compatibility groups (*nit* mutant complementation) may not fully measure the potential for anastamosis and recombination. He et al. (15) suggested that the 2-Mb chromosome may lack incompatibility loci and, therefore, may escape processes that constrain recombination between the genomes of biotypes A and B.

Two models for transfer of the 2-Mb chromosome have been proposed (15) (Fig. 2). In both models a transient or slow-growing heterokaryon is formed between the biotypes. In Model 1 (Fig. 2) the nuclei fuse, and all biotype A chromosomes are lost with the exception of the 2-Mb chromosome. Alternatively, in Model 2 a mechanism may exist for the specific transfer of the 2-Mb chromosome between nuclei in a transient heterokaryon. Direct experimental proof for the formation of an inter-biotype heterokaryon was obtained by He (13). In this study,

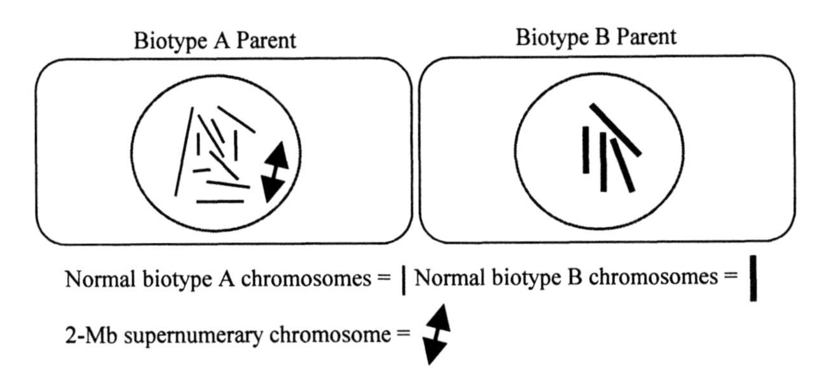

Model 1. Nuclear fusion: An inter-biotype heterokaryon is formed, nuclear fusion occurs, and all biotype A chromosomes except the 2-Mb supernumerary chromosome are unstable.

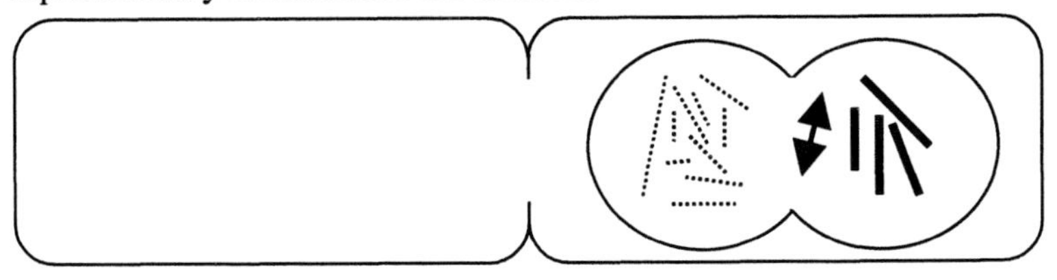

Model 2. Inter-nuclear transfer: A heterokaryon is formed, and the 2-Mb chromosome moves from the biotype A nucleus to the biotype B nucleus.

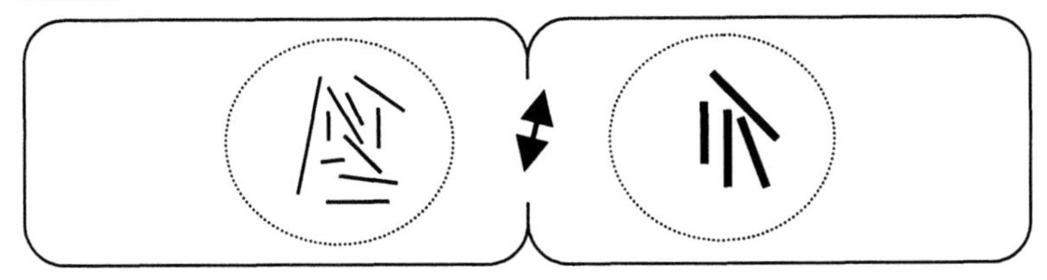

Recombinant progeny: Genotypes carrying the 2-Mb chromosome in a biotype B genetic background have been detected in the field and laboratory (15,26).

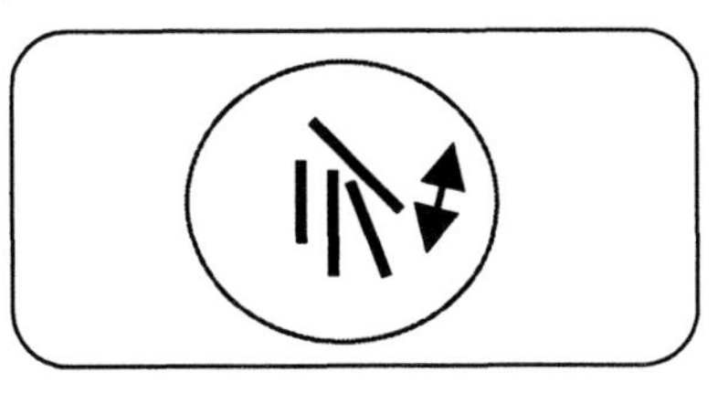

Fig. 2. Two possible models for the transfer of the 2-Mb supernumerary chromosome from biotype A to B (15).

hygromycin-resistant and phleomycin-resistant cultures of biotypes A and B, respectively, were paired and subsequently subcultured on media containing both antibiotics by the transfer of hyphal tips. One colony grew very slowly on both antibiotics and contained the molecular markers for both biotypes A and B. This colony gave rise only to conidia that were either biotype A or B. This slow-growing inter-biotype heterokaryon is a potential intermediate in the chromosome transfer process, but its poor fitness in culture suggests that it would not survive in nature.

The 2-Mb supernumerary chromosome of *C. gloeosporioides* which infects *Stylosanthes* may represent a mobile genetic entity which is compatible with different genetic backgrounds of *C. gloeosporioides*. The function of this chromosome is now under study by analyzing the genes that it encodes. So far, little laboratory evidence suggests that this chromosome has any major role in either pathogenicity or virulence on *Stylosanthes* (15,26). The observation that Bx isolates are maintained in the field (26) suggests that the 2-Mb chromosome may provide some selective advantage in the natural environment of the fungus.

Horizontal Transfer Processes in the Evolution of *C. gloeosporioides*

In *C. gloeosporioides,* we have identified several DNA elements that appear to have been introduced into the lineage since the separation of biotype B and biotype A progenitors (23). The transferred DNA elements include the retrotransposon *Cg*T1 (14) and the 1.2- and 2-Mb supernumerary chromosomes (15,25,26). The isolate Bx that was detected in 1992 carries all of these elements and appears to be the most recent genotype in this lineage in Australia. Although the time frame for the events leading to the formation of Bx is not known precisely, the 2-Mb chromosome from A to B was most probably transferred in the past 30 years when the respective hosts for these two biotypes were grown extensively in northern Australia.

Pathogenicity Processes on *Stylosanthes*

The ability of fungal phytopathogens to infect their hosts depends on their capacity to obtain adequate nutrients while avoiding or negating the host defenses. Genes that are essential for fungal pathogens to infect their hosts have been termed pathogenicity genes (30,36). There is considerable genetic diversity within the genus *Colletotrichum* and considerable variation in the modes of pathogenesis of the species and biotypes observed on specific hosts (1). Perhaps, because of this diversity, no typical experimental model host-pathogen interaction has emerged for this genus. The histopathology of infection of *C. gloeosporioides* on *Stylosanthes* has been described (29,37,44). Conidia germinate to form melanized appressoria which penetrate an epidermal cell within 24 h after inoculation. An intracellular vesicle or microcolony then forms within the epidermal cell, and after 48 h intracellular and intercellular hyphae colonize the mesophyll tissues. In biotype B, subcuticular and superficial hyphae form and spread rapidly to initiate new infection sites elsewhere on the leaf, which leads to the blight symptom characteristic of Type B anthracnose. A fungus-specific DNA probe from the 25S rRNA of *C. gloeosporioides* has been identified (39) and can be used to quantify the development of the fungus following infection of *S. guianensis* without interference from host rRNA (Fig. 3). Most of the fungal development occurs between 2 and 4 days after inoculation. This DNA probe should have wide application in quantifying *Colletotrichum* growth during infection.

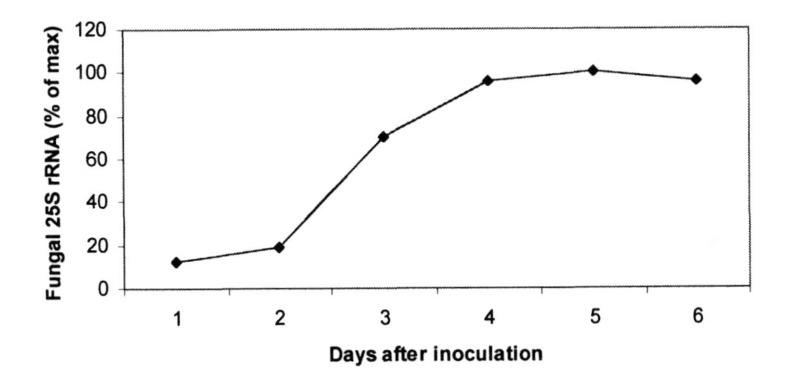

Fig. 3. Measurement of fungal growth in leaves of *Stylosanthes guianensis* following inoculation with *Colletotrichum gloeosporioides* by measuring the abundance of fungal 25S rRNA. Total leaf RNA was extracted at various days after inoculation and 20µg blotted onto nylon membranes and hybridized to ^{32}P-labeled clone pCgRL1 which is specific for fungal 25S rRNA (39). The amount of hybridizing probe was quantified using a phosphorimager.

Pathogenicity Genes in *Colletotrichum*

Recently, some genes that may have important roles in pathogenicity have been cloned from a few pathogenic species within the genus *Colletotrichum*. In a few instances, definitive roles for these genes in pathogenicity have been shown using gene disruption or mutant complementation. In general, two approaches have been used to identify pathogenicity genes in phytopathogenic fungi. Examples of these approaches are detailed in other chapters in this volume.

1. The first approach is to first identify mutants impaired in their ability to infect normally susceptible host plants. If the nonpathogenic mutants are derived via either chemical or physical processes then the gene can subsequently be isolated by complementation of the mutant using a genomic library of the wild type in a fungal transformation vector (32,33). Alternatively, if the mutation is derived by insertional mutagenesis with a fungal transformation vector then the wild-type gene can be cloned using the flanking sequences at the insertion site as a DNA probe (12).
2. The second approach is to identify a candidate gene, then test the role of this gene by targeted gene disruption. Candidate genes that have been isolated from *Colletotrichum* include (a) genes expressed specifically in appressoria (16), (b) genes expressed in intracellular hyphae (31), and (c) genes encoding a range of extracellular hydrolytic enzymes, e.g. proteinases (9), pectinases (6,43,48) and (d) genes induced during growth *in planta* (39).

Gene Expression during Infection of *Stylosanthes*

The overall picture that is emerging from these studies is that during the infection process a coordinated profile of expression of specific genes is induced in the pathogen. For example, Stephenson et al. (39) described the application of a differential hybridization technique for isolating candidate pathogenicity genes in biotype B of *C. gloeosporioides*. In this approach, cDNAs were selected on the basis of high expression during leaf infection of *S. guianensis* compared to expression during axenic culture in rich medium. Among the six distinct cDNAs isolated were clones that had homology to eukaryote 1β elongation factors and to glutamine synthetase (GS) (39). Subsequent analysis of the expression of the GS gene indicated that transcripts of this gene were three to six times more abundant relative to fungal 25S rRNA in the fungus when it was growing in the host than when it was growing in a rich medium. The GS gene was also induced to a similar level when the fungus was grown in media lacking nitrogen. In most prokaryote and eukaryote cell types, an abundance of GS is a biochemical marker for conditions of nitrogen deficiency (27,46). Recent studies of a variety of plant pathogens have also suggested that relative nutrient deprivation occurs *in planta* (10,19,34,41,45). Taken together, all these observations suggest that *C. gloeosporioides* may be subjected to nitrogen limitation during the early stages of infection of *S. guianensis.*

This nitrogen deprivation may act as a signal for the expression of genes necessary for pathogenicity. In a recent test of this hypothesis, eight cDNAs were isolated from *C. gloeosporioides* that had been starved for nitrogen in axenic culture (40). One of these cDNAs corresponds to a gene, termed *CgDN3,* that most probably encodes a secreted basic peptide of approximately 56 amino acids with no homology to any known protein. This gene is expressed at early stages of infection with transcripts detectable at 24 hr after inoculation using RT-PCR techniques. The gene is also strongly induced in nitrogen-starved mycelium. Importantly, disruption of the *CgDN3* gene leads to a loss of pathogenicity and the production of a hypersensitive response in the inoculated host. Recent studies of the transformants mutated in the *CgDN3* gene have shown that these strains can colonize the host tissue if placed on wounds (J. Hatfield, personal communication). This suggests that the *CgDN3* gene is either a molecular suppresser of a hypersensitive reaction during appressorial penetration of the host epidermis, or that it is involved in a developmental program at primary infection that is essential to avoid triggering the host response. Current studies are focused on determining the site of expression of the *CgDN3* gene during primary penetration. Because the *CgDN3* gene represents a novel pathogenicity gene for fungal phytopathogens, studies of the mode of action of its product may provide clues as to how hemi-biotrophic fungal pathogens suppress or avoid hypersensitivity during infection.

Conclusions

The interaction between *Stylosanthes* and *C. gloeosporioides* has provided a wealth of information on processes leading to genotypic variation in this fungus. Although this fungus is primarily clonal in Australia, there is now unequivocal evidence that some genetic recombination occurs, albeit infrequently, between genotypes that otherwise appear to be genetically isolated. The extent of this genetic recombination in the field and the impact it has on the success of the pathogen will now require further study. Recent research on the processes of infection has shown that some genes are induced at early stages of the infection process. Some of these genes, e.g., glutamine synthetase, may have roles in fungal nutrition, while others may have roles directly effecting pathogenesis. One pathogenicity gene has been identified in *C. gloeosporioides*, and studies of mutants of this gene suggest that it has a role in either suppressing or avoiding a host hypersensitive response during the primary infection process.

Acknowledgements

We are grateful to Prof. John Irwin and Dr. Don Cameron for their introduction and help with the stylo anthracnose system, to Drs. Andrew Masel and Sukumar Chakraborty for their interest in this work over the years, and to CIAT, The Crawford Fund, The Australian Research Council and the University of Queensland for funding.

Literature Cited

1. Bailey, J.A., O'Connell, R.J., Pring, R.J., and Nash, C. 1992. Infection strategies of *Colletotrichum* species. Pages 88-120 in: *Colletotrichum* Biology, Pathology and Control. J.A. Bailey and M.J. Jeger, eds. CAB International, Wallingford, U.K.
2. Braithwaite, K.S., Irwin, J.A.G., and Manners, J.M. 1990a. Ribosomal DNA as a molecular taxonomic marker for the group species *Colletotrichum gloeosporioides*. Aust. Syst. Bot. 3:733-738.
3. Braithwaite, K.S., Irwin, J.A.G., and Manners, J.M. 1990b. Restriction fragment length polymorphisms in *Colletotrichum gloeosporioides* infecting *Stylosanthes* spp. in Australia. Mycol. Res. 94:1129-1137.
4. Braithwaite, K.S. and Manners, J.M. 1989. Human hypervariable minisatellite probes detect DNA polymorphisms in the fungus *Colletotrichum gloeosporioides*. Curr. Genet. 16:473-475.
5. Cameron, D.F., Trevorrow, R.M., and Liu, C.J. 1997. Approaches to breeding for anthracnose resistance in *Stylosanthes* in Australia. Trop. Grassl. 31:424-429.
6. Centis, S., Dumas, B., Fournier, J., Marolda, M., and Esquerre-Tugaye, M-T. 1996. Isolation and sequence analysis of *Clpg1*, a gene coding for an endopolygalacturonase of the phytopathogenic fungus *Colletotrichum lindemuthianum*. Gene 170:125-129.

7. Chakraborty, S. 1997. Advances in research on *Stylosanthes* anthracnose epidemiology in Australia. Trop. Grassl. 31:445-453.
8. Chakraborty, S., Perrott, R., Ellis, N., and Thomas, M.R. 1999. New aggressive *Colletotrichum gloeosporioides* strains on *Stylosanthes scabra* detected by virulence and DNA analysis. Plant Dis. 83:333-339.
   Clark, S.J., Templeton, M.D., and Sullivan, P.A. 1997. A secreted aspartic proteinase from *Glomerella cingulata:* purification of the enzyme and molecular cloning of the cDNA. Microbiol. 143:1395-1403.
9. Coleman, M., Henricot, B., Arnau, J., and Oliver, R.P. 1997. Starvation-induced genes of the tomato pathogen *Cladosporium fulvum* are also induced during growth in planta. Mol. Plant-Microbe Interact. 10:1106-1109.
10. Covert, S.F. 1998. Supernumerary chromosomes in filamentous fungi. Curr. Genet. 33:311-319.
11. Dufresne, M., Bailey, J.A., Dron, M., and Langin, T. 1998. *clk1,* a serine/threonine protein kinase-encoding gene, is involved in pathogenicity of *Colletotrichum gloeosporioides* on common bean. Mol. Plant-Microbe Interact. 11:99-108.
12. He, C. 1997. Genetic variation in the fungal plant pathogen *Colletotrichum gloeosporioides*. Ph.D. Thesis. The University of Queensland, Australia.
13. He, C., Nourse, J.P., Kelemu, S., Irwin, J.A.G., and Manners, J.M. 1996. *Cg*T1: A non-LTR retrotransposon with restricted distribution in the fungal phytopathogen *Colletotrichum gloeosporioides*. Mol. Gen. Genet. 252:320-331.
14. He, C., Rusu, A.G., Poplawski, A., Irwin, J.A.G., and Manners, J.M. 1998. Transfer of a supernumerary chromosome between vegetatively incompatible biotypes of the phytopathogen *Colletotrichum gloeosporioides.* Genetics 150:1459-1466.
15. Hwang, C-S., Flaishman, M.A., and Kolattukudy, P.E. 1995. Cloning of a gene expressed during appressorium formation by *Colletotrichum gloeosporioides* and a marked decrease in virulence by disruption of this gene. Plant Cell 7:183-193.
16. Irwin, J.A.G. and Cameron, D.F. 1978. Two diseases in *Stylosanthes* spp. caused by *Colletotrichum gloeosporioides* in Australia, and pathogenic specialisation within one of the causal organisms. Aust. J. Ag. Res. 29:305-317.
17. Irwin, J.A.G., Cameron, D.F., Davis, R.D., and Lenne, J. 1986. Anthracnose problems with *Stylosanthes*. Trop. Grassl. Soc. Occl. Publ. 3:38-46.

18. Jelitto, T.C., Page, H.A., and Read, N.D. 1994. Role of external signals in regulating the pre-penetration phase of infection by the rice blast fungus, *Magnaporthe grisea.* Planta 194:471-477.
19. Levy, M., Romao, J., Marchetti, M.A., and Hamer, J.E. 1991. DNA fingerprinting with a dispersed repeat sequence resolves pathogenic diversity in the rice blast fungus. Plant Cell 3:95-102.
20. Maclean, D.J., Braithwaite, K.S., Manners, J.M., and Irwin, J.A.G. 1993. How do we identify and classify fungal plant pathogens in the era of DNA analysis? Adv. Plant Pathol. 10:207-244.
21. Manners, J.M., Masel, A., Braithwaite, K.S., and Irwin, J.A.G. 1992. Molecular analysis of *Colletotrichum gloeosporioides* pathogenic on the tropical pasture legume *Stylosanthes*. Pages 250-268 in: *Colletotrichum* Biology, Pathology and Control. J.A. Bailey and M.J. Jeger, eds. CAB International, Wallingford, U.K.
22. Manners, J.M. and He, C. 1997. Molecular approaches to studies of *Colletotrichum gloeosporioides* causing anthracnose of *Stylosanthes* in Australia. Trop. Grassl. 31:435-445.
23. Masel, A., Braithwaite, K.S., Irwin, J.A.G., and Manners, J.M. 1990. Highly variable molecular karyotypes in the plant pathogen *Colletotrichum gloeosporioides*. Curr. Genet. 18:81-86.
24. Masel, A., Irwin, J.A.G., and Manners, J.M. (1993) DNA deletion or addition is associated with a major karyotype polymorphism in the phytopathogen *Colletotrichum gloeosporioides*. Mol. Gen. Genet. 237:73-80.
25. Masel, A., He, C., Poplawski, A.M., Irwin, J.A.G., and Manners, J.M. 1996. Molecular evidence for chromosome transfer between biotypes of *Colletotrichum gloeosporioides*. Mol. Plant-Microbe Interact. 9:339-348.
26. Mora, Y., Chavez, J., and Mora, J. 1980. Regulation of *Neurospora crassa* glutamine synthetase by the carbon and nitrogen source. J. Gen. Microbiol. 118:455-463.
27. Ogle, H.J., Irwin, J.A.G., and Cameron, D.F. 1986. Biology of *Colletotrichum gloeosporioides* isolates from tropical legumes. Aust. J. Bot. 34: 281-292.
28. Ogle, H.J., Gowenlock, D.H., and Irwin, J.A.G. 1990. Infection of *Stylosanthes guianensis* and *S. scabra* by *Colletotrichum gloeosporioides.* Phytopathol. 80: 837-842.
29. Oliver, R. and Osbourn, A. 1995. Molecular dissection of fungal phytopathogenicity. Microbiol. 141:1-9.
30. Perfect, S.E., O'Connell, R.J., Green, E.F., Doering-Saad, C., and Green, J.R. 1998. Expression cloning of a fungal proline-rich glycoprotein specific to the biotrophic interface formed in the *Colletotrichum*-bean interaction. Plant J. 15:273-279.

31. Perpetua, N.S., Kubo, Y., Okuno, T., and Furusawa, I. 1994. Restoration of pathogenicity of a penetration-deficient mutant of *Colletotrichum lagenarium* by DNA complementation. Curr. Genet. 25:41-46.
32. Perpetua, N.S., Kubo, Y., Yasuda, N., Takano, Y., and Furusawa, I. 1996. Cloning and characterisation of a melanin biosynthetic *THR1* reductase gene essential for appressorial penetration of *Colletotrichum lagenarium.* Mol. Plant-Microbe Interact. 9:323-329.
33. Pieterse, C.M.J., Derksen, A.M.C.E., and Govers, F. 1994. Expression of the *Phytophthora infestans ipiB* and *ipiO* genes in planta and in vitro. Mol. Gen. Genet. 244:269-277.
34. Poplawski, A.M., He, C., Irwin, J.A.G., and Manners, J.M. 1990. Transfer of an autonomously replicating vector between vegetatively incompatible biotypes of *Colletotrichum gloeosporioides.* Curr. Genet. 32:66-72.
35. Schafer, W. 1994. Molecular mechanisms of fungal pathogenicity to plants. Annu. Rev. Phytopathol. 32:461-477.
36. Sharp, D., Braithwaite, K.S., Irwin, J.A.G., and Manners, J.M. 1990. Biochemical and cytochemical responses of *Stylosanthes guianensis* to infection by *Colletotrichum gloeosporioides*: association of callose deposition with resistance. Can. J. Bot. 68:505-511.
37. Sherrif, C., Whelan, M.J., Arnold, G.M., Lafay, J-F., Brygoo, Y., and Bailey, J.A. 1994. Ribosomal DNA sequence analysis reveals new species groupings in the genus *Colletotrichum.* Exp. Mycol. 18:121-138.
38. Stephenson, S.-A., Green, J.R., Manners, J.M., and Maclean, D.J. 1997. Cloning and characterisation of glutamine synthetase from *Colletotrichum gloeosporioides.* Curr. Genet. 31:447-454.
39. Stephenson, S.-A. 1998. Cloning and characterisation of pathogenicity genes from *Colletotrichum gloeosporioides.* Ph.D. thesis, The University of Queensland, Australia.
40. Talbot, N.J., Ebbole, D.J., and Hamer, J.E. 1993. Identification and characterisation of *MPG1,* a gene involved in pathogenicity from the rice blast fungus, *Magnaporthe grisea.* Plant Cell 5:1575-1590.
41. Tateno, Y. 1994. Evolution of glutamine synthetase is in accordance with the neutral theory of molecular evolution. Jap. J. Genet. 69:489-502.
42. Templeton, M.D., Sharrock, R.K., Bowwen, J.K., Crowhurst, R.N., and Rikkerink, E.H.A. 1994. The pectin lyase encoding gene *(pnl)* family from *Glomerella cingulata,* characterisation of pnlA and its expression in yeast. Gene 142:141-148.

43. Trevorrow, P.E., Irwin, J.A.G., and Cameron, D.F. 1988. Histopathology of compatible and incompatible interactions between Type A pathotypes of *Colletotrichum gloeosporioides* and *Stylosanthes scabra*. Trans. Brit. Mycol. Soc. 90:421-429.
44. Van den Ackerveken, G., Dunn, R., Cozijnsen, A., Vossen, J., van den Broek, H., and de Wit, P. 1994. Nitrogen limitation induces expression of the avirulence gene *avr9* in the tomato pathogen *Cladosporium fulvum.* Mol. Gen. Genet. 243:277-285.
45. Vichido, I., Mora, Y., Quinto, C., Palacios, R., and Mora, J. 1979. Nitrogen regulation of glutamine synthetase in *Neurospora crassa.* J. Gen. Microbiol. 106:251-259.
46. Vinijsanun, T., Irwin, J.A.G., and Cameron, D.F. 1987. Host range of three strains of *Colletotrichum gloeosporioides* from tropical pasture legumes, and comparative histological studies of interactions between Type B disease producing strains and *Stylosanthes scabra* (non-host) and *S. guianensis* (host). Aust. J. Bot. 35:665-677.
47. Wattad, C., Kobiler, D., Dinoor, A., and Prusky, D. 1997. Pectate lyase of *Colletotrichum gloeosporioides* attacking avocado fruits: cDNA cloning and involvement in pathogenicity. Physiol. Mol. Plant Pathol. 50:197-212.

Chapter 12

The Endopolygalacturonases of *Colletotrichum lindemuthianum*: Molecular Characterization, Gene Expression, and Elicitor Activity

Bernard Dumas, Georges Boudart, Sylvie Centis, and Marie-Thérèse Esquerré-Tugayé

Pectin, a methylated heteropolymer containing α-1,4-linked galacturonic acid, is one of the predominant polysaccharides of the plant cell wall (7). Since this material constitutes a mechanical barrier as well as a carbon source, many phytopathogenic and saprophytic microorganisms secrete pectolytic enzymes such as polygalacturonase, pectin methyl esterase, pectate lyase, or pectin lyase. In filamentous fungi, pectolytic enzymes are rapidly secreted when these microorganisms are grown *in vitro* on pectin. Their action causes tissue maceration in the host plant, thus allowing the fungal parasite to spread inter-, and intracellularly. These enzymes play a dual role in pathogenicity by inducing cell wall degradation (11,12) and defense gene expression in the host plant (4,5,14).

Colletotrichum lindemuthianum, a fungal pathogen causing anthracnose on bean seedlings, secretes an endopolygalacturonase (endoPG) when grown on pectin-containing media. This enzyme cleaves α-(O)-D-galacturonosyl bonds of homogalacturonan and pectic polymers producing mono-, di-, and trigalacturonic acid as final hydrolysis products. EndoPG is partly inhibited during the parasitic stage of the fungus by the bean cell wall polygalacturonase inhibitory protein (PGIP) (1,18,20). We recently reported that the endoPG purified from *C. lindemuthianum* race β elicits the biosynthesis of PR (pathogenesis-related) proteins in bean cuttings in a cultivar-specific manner (17). In this chapter, we present recent advances on the molecular characterization and expression of two endoPG genes and the differential elicitor activity of pectic fragments solubilized from susceptible or resistant bean seedlings by pure endoPG.

Molecular Characterization of Endopolygalacturonases

ISOLATION OF ENDOPOLYGALACTURONASE GENOMIC CLONES

An oligonucleotide primer based on the N-terminal amino acid sequence of *C. lindemuthianum* race β endoPG and an antisense primer designed from an internal amino acid sequence well-conserved among endoPGs were synthesized. These two primers were used to amplify by PCR a genomic DNA fragment encoding a portion of endoPG (8). This fragment was used to screen a partial genomic library of *C. lindemuthianum*. Two different types of clones, designated *clpg1* and *clpg2*, were isolated and characterized.

CHARACTERIZATION OF *CLPG1* AND *CLPG2*

From the cloned genomic DNA, 1700 nucleotides (nt) were sequenced for *clpg1* (8). Analysis of the sequence revealed an open reading frame of 1152 bp interrupted by an intron of 70 bp (Fig. 1). The 5'-noncoding region of the clone contained a CCAAT sequence which is often observed in fungal promoters. Nucleotide sequencing was done on 2020 nt for *clpg2* (9). This sequence contained an open reading frame of 1098 bp interrupted by one intron of 53 bp (Fig. 1). Two CCAAT sequences were detected in the 5'-noncoding sequence (Fig. 1). The proteins encoded by *clpg1* and *clpg2* comprise 363 and 365 amino acids, respectively. The N-terminal amino acid sequence obtained by microsequencing the endoPG purified from axenic culture of *C. lindemuthianum* race β (16) corresponds to amino acids 27 to 62 of CLPG1 showing that *clpg1* encodes the major endoPG secreted by *C. lindemuthianum* during saprophytic growth.

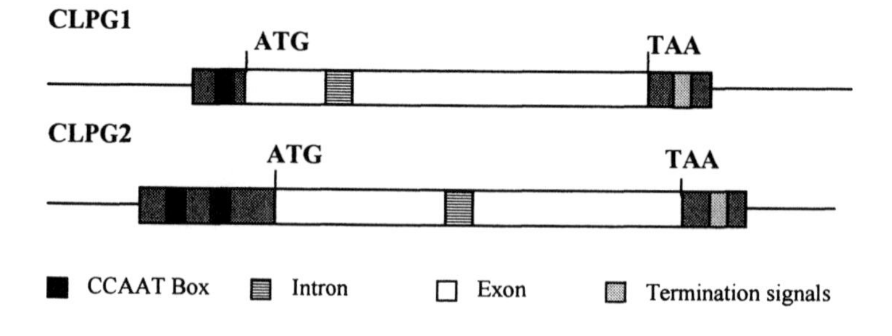

Fig. 1. Structural organization of *clpg1* and *clpg2*.

Table 1. Comparison of CLPG1 and CLPG2 amino acid sequences with those of some fungal endoPG

Species	EndoPGs	CLPG1[a]	CLPG2[a]
Cryphonectria parasitica	ENPG1	66%	53%
Cochliobolus carbonum	CCLPGN1	63%	59%
Sclerotinia sclerotiorum	SCEEPG	66%	57%
Aspergillus niger	ANPG	53%	55%
Fusarium moniliforme	FSOWPGA	37%	37%

[a] Percentages of identity between amino acid sequences of CLPG1 and CLPG2 with that of other fungal endoPGs were determined for most favorable alignment in each case by the CLUSTAL software program (IntelliGenetics)

Sequence alignment of the amino acids deduced from *clpg1* and *clpg2* revealed 61% amino acid identity. Comparison with different fungal endoPG showed amino acid identities ranging from 66% to 37% (Table 1). The most striking feature of CLPG2 is the amino acid sequence surrounding the conserved His 224 residue which is essential for the catalytic activity of endopolygalacturonase (3,6,19) and is conserved in all polygalacturonases sequenced so far in plants, fungi, and bacteria. The consensus sequence of this site is "GHG" (Fig. 2), whereas in CLPG2, a serine at position 223 replaces the glycine which precedes histidine. To determine whether this

```
    CLPG1     TGGTCSGGHGLSIGSVGG-RKDNVVKSVSITNSKIINSDNGVRIKTVAGA  264
    COCCA     TNCQCSGGHGVSIGSVGG-RKDNTVKGVVVSGTTIANSDNGVRIKTISGA  265
    CCLPGN1   TNCQCSGGHGVSIGSVGG-RKDNTVKGVVVSGTTIANSDNGVRIKTISGA  265
    SCEEPG    TGGTCSGGHGLSIGSVGG-RSDNVVSDVIIESSTVKNSANGVRIKTVLGA  284
1   CLPG2     ENGYCYGSHGLSIGSVGG-RTSNTVKDIIIRDSTIEKADNGIRIKTIAKK  264
    PG5       SGGVCSGGHGLSVGSVGG-RDDNIVQTVNFENSEIKNSQNGVRIKTISGD  286
    FSOWPG    SNMYCSGGHGLSIGSVGG-KSDNVVDGVQFLSSQVVNSQNGCRIKSNSGA  274
    AVOPOLYG  TNITCGPGHGISIGSLGDRNSEAHVSGVLVDGGNLFDTTNGLRIKTWQGG  342
    TOMPOL    TNITCGPGHGISIGSLGSGNSEAYVSNVTVNEAKIIGAENGVRIKTWQGG  334
    CHOUXPOL  EKVVCGPGHGISVGSLGRYGWEQDVTDITVKNCTLEGTSNGLRIKTWPSA  274
2   LUZERPOL  QGVNCGPGHGLSVGSLGKFTTEENVEGITVKNCTLTATDNGVRIKTWPDA  283
    TABACPOL  TRVTCGPGHGISVGSLGGNPDEKPVVGVFVSNCTVTNTDNGVRIKTWPAS  266
    MAISPOL   TGVTCGPGHGISIGSLGRYKDEKDVTDINVKDCTLKKTMFGVRIKAYEDA  297
3   ERWPLYGL  LHNEFGTGHGMSIGSET-MGVYNVTVDDLVMTG----TTNGLRIKS--DK  311
    PSEUDOPO  AHNHFYYGHGLSIGSETNTGVSNMLVTDLTMDGNDSSAGNGLRIKS--DA  370
                   .**.*.** .         .     .    .   * ***.
```

Fig. 2. Alignment of some endoPG amino acid sequences from fungi [1], plants [2], and bacteria [3]. The conserved region comprising the His residue involved in catalytic activity of endoPG is indicated in bold letters.

feature is encountered in the CLPG2 sequences of different races of *C. lindemuthianum*, oligoprimers located at the 5' and 3' end of *clpg2* were synthesized and used to amplify a genomic DNA fragment of *C. lindemuthianum* avirulent race (α) and virulent race (κ) toward *Phaseolus vulgaris* cv. Processor carrying the *ARE* resistance gene. Alignment of the deduced amino acid sequence of the potential catalytic site revealed that in the CLPG2 sequence of *C. lindemuthianum* races α and κ, a proline residue at position 223 replaces the serine found in the gene of race β. Hence, this region which is usually extremely conserved among plant, bacterial, and fungal endoPGs is heterogeneous in the *clpg2* genes. The production of CLPG2 in a heterologous system is in progress to determine whether this amino acid change modifies the catalytic activity of this protein.

Expression of Endopolygalacturonase Genes

EndoPG Gene Expression in Axenic Cultures of *C. lindemuthianum*

EndoPG gene expression was studied from mycelium obtained on a medium containing pectin as a sole carbon source or a mixture of pectin and glucose. Northern blot analysis of total RNA was conducted with specific probes for *clpg1* or *clpg2*. Both genes are transiently and rapidly induced, as soon as 12 hr of culture on pectin (9) and partly repressed by glucose. However, over a longer period of culturing (3 to 6 d), only *clpg1* is expressed on pectin whereas *clpg2* remains silent. The expression pattern of *clpg1* is consistent with the finding that only one form of endoPG, whose amino acid sequence is identical to that of CLPG1 is present in the culture medium.

The effect of different sugar monomers in the apoplastic compartment of plant cells on endoPG gene expression was studied. Among them, two neutral sugars, arabinose and rhamnose, were found to be strong inducers of endoPG (15). Northern blot experiments using specific probes for each endoPG gene *clpg1* and *clpg2*, showed that only *clpg1*, but not *clpg2,* was induced by arabinose and rhamnose.

Expression of EndoPG Genes During Parasitic Growth of *C. lindemuthianum*

Race β-inoculated susceptible bean seedlings were used for these studies. Immunolocalization of endoPG in ultrathin sections of infected bean hypocotyls using antibodies against CLPG1 were performed 3 days after inoculation. At this stage, the necrotrophic phase of the disease has started. Intense immunogold labelling of the fungal cell surface was observed showing the presence of the protein at the interface with the host plant (9). To identify the endoPG gene expressed at this time of infection, total RNA extracted from infected tissues was analyzed by RT-PCR using oligonucleotides specific for *clpg1* or *clpg2*. A signal was detected only for *clpg1* but not for *clpg2*. To further detect the possible expression of *clpg2* during early stages of pathogenesis, RT-PCR analysis was performed with total RNA extracted 24 hr after inoculation. Strong expression of *clpg2* was detected, suggesting that this gene is expressed *in planta* rapidly and transiently in a manner similar to that observed *in vitro* (B. Dumas, S. Centis and M.-T. Esquerré-Tugayé, unpublished).

USE OF GREEN FLUORESCENT PROTEIN (GFP) TO STUDY *CLPG2* REGULATION

The 5'-noncoding region of *clpg2* was fused to the coding sequence of a gene encoding a modified version of GFP (10). The construct was introduced into the fungal genome and the expression of GFP was detected by measuring the fluorescence of protein extracts from the mycelium grown *in vitro* in the presence of various carbon sources (B. Dumas, unpublished). Among them, pectin was the best inducer, and glucose had only a limited repression effect, showing that the promoter fragment used conferred pectin inducibility to the marker gene. Fluorescence microscopy revealed that GFP was expressed at early stages of germination (Fig. 3).

To look for the expression of GFP during infection of the host, bean hypocotyls of the susceptible Early Wax cultivar were inoculated with the GFP-expressing strain. The epidermis was peeled off 24 hr after inoculation and analyzed by fluorescence microscopy. As shown in Fig. 4, fluorescent germ tubes and swelling appressoria were detected at the surface of infected plant tissue, confirming that transcriptional activation of *clpg2* occurred rapidly when the fungus entered its parasitic stage. At later stages, fluorescence of the germ tube decreased, whereas fluorescence increased in the appressorium, possibly reflecting the migration of the cytoplasm into this swelling structure. At the onset of necrosis, fluorescence was no longer detected, whereas after prolonged incubation (15 d after inoculation) fluorescent hyphae were visible.

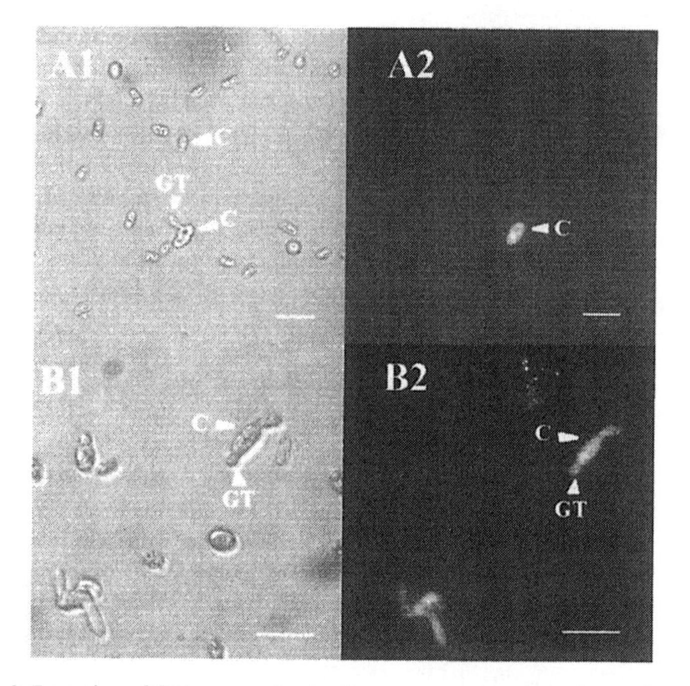

Fig. 3. Detection of GFP expression by fluorescence microscopy in axenic culture. Conidia of a *clpg2*-GFP-expressing strain of *C. lindemuthianum* were allowed to germinate for 12 hr (A1-A2) or 24 hr (B1-B2) on pectin medium cleared by filtration, then examined by visible (A1-B1) or fluorescent light (B1-B2) microscopy. Bar = 20 μm. C = conidia, GT = germ tube.

Elicitor Activity of EndoPGs

Differential Elicitation of Defense Responses by Pectic Fragments Solubilized from Bean Cell Walls by EndoPG

A bioassay allowing absorption of pure endoPG or pectic fragments from the plant cell walls was designed to measure the elicitor effect of the enzyme (17) on two near-isogenic lines of *P. vulgaris* cv. Processor (susceptible or resistant to race β of *C. lindemuthianum* [2]). In response to endoPG, β-1,3-glucanase activity is induced early in resistant seedlings, but only in a delayed manner in susceptible ones. Induction is very efficient since pmole amounts of endoPG per seedling are sufficient to induce the defense reaction at the whole plant level. Thus, endoPG mimics the effect of the fungus itself on defense induction (13). Since the differential effect of endoPG might

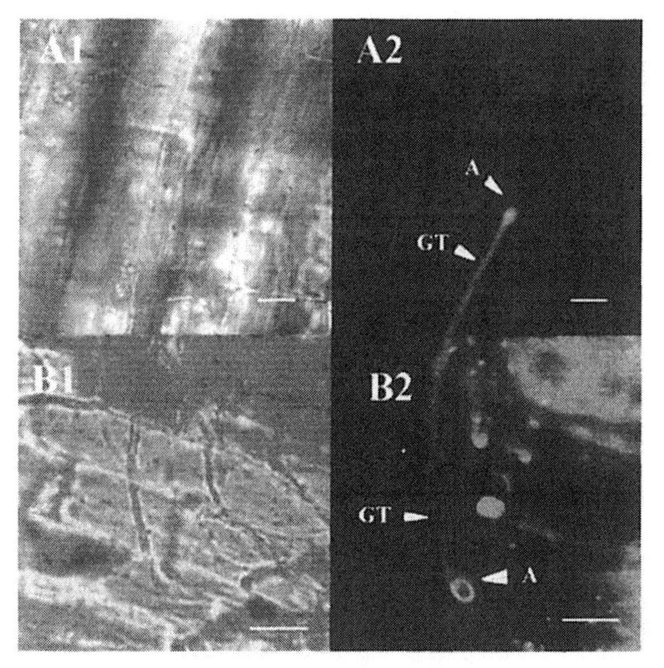

Fig. 4. Detection of GFP expression during early stages of pathogenesis. Bean hypocotyls were inoculated with a conidial supension of a *clpg2*-GFP-expressing strain of *C. lindemuthianum*. Samples were assayed 24 h after inoculation and analyzed by visible (A1-B1) or fluorescent light (B1-B2) microscopy. Two different stages of appressoria development are presented in A and B. Bar = 20 µm. A = appressorium, C = conidia.

result either from the protein itself of from the nature of the fragment it releases from the cell walls, elicitor activity of pectic fragments generated through the action of endoPG on susceptible or resistant bean cell walls was analyzed (5). The solubilized fragments were separated according to their charge and size. Their sugar composition revealed that neutral sugars were more abundant in the fragments released from the resistant plant than from the susceptible one suggesting different chemical compositions. The fragments solubilized from the resistant plant induced an increase of PR proteins (β-1,3-glucanases and chitinases) both at the transcript and enzyme activity level (5) when challenged on resistant or susceptible bean seedlings. On the other hand, pectic fragments released from susceptible bean cell walls only weakly elicited defense reactions in both cultivars. Thus, endoPG-released pectic fragments showed the same ability as the pure endoPG to discriminate resistant and susceptible plants.

Conclusion

C. lindemuthianum produces endoPG both in axenic culture and during pathogenesis. Molecular analysis of endoPG genes allowed the characterization of two endoPG genes, *clpg1* and *clpg2* which are differentially expressed. *Clpg1* encodes the major endoPG isoform which is secreted in axenic cultures of the fungus, whose expression is induced by pectin and the two neutral sugars, arabinose and rhamnose. Moreover, detection of *clpg1* mRNA during the necrotrophic stage of pathogenesis suggests that this protein plays a role in host cell wall degradation. On the other hand, CLPG1 has the ability to elicit plant defense responses in a cultivar-specific manner. Elicitor activity is probably mediated through the release of active pectic fragments from the plant cell wall, but a direct perception of CLPG1 by the plant cell should not be excluded. The production of a mutated endoPG, devoid of enzymatic activity, produced in a heterologous host is in progress to clarify this point. In addition, mutant β strains obtained by disruption of *clpg1* should prove useful to study the involvement of endogenous elicitor signals in the outcome of the bean-*Colletotrichum* interaction.

The role of the second endoPG gene, *clpg2*, is still unclear. Analysis by isoelectric focusing and western blotting with antibodies against endoPG during *in vitro* culture did not allowed identification of the protein product of *clpg2* even when the transcript was detected. Therefore, CLPG2 is probaably not secreted in the culture medium and could be associated with the fungal cell wall. The expression pattern of *clpg2*, as determined by northern blotting, RT-PCR analysis, and GFP-gene fusion, showed that this gene is rapidly and transiently induced by pectin during appressoria development. This induction suggests that CLPG2 might participate in the very early stages of interaction with the host plant. Future work will be aimed at unravelling the roles that *clpg1* and *clpg2* play in host degradation and cell signalling.

Literature Cited

1. Albersheim, P. and Anderson, A.J. 1971. Proteins from the plant cell walls inhibit polygalacturonases secreted by plant pathogens. Proc. Natl. Acad. Sci. USA 68:1815-181.
2. Bannerot, H. 1965. Résultats de l'infection d'une collection de haricots par six races physiologiques d'anthracnose. Ann. Amélior. Plantes 15:201-222.
3. Benen, J.A.E., Kester, H.C.M., Parenicova, L., and Visser, J. 1996. Kinetics and mode of action of *Aspergillus niger* polygalacturonases.

Pages 221-230 in: Pectins and Pectinases. A.G.J.Voragen and J. Visser, eds. Elsevier, Amsterdam.

4. Boudart, G., Déchamp-Guillaume, G., Lafitte, C., Ricart, G., Barthe, J.P., Mazau, D., and Esquerré-Tugayé, M.-.T. 1995. Elicitors and suppressors of hydroxyproline-rich glycoprotein accumulation are solubilized from plant cell walls by endopolygalacturonase. Eur. J. Biochem. 232:449-457.
5. Boudart, G., Lafitte, C., Barthe, J.P., Frasez, D., and Esquerré-Tugayé, M.T. 1998. Differential elicitation of defense responses by pectic fragments in bean seedlings. Planta 206:86-94.
6. Caprari, C., Mattei, B., Basile, M.L., Salvi, G., Crescenzi, V., de Lorenzo, G., and Cervone, F. 1996. Mutagenesis of endopolygalacturonase from *Fusarium moniliforme*: Histidine residue 234 is critical for enzymatic and macerating activities and not for binding to polygalacturonase-inhibiting protein (PGIP). Mol. Plant-Microbe Interact. 9:617-624.
7. Carpita, N.C. and Gibeaut, D.M. 1993. Structural models of primary cell walls in flowering plants: Consistency of molecular structure with the physical properties of the walls during growth. Plant J. 31:1-30.
8. Centis, S., Dumas, B., Fournier, J., Marolda, M., and Esquerré-Tugayé, M.-.T. 1996. Isolation and sequence analysis of *Clpg1*, a gene coding for an endopolygalacturonase of the phytopathogenic fungus *Colletotrichum lindemuthianum*. Gene 170:125-129.
9. Centis, S., Guillas, I., Séjalon, N., Esquerré-Tugayé, M.-.T., and Dumas, B. 1997. Endopolygalacturonase genes from *Colletotrichum lindemuthianum* : Cloning of *CLPG2* and comparison of its expression to that of *CLPG1* during saprophytic and parasitic growth of the fungus. Mol. Plant-Microbe Interact. 10:769-775.
10. Chiu, W., Niwa, Y., Zeng, W., Hirano, T., Kobayashi, H., and Sheen, J. 1996. Engineered GFP as a vital reporter in plants. Curr. Biol. 6:325-330.
11. Collmer, A. and Keen, N.T. 1986. The role of pectic enzymes during pathogenesis. Annu. Rev. Phytopathol. 24:1389-1409.
12. Cooper, R.M. and Wood, R.K.S. 1975. Regulation of synthesis of cell wall degrading enzymes by *Verticillium albo-atrum* and *Fusarium oxysporum* f. sp. *lycopersici*. Physiol. Plant Pathol. 5:135-156.
13. Daugrois, J.H., Lafitte, C., Barthe, J.P., and Touzé, A. 1990. Induction of β-1,3-glucanases and chitinase activity in compatible and incompatible interactions between *Colletotrichum lindemuthianum* and bean cultivars. J. Phytopathol. 130:225-234.
14. Hahn, M.G., Darvill, A.G., and Albersheim, P. 1981. Host pathogen interaction, XIX. The endogenous elicitor, a fragment of a plant cell wall polysaccharide that elicits phytoalexins accumulation in soybeans. Plant Physiol. 68:1161-1169.

15. Hugouvieux, V., Centis, S., Lafitte, C., and Esquerré-Tugayé, M.-.T. 1997. Induction by α-L-arabinose and α-L-rhamnose of endopolygalacturonase gene expression in *Colletotrichum lindemuthianum*. Appl. Environ. Microbiol. 63:2287-2292.
16. Keon, J.P.R., Waksman, G., and Bailey, J.A. 1990. A comparison of the biochemical and physiological properties of polygalacturonase from two races of *Colletotrichum lindemuthianum*. Physiol. Mol. Plant Pathol. 37:193-206.
17. Lafitte, C., Barthe, J.P., Gansel, X., Déchamp-Guillaume, G., Faucher, C., Mazau, D., and Esquerre-Tugaye, M.-.T. 1993. Differential induction by endopolygalacturonase of β-1,3 glucanases in *Phaseolus vulgaris* isolines susceptible and resistant to *Colletotrichum lindemuthianum* race β. Mol. Plant-Microbe Interact. 6:628-634.
18. Lafitte, C., Barthe, J.P., Montillet, J.L., and Touzé, A. 1984. Glycoprotein inhibitors of *Colletotrichum lindemuthianum* endopolygalacturonase in near isogenic lines of *Phaseolus vulgaris* resistant and susceptible to anthracnose. Physiol. Plant Pathol. 25:39-53.
19. Rexova-Benkova, L. and Mrackova, M. 1978. Active groups of extracellular endo-D-galacturonase of *Aspergillus niger* derived from pH effect on kinetic data. Biochim. Biophys. Acta 523:162-169.
20. Wijesundera, R.L.C., Bailey, J.A., Byrde, R.J.W., and Fielding, A.H. 1989. Cell wall degrading enzymes of *Colletotrichum lindemuthianum*: their role in the development of bean anthracnose. Physiol. Mol. Plant Pathol. 34:403-413.

Chapter 13

Signal Exchange During *Colletotrichum trifolii*-Alfalfa Interactions

Marty Dickman

The ability to respond to developmental and environmental cues is a characteristic that extends from single cells to complex multicellular organisms. Recent findings show unexpected similarities in the fundamental properties by which plant, animal, and human hosts respond to environmental stimuli. Such situations generally involve activation of signaling molecules and pathways, ultimately leading to changes in gene expression and defined cellular responses. In many cases the structure and, to some extent, the function of signaling molecules have been conserved during evolution. We are using this premise in studying the molecular mechanisms of disease development in the *Colletotrichum trifolii*-alfalfa interaction. *C. trifolii* (and all *Colletotrichum* species) represent, to varying degrees, an experimental compromise between the sophisticated genetics available to saprophytic unicellular eukaryotes such as yeast and the complex multicellular "higher" organisms. Pathogenicity of this fungus depends on a precisely orchestrated sequence of developmental transitions including conidial attachment, germination of the conidium to form a germ tube, differentiation of the germ tube into a specialized infection structure (appressorium), penetration of the plant cell by a penetration peg, biotrophic hyphal growth and nutrient assimilation within plant tissue, and eventual differentiation of hyphal tips into asexual conidia which rupture through the plant via acervuli. Using a "comparative pathobiology" approach, we are characterizing the stimulus responses controlling these pathways. Our assumption is that the crucial aspects of disease development (as well as host resistance) operate through signaling molecules. In the following chapter, this concept is illustrated with recent data from both the plant (alfalfa) and the fungus (*C. trifolii* Bain).

The mechanisms of recognition and adaptive responses to environmental stimuli are well understood in some systems, particularly in vertebrates, and appear to have commonalities in all eukaryotes (8,37,38,58). An emerging body of evidence indicates that a number of components employed in host-pathogen interactions are shared, either conceptually and/or mechanistically by mammals, plants, and their pathogens. This idea is clearly illustrated in host-bacterial interactions where pathogenic bacteria deliver virulence factors by a common means to their respective plant and animal hosts. While these virulence proteins generally differ, the mechanism to deliver such molecules and thus initiate disease is remarkably conserved, to the degree that genes can be functionally interchanged (1). An intriguing question arises as to whether fungi, particularly biotrophs, have analogous secretion systems. Similarly, the explosion in the cloning of R genes has not only revealed structural similarity between the proteins irrespective of the nature of the stimulus (pathogen), but also has made it clear that these genes are functionally similar and are likely to be involved with signal transduction pathways. In addition, a number of these genes have homologs in the animal kingdom. For example, the first specific R gene identified was a serine/threonine kinase (73). The N gene from tobacco has a similarity to the *Drosophila Toll* gene and the mammalian interleukin-1 receptor, both of which are involved in stress responses. Similar examples have been and continue to be reported (3). Thus, comparing the *C. trifolii*-alfalfa system with the more thoroughly characterized stimulus perception and signal delivery systems in the animal world is probably constructive. Examination of specific molecules and cell communication pathways in relatively well-developed animal host-pathogen/stress combinations may provide insight into analogous molecules and pathways which govern fungal-plant interactions. Dissection of signal transduction pathways will provide not only a means for understanding fungal pathogenesis, but also will identify new potential targets for disease control.

The interactions of fungi and their hosts provide particularly interesting examples of fungal responses to signals derived from the host in order to achieve infection. In nearly all animal cell systems studied to date, the receptor for an external signal is located in the plasma membrane, and signal exchange is mediated by second messengers. Recent studies have provided evidence that similar signaling pathways are also operative in *C. trifolii* and could provide the speed, specificity, and subtlety required for the events which lead to compatibility or incompatibility (26). The overall challenge now is to determine the *function* of these proteins in transducing signals. Even in mammalian systems, the functions of many kinases and G-proteins are unknown, and there are many functions for which a specific molecule has yet to be assigned (38,94). An additional complication is that, whereas signal transduction pathways in animal systems are mostly concerned with the regulation of basic cellular processes and integration of metabolic activities within multicellular organisms, we are considering signaling processes that probably evolved during the interaction of two different organisms. We have taken such a comparative biological approach for analyzing the molecular basis for communication between plant and fungus to understand mechanisms underlying host specificity and disease development with *C. trifolii*, the causal agent of alfalfa (*Medicago sativa*) anthracnose.

Anthracnose of alfalfa is one of the most destructive diseases of alfalfa in the United States and many other regions of the world (28). The disease is typified by stem or crown infections which usually kill the plant or lead to rapid deterioration of established stands. Anthracnose has been recognized as a serious alfalfa pathogen only within the last 20 years (26). As production and field costs rise (alfalfa is an important legume for hay and pasture), crop losses due to anthracnose have become a major concern, particularly in the southeast United States where the industry is rapidly expanding and the disease is most severe. Because fungicides are not economically feasible, considerable effort has been and still is being expended to breed and develop resistant alfalfa varieties (25,85). The use of resistant cultivars is now the principal means of control; however, breeding for resistance cannot always keep pace with the evolution of new pathogen races.

When a spore of *C. trifolii* encounters an alfalfa leaf surface, the spore attaches and undergoes a precise developmental sequence of events prior to attempted colonization. This process begins with adherence to the host cuticle and continues with spore germination, germ tube formation, and differentiation into the appressorium. Mutants of *C. trifolii* and other fungal phytopathogens, particularly *Colletotrichum* spp. which are defective in appressorium formation are nonpathogenic (26).

Since morphological transitions from conidial gemination to mature appressoria formation are essential for colonization by *C. trifolii* (and virtually all other *Colletotrichum* spp.), studying the regulation of infection-related morphogenesis is a key step in understanding disease development in the *C. trifolii*-alfalfa pathosystem. From a practical standpoint, traditional breeding for anthracnose resistance is not only time consuming but importantly, has not been durable, thus new and effective control strategies are required. Enumeration of genes and proteins required for prepenetration development will identify potential targets and possible pathways for disease control.

We have studied this interaction by two approaches (26). One is to isolate "important" genes making no assumption about their products and then identifying their biochemical roles by molecular analysis (e.g., sequence, expression pattern). Toward this end, we are using differential display of transcripts from infected host (alfalfa) tissue to isolate fungal genes expressed after penetration which are specifically induced during disease development. Alternatively, known genes encoding signal transducers from other organisms serve as heterologous probes or primers. By these approaches, we have isolated several fungal genes which are maximally expressed during prepenetration morphogenesis.

Isolation of Differentially Expressed Alfalfa cDNAs

To isolate alfalfa or *C. trifolii* genes which function during disease development, mRNA differential display was performed. mRNA differential display is a PCR-based technique for identifying differentially expressed genes by comparing gel migration patterns of randomly amplified mRNA (cDNA) fragments from two or more cell samples (68,69). Those cDNA fragments "differentially displayed" in one cell sample are recovered from the gel and used as probes to isolate the corresponding full-length differentially expressed genes from cDNA or genomic libraries (68,69). This approach was used for a three-way comparison of gene expression with RNA from *C. trifolii* race 1 mycelium grown in culture, healthy alfalfa leaves (cv. Saranac), and alfalfa leaves infected with *C. trifolii* race 1 for 48 hours. At this time, early disease symptoms were visible and mycelia were observed microscopically in and around the inoculation site. Initially, the mRNA differential display protocol described by Liang et al. (69) was followed using the $T_{12}MN$ primer in the PCR reactions. This modification increases primer sequence complexity and, therefore, decreases the number of bands displayed on the gel (42).

The hypothesis for this experiment was that plant and fungal genes important for disease establishment will be induced only when the plant and fungus interact in a compatible interaction. No assumptions were made concerning specific gene functions. This "black box" approach for gene cloning led to the identification of a novel alfalfa gene, SRG1, which was induced in response to infection with *C. trifolii* (104).

A modified differential display procedure was performed using 30 total primer set combinations per RNA sample. Seventeen differentially displayed cDNA tags were isolated from the infected leaf samples and tested for their differential expression in northern blots. Of these tags, three corresponded to differentially expressed genes (see below), four did not detect any signal, and the rest were false positives that detected either signals in healthy and infected alfalfa RNA or rRNAs in all three samples. The three differentially expressed tags were isolated from different pools of infected leaf cDNAs. However, these tags appeared to correspond to either the same gene or members of a gene family because they displayed a number of similarities. For example, all three tags were amplified using the same random decamers (E3 and E5), the size of each tag was 300 bp, and in northern blots, each detected an ~1 kb transcript in infected leaves with no signal in healthy leaves or in fungal mycelia. Due to these similarities, only one tag, tag G, was analyzed further.

RNA from infected plants is a mixture of plant and fungal RNA. Southern blot analysis showed tag G hybridized to DNA from alfalfa but not C. *trifolii*. Accordingly, tag G was used to screen a fungal elictor-induced alfalfa cDNA library. Two clones of different sizes (~700 bp and ~300 bp) were recovered. The inserts, named SRG1 and SRG2 for stress response genes, served as probes on northern blots to confirm differential expression. Like tag G, both SRG1 and SRG2 hybridized to an ~1 kb mRNA in infected leaves. SRG1 detected a faint signal in healthy leaves after prolonged exposure of the blot. These results indicated that SRG1 and SRG2 correspond to tag G and are alfalfa cDNAs induced by infection with *C. trifolii.*

SRG1 and the SRG2 partial sequence are 96% identical at the nucleotide level. Only one difference in their deduced amino acid sequences was revealed, an amino acid replacement (T to I) at position 113 in the SRG1 sequence. Assuming the sequence similarity between the SRG clones is maintained in the 5'-coding region of SRG2, these cDNAs are likely members of a gene family.

Comparison of SRGl with proteins in GenBank revealed similarities to several plant defense-related proteins (106), including the disease resistance-response proteins (DRRG49-a, DRRG49-b and DRRG49-c/RH2; 18,34,82; 79-82 % identical), the abscisic acid-responsive proteins from pea (ABR17 and ABR18; 52; 59 and 64% identical, respectively), the stress-induced proteins from soybean (SAM22 and H4; 21; 66% identical), and the pathogenesis-related proteins from chickpea (capRI; 64% identical), kidney bean (PvPR1 and PvPR2; 106; 63% identical), potato (STH-2 and STH-21; 74; 46 and 52% identical, respectively), and alfalfa (the PR10 partial sequence; 31; 86% identical). SRG1 is also 50 to 59% identical to the various isoforms of the major pollen allergens from birch (Bet v I; 13,95), hazel (Cor a I, 12) alder (Aln g I, 11), and hornbeam (Car b I; 64). Though the entire sequence of PR10 is not known, it is tempting to speculate that PR10, SRG1, and SRG2 are members of a gene family in alfalfa.

The expression patterns of some plant defense-related genes have been reported to correlate with disease resistance. In peas DRRG49-a (pI49) expression dramatically increases within 6 hr and again at 12-24 hr after inoculation with *Fusarium solani* f.sp. *phaseoli* (avirulent fungal pathogen), but not after inoculation with *F. solani* f.sp. *pisi* (virulent fungal pathogen) (33). In alfalfa, transcripts hybridizing to the PR10 PCR product are detected 30 hr after leaves are infiltrated with *Pseudomonas syringae* pv. *pisi* (avirulent bacterial pathogen) but not after infiltration with *Xanthomonas campestris* pv. *alfalfa* (31). In contrast, pea genes have been identified that are induced after challenge with *F. solani* f. sp. *phaseoli,* but whose expression patterns do not correlate with the resistance response (34).

To test whether accumulation of SRG1-like transcripts in alfalfa correlates with the resistance response to *C. trifolii,* accumulation of SRG1-hybridizing transcripts was followed over time in leaves (cv. Arc) inoculated with either C. *trifolii* race 1 (avirulent) or race 2 (virulent). If

SRG1 transcript accumulation correlates with resistance to *C. trifolii,* one would expect SRG1 transcripts to accumulate to a maximum level at an earlier time in race 1- versus race 2-infected leaves. Importantly, normalization of the RNA signals showed that at 2 hr SRG1 transcript accumulation increased approximately four-fold in the fungal-treated samples and three-fold in the surfactant-treated controls. While accumulation in control leaves remained relatively constant until 72 h, accumulation in race 1- and race 2-infected leaves increased to maximum levels at 8 hr (~ eight-fold induction) and 4 hr (~ seven-fold induction), respectively. The relatively high accumulation levels in the fungal-treated samples were maintained until 72 h, when accumulation dropped to five-fold in both samples. Since the early time points are likely key in determining resistance or susceptibility, and SRG1 transcripts accumulated to high levels in response to both fungal races, but reached a maximum earlier in response to the virulent fungal race, the SRG1-like genes do not appear specific for the resistance response to *C. trifolii.* Instead, SRG1 transcripts likely accumulated in response to the stress associated with inoculation. SRG1 transcript accumulation in control leaves treated with surfactant is consistent with this hypothesis.

To further define the conditions that cause accumulation of SRG1-like transcripts in alfalfa, northern blots were performed on RNA from senescent, UV-treated or mechanically wounded Arc leaves. Neither senescence nor UV treatment caused SRG1 transcript accumulation. However, mechanical wounding caused a high level of SRG1 transcript accumulation. Like SRG1, potato STH-2 (74), soybean SAM22 (21), and asparagus AoPR1 (107) are induced by wounding.

Taken together, the northern analyses indicate that external stresses, including inoculation, surfactant, and mechanical wounding, cause accumulation of SRG1-like transcripts in alfalfa leaves. The SRG1-like genes may possibly be induced by breaches in the leaf cell wall, since microscopic studies of the alfalfa-*C. trifolii* interaction revealed that both virulent and avirulent races of the fungus penetrate the plant cuticle (79,80).

Though expression of PR10, an alfalfa gene potentially closely related to SRG1, correlates with the resistance response to a bacterial pathogen (31), SRG1 transcript accumulation did not appear to correlate with the resistance response to *C. trifolii.* These results, however, do not preclude the possibility that genes closely related to SRG1 (e.g., PR10) are specific for pathogen attack. The SRG1 probe used in the expression analyses likely hybridized to transcripts of many closely related alfalfa genes, some of which may be specific for a resistance response, and thus possibly masked their differential expression. The differences in PR10 and SRG1 expression may also reflect differences in alfalfa's resistance to bacterial versus fungal pathogens or because PR10 expression was compared in alfalfa treated with two different bacterial species, whereas SRG1 transcript accumulation was compared in alfalfa treated with two closely related fungal races.

Forty-eight hours after alfalfa plants were inoculated with virulent *C. trifolii* spores, fungal mycelia were observed microscopically in and around the inoculation site. Our initial hypothesis was that both induced plant and fungal genes could be isolated from infected leaf tissue using mRNA differential display. However, aside from possible fungal rRNA tags, no fungal genes (either induced or false-positives) were identified. Although this result may be explained by a number of reasons, perhaps most important is that only a small amount of the RNA mass in the infected leaf sample is of fungal origin. Therefore, plant genes were far more likely to be amplified in the PCR reactions.

We subsequently changed our approach by taking advantage of the fact that a number of plant-induced fungal genes are also induced by conditions of starvation (e.g., 97) *C. trifolii* was grown *in vitro* under nutrient-limiting conditions (carbon and/or nitrogen) as well as in rich media. Differential display was performed and mRNAs were isolated that were specifically induced during nutrient limitation. These transcripts were labeled and used as probes in northern blots containing RNA from inoculated alfalfa tissue (24 hr). Fungal transcripts that were only induced in inoculated tissue were further analyzed. Two clones were sequenced and analyzed. One (DD-9) contained a 200-bp region of Alu repeats and is still being characterized. The other (DD-60) has been completely sequenced. DD-60 has a basic region, followed by the so-called "leucine repeat" of the B-zip class of DNA-binding proteins (63). This "myc-like" amphipathic helix (Ct-LR1) is characterized by the presence of leucine residues at every seventh position. Moreover, a second leucine repeat of the coiled-coil class (Ct-LR2) was found immediately adjacent to Ct-LR1. Two leucine repeats aligned in a parallel fashion can result in the formation of homo- or heterodimers and are found in a number of mammalian and yeast transcription factors. Data base analysis has revealed DD-60 to be most like the mammalian transcription factor AP-4 (50). A helical wheel analysis confirms the amphipathic helix of DD-60, and intriguingly, all charged and uncharged groups (Q, E, R, E) align on the same face (Niefeldt and Dickman, unpublished).

RAS

Ras proteins are small (21-24 kd), monomeric GTP-binding proteins that transduce signals for growth and differentiation in eukaryotic organisms. The *ras* genes were first identified as oncogenes of the Harvey and Kirsten strains of rat sarcoma viruses (29). Cellular *ras* genes were isolated later, after the discovery in 1982 that certain human tumors harbor mutated *ras* alleles capable of transforming mouse NIH 3T3 cells in gene transfer assays (24,87; see ahead). Ras has since been intensely studied, and genes encoding *ras* homologs have been identified in evolutionarily diverse organisms. In mammals, the importance of Ras in regulating growth is underscored by the observation that activating mutations in *ras,* together with inactivation of the tumor suppressors p53 and pl6, are the most prevalent mutations which lead to human tumors (46,48).

Ras proteins are synthesized as cytosolic precursors which must undergo post-translational modification, including farnesylation and palmitoylation, for proper membrane localization and function (17,20). The signal for post-translational changes is the conserved CAAX motif (C, cysteine; A, an aliphatic amino acid; X, usually methionine or serine) at the C-terminus of all Ras proteins. Addition of phenyl groups renders Ras hydrophobic and promotes association with the plasma membrane, where it interacts with various regulators and effectors. Once localized, Ras serves as a molecular switch, coupling activated membrane receptors to downstream signaling molecules by alternating between GTP-bound (active) and GDP-bound (inactive) conformations (10,71). Ras-GTP associates with and stimulates target effectors, whereas Ras-GDP cannot. The rate of conversion between the active and inactive conformations is modulated by the weak intrinsic GTPase activity of Ras and two kinds of regulatory proteins, guanine nucleotide exchange factors (GEFs) and GTPase activating proteins (GAPs) (9). GEFs activate Ras by increasing the dissociation rate of guanine nucleotides, allowing GTP, which is more prevalent in the cell, to bind Ras. GAPs inactivate Ras by dramatically increasing the slow intrinsic rate of GTP hydrolysis. Dominant activating (oncogenic) mutations occur at amino acids critical for guanine nucleotide coordination (71,90). Oncogenic Ras mutants have an impaired intrinsic GTPase and are insensitive to GAPs, and are thus unable to switch off transmitted signals (90). In animals, flies, worms, and fission yeast, Ras regulates conserved mitogen-activated protein kinase pathways (MAPK) that convey signals from the plasma membrane to the nucleus.

That *ras* genes activated by point mutations confer aspects of an oncogenic phenotype upon cultured cells implies that their encoded proteins function in the transduction of signals that regulate cell proliferation. Other pathways are also probably affected, since *ras* is expressed in non-dividing cells and activated *ras* genes can induce neuronal differentiation (6). In metazoans and fission yeast, ras mediates two different pathways-one via the MAPK pathway as mentioned and the other via the Rho (G protein) family of proteins. In fission yeast, these pathways are involved with mating and morphogenesis, respectively. In contrast, the two redundant *ras* genes in budding yeast (*S. cerevisiae*) are not linked to any of these pathways. Rasl activates adenylate cyclase, leading to synthesis of cAMP, and Ras2 has recently been shown to be involved in inducing filamentous growth (78). A *ras* homolog has also been described in *Aspergillus nidulans* (103). This essential gene was found to regulate developmental decisions revolving around conidiation. Constitutively active Ras led to failure of cells to initiate the first step in morphogenesis, production of a germ tube. Low levels of Ras expression resulted in premature development of conidia. A gene disruption of *ras* was lethal.

C. trifolii Ras

Using degenerate primers and PCR, we have recently isolated and cloned the unique *ras* cDNA from *C. trifolii* (104). The *C. trifolii* Ras protein (CT-Ras) has considerable sequence identity with *ras* genes from other fungi; interestingly, CT-Ras is as similar to the budding yeast Ras as it is to human Ras.

Because at the time these studies were initiated we were unable to perform targeted disruptions in *C. trifolii* (this has now changed), we sought alternative means to functionally analyze the putative *ras* gene. In *S. cerevisiae,* disruption of either Ras gene alone does not affect phenotype, but disruption of both *ras* genes is lethal (56); therefore, we utilized a novel approach using a well-characterized animal cell line (NIH 3T3).

Oncogenes in mammals have long been analyzed by similar gene transfer techniques. In particular, *ras* oncogenes have been extensively characterized by transfection of mouse NIH 3T3 cells, a fibroblast cell line extremely sensitive to perturbations in Ras activity. Expression of a functional *ras* oncogene (from any organism) in NIH 3T3 cells produces a malignant (transformed) phenotype, which is characterized by alterations in cell morphology, increased growth rate, reduced serum dependence, loss of density-dependent growth inhibition, anchorage-independent growth, changes in gene expression, and the ability to form tumors in experimental animals. Among these properties, the most reliable determinant of cellular transformation is tumor formation *in vivo.*

Using standard recombinant DNA technology and PCR, site-directed mutations in the C. *trifolii ras* were made such that the *ras* gene would be constitutively active, based upon what is known in human *ras* genes and the fact that there is a high degree of conservation at the sequence level.

Thus, two known and separate activating mutations in C. *trifolii ras* gene were made using PCR:

1. CT-Ras, $G^{17} \rightarrow V^{17}$; equivalent to $G^{12} \rightarrow V^{12}$ in animal tumors (91).
2. CT-Ras, $Q^{66} \rightarrow L^{66}$; equivalent to $Q^{61} \rightarrow L^{61}$ in animal tumors (23).

These constructs along with the wild-type *C. trifolii ras* cDNA were cloned into a mammalian expression vector. NIH 3T3 cells (mouse fibroblasts) were transfected, selected on geneticin media (mock transfected cells were dead in 2 weeks), and assayed for a transformed phenotype, using three standard criteria (morphology, soft agar assay, tumor formation).

1. Morphology. Transformed cells grow in a cluster of spindle shapes that are highly refractile when viewed under the microscope. Untransformed cells are flat and non-refractile (19).
2. Soft Agar Assay. This is the most commonly used assay for Ras-transformed cells (19). Untransformed cells require a solid substrate for attachment as well as a rich nutritional medium for growth (8% fetal calf serum). Transformed cells exhibit anchorage-independent growth (they will readily proliferate in liquid culture suspension or when placed in semi-solid medium). Normal cells are anchorage-dependent. Transformed cells will also grow in low amounts of serum (3%) at concentrations which will not support growth of normal cells.

Cells exhibiting the transformed phenotype based on morphology contained CT-Ras. CT-Ras can easily be distinguished from the endogenous animal Ras based on size difference. (CT-Ras encodes a 25-kd protein; mammalian is 21 kd, data not shown). Such cells were used in the soft agar assay procedure (anchorage-independent growth), and after 4 weeks on 0.3% agar and 3% fetal calf serum only transformed cells will grow; non- or mock-transfected NIH 3T3 cells did not grow, nor did NIH 3T3 cells transfected with the wild-type *C. trifolii CT-ras* cDNA. While there were no qualitative differences, the $G^{17}{\rightarrow}V^{17}$ mutation consistently yielded increased rates of transformation and large tumors, relative to the $Q^{66}{\rightarrow}L^{66}$ mutation.

3. Tumor formation. Lastly, the proliferating cells were injected into mice and tumors rapidly developed.

Two important points need to be made:

1. Clearly, the *C. trifolii ras* gene has the genetic capability to function as a *bona fide* oncogene (via a single base change).
2. Mammalian cell lines, which are very well characterized and relatively easy to manipulate, can be used to ask and address questions about gene function in filamentous fungi.

We have taken both constitutively activated *ras* gene constructs, transcriptionally fused them to the *Aspergillus* glyceraldehyde dehydrogenase promoter (gdp) (89), and inserted the constructs, which also contained a hygromycin-resistance gene for selection, into wild-type *C. trifolii* to evaluate the consequences of "oncogenic" expression. Our preliminary findings indicated high constitutive levels of oncogenic CT-ras expression, as determined by RNA blots (data not shown). On nutritionally rich media (e.g., potato dextrose agar), these mutants are phenotypically identical to wild type. However, on minimal medium, the CT-Ras (G 17V) and CT-Ras (Q66L) isolates exhibited an unusual growth phenotype that was not observed in control strains, including transformants containing an additional copy of the wild-type gene. Hyphal growth normally proceeds by both polarized linear extension of hyphal tips and by branching.

"Oncogenic" *C. trifolii* has lost the ability to grow in a polar manner, and hyphae that branch became curled and distorted, indicating a defect in orientation of hyphal tip growth. The hyphal distortions increased in severity with time, although the growth was abundant; these mutants actually had greater mass than the wild type under the same conditions. The wild type in minimal medium sporulates profusely. This is not surprising since there is little nutrition to support vegetative growth. The mutants, however, are unable to sporulate, or at best sporulate at considerably reduced levels, though the fungus appears to be attempting to sporulate (104). Moreover, when an aqueous spore suspension is placed on a glass slide, which should induce appressorium formation (108), these mutants were unable to differentiate. Thus, in conditions of nutrient limitation, when an oncogene in *C. trifolii* is expressed, there is uncontrolled, rapid vegetative growth, a loss of polarity, and an inability to undergo normal development.

Our results are consistent with a model in which CT-Ras regulates a signal transduction pathway that senses and responds to nutrients. The fact that CT-Ras regulates fungal cellular responses to the nutrient environment is reminiscent of the situation in mammalian cells (e.g., NIH 3T3 cells) where Ras is involved in transducing growth factor signals. When these growth factors are removed (a nutrient-limiting situation), wild-type cells arrest growth or differentiate, whereas cells expressing constitutively active Ras continue cell division, and oncogenesis or apoptosis occurs. Under nutrient-limiting conditions in *C. trifolii,* wild-type cells arrest vegetative growth (hyphal elongation) and differentiate (conidiate), whereas cells expressing constitutively active CT-Ras continue vegetative growth, do not differentiate, and, importantly, are impaired in polarized, asymmetric growth.

Cell polarity is important for many processes in a wide range of cells. For example, it is required for transmission of a nerve impulse, the development of a fertilized egg, transport of molecules across epithelial cell layers, crawling of a fibroblast cell, and the growth of fungi. Since fungal hyphae have many of the basic morphological and structural elements found in neurons, as well as transmembrane ion currents believed to be involved in branching, hyphal growth may share some of the elements involved in neurite extension in the developing nervous system (57).

In all eukaryotes examined except *S. cerevisiae,* Ras links receptor tyrosine kinases to kinases in pathways central to growth control (71). Using heterologous probes and anti-phospho-tyrosine antibodies, we have been unable to demonstrate the existence of tyrosine kinases in C. *trifolii* (unpublished). Identifying components upstream of CT-Ras will, therefore, be particularly interesting. Because we cannot exclude the possibility that CT-Ras may be a component of the cAMP/PKA pathway as in *S. cerevisiae* (71), we have recently cloned *C. trifolii* genes encoding the regulatory and catalytic subunits of PKA (112,113) so we can directly address this hypothesis.

Although the molecular contexts of pathogenic filamentous fungi and mouse NIH 3T3 fibroblast cells are clearly distinct in both cell types, Ras is an important sensor and is responsible for external stimuli which activate the appropriate signal response pathway in both cell types.

Protein Phosphorylation

A common mechanism by which eukaryotes (and prokaryotes) respond to external and internal stimuli is reversible phosphorylation of substrate proteins. Protein phosphorylation/dephosphorylation is one of the principal mechanisms which regulate signal transduction pathways in eukaryotes (55). Thus, protein kinases have been extensively studied and found to be involved in virtually every aspect of cellular events (44). Humans are estimated to have over 2000 protein kinases (51). Although a number of fundamental processes in fungi such as the cell cycle, transcription, and mating require protein phosphorylation, the analysis of protein kinases and phosphatases in filamentous fungi is in its infancy; however, it is already apparent that kinases and phosphatases are likely important mediators of fungal growth and development. For a comprehensive review of protein kinases (and phosphatases) in filamentous fungi, see Dickman and Yarden (27).

Of the various protein kinase pathways known, the MAP (mitogen activated protein) kinase cascade is known to be conserved throughout eukaryotes and to act by converting receptor signals into a variety of specific responses. The cascade consists of a MAP kinase kinase kinase (MAPKKK), a MAP kinase kinase (MAPKK), and a MAP kinase. The MAP kinase is activated through threonine/tyrosine phosphorylation catalyzed by the dual-specificity MAPKK, which in turn is activated through serine phosphorylation catalyzed by MAPKKK. Recent evaluation of the MAP kinase cascade arrangement suggests that it is particularly suited to mediating processes where rapid switching from one discrete response to another is required (see Hirt, 47; Dickman and Yarden, 27; for more comprehensive reviews).

MAP kinases have recently been shown to be important for fungal development and pathogenicity (110) as well as plant defense (70). We, therefore, have initiated studies to determine whether MAP kinase activity is altered during *C. trifolii* challenge of alfalfa. In collaboration with H. Hirt, who has developed antibodies to alfalfa MAPK, we have shown that MAP kinase activity is significantly enhanced during the incompatible but not the compatible interaction (M. Dickman and H. Hirt, unpublished). Thus,MAP kinase may play a pivotal role in determining alfalfa resistance.

PKA

As one of the first protein kinases to be purified (105), cyclic AMP-dependent protein kinase (PKA) was the first protein kinase to be sequenced and crystallized (60,61,92). PKA has been found in all eukaryotes studied except plants (99) and remains the main (if not the only) receptor for cAMP, the first documented second messenger. Intracellular cAMP levels are regulated by two enzymes. Adenylate cyclase synthesizes cAMP from ATP, and cyclic AMP phosphodiesterase degrades cAMP to AMP. With the exception of PKA in *Dictyostelium discoideum*, PKA is a tetrameric protein, consisting of two catalytic (C) and two regulatory (R) subunits in the inactive holoenzyme. The conformation of the holoenzyme changes upon binding of cAMP to the R subunits, resulting in the release of the two active C subunits (99).

While numerous studies have been performed in understanding the wide range of PKA function in mammals and yeasts, only recently has PKA been studied and functionally analyzed in phytopahogenic fungi (62). Of particular interest is a report that cAMP induces appressoria formation by *Magnaporthe grisea* and a mutant defective in appressoria initiation can be restored to wild type by addition of cAMP (66). Under conditions in which *M. grisea* conidia normally do not form appressoria, exogenously supplied cAMP, cAMP analogs, cAMP phosphodiesterase inhibitors, and adenylate cyclase activators induced appressoria development. The PKA catalytic subunit gene was cloned and disrupted; the knockout mutant of the catalytic subunit grew normally in rich media, but appressoria development was delayed. Host tissue was not penetrated (7,109). In addition, the α-factor of *S. cerevisiae* inhibited *M. grisea* appressoria development, but 10 mM cAMP in the media reversed the inhibition, indicating that the α-factor of yeast or the pheromone from *M. grisea* blocks signaling events leading to appressoria formation in the opposite mating type. Thus, the cAMP/PKA pathway might act downstream of pheromone signaling receptors (7). In both budding yeast and fission yeast as well as *Ustilago maydis* (described later), pheromone receptors are coupled to heterotrimeric G-proteins. Since adenylate cyclase is coupled to and activated by G-proteins in all eukaryotes (except for budding yeast), pheromone inhibition of appressoria development further demonstrated the relationship of the cAMP/PKA pathway with pathogenicity-related development.

Pheromone-induced mating events and pathogenicity-related dimorphism of *U. maydis,* the fungus causing corn smut, has been studied in great detail at the molecular level (e.g., 40). Mating type genes are directly involved in the morphological switches between budding and filamentous growth in *U. maydis.* A constitutively filamentous mutant proved to be defective in adenylate cyclase activity. This phenotype was suppressed by a mutation in the regulatory (R) subunit of PKA or exogenous cAMP.

Mutation in the R subunit of PKA also affected cytokinesis. Taken together, these results indicate that the cAMP/PKA pathway influences *U. maydis* morphogenesis, which is required for pathogenesis (40).

The PKA gene from *M. grisea* is actually required for appressoria penetration, and, as mentioned previously, only delays formation of appressoria. Thus, PKA knockout mutants are infectious if a wound is present in the plant (109). Interestingly, Xu and Hamer (109) have identified a MAP (mitogen activated protein) kinase in *M. grisea* that when replaced by a defective copy can no longer differentiate appressoria *and* are also nonpathogenic even if a wound is present. Our previous studies have shown that PKA regulated conidial gemination and appressoria formation in *C. trifolii* (111). To analyze PKA function directly, we have isolated and cloned the C and R subunit genes of PKA (112,113). The deduced protein was similar to other PKA C subunits. Similarities were highest between CT-PKAC protein and the C subunits of PKA in other fungi (72,75,86,101). We functionally analyzed the *C. trifolii* PKA C gene by transforming it into an *S. pombe* PKA mutant (112,113). Expression of Ct-PKAC suppressed the defect in the *S. pombe* mutant. These results strongly suggest that Ct-PKAC encodes a functional PKA C subunit gene. An interesting feature of CT-PKAC protein is the presence of 26 consecutive glutamine residues near its N-terminus. Other fungal serine/threonine kinases that contain glutamine-rich regions are the C subunit of PKA in *M. grisea* and TB3 in *C. trifolii* (16,75). Glutamine-rich regions are important in protein-protein interactions, and are found in proteins which activate transcription (30,36,98).

We also functionally analyzed the Ct-PKAR gene (112). This analysis was particularly important since neither overexpression nor disruption of this gene visibly altered the phenotype. We used the *Neurospora crassa mcb* mutant which has a temperature sensitive mutation in the R subunit of PKA (14). When *mcb* protoplasts were transformed with Ct-PKAR, functional complementation was achieved (112). Heterologous hosts, with well-characterized mutations can be useful tools for analyzing gene function in less well-characterized fungi like *C. trifolii.*

The generation of CT-PKAC replacement mutants showed that this fungus is amenable to targeted gene disruption, and the methodology is important as a framework in the characterization of other genes cloned in *C. trifolii.* The intentionally inactivated mutants could be distinguished from other ectopic integration transformants by slow growth rates, and phenotypes were confirmed by Southern analysis. *Ct-PKAC* transcripts were also absent in these mutants. Three PKA mutants were considerably retarded in growth regardless of nutritional conditions, but the timing of conidia germination and initiation of appressoria development were similar but slower than the wild type. In both budding yeast and fission yeast, the loss of PKA activity delayed spore germination (72,101). Appressoria development was delayed in the rice blast pathogen *M. grisea* when the C subunit of PKA was disrupted (75,109). Our data is consistent with the reduced growth in PKA-deficient mutants of other organisms.

This growth defect might be explained by the inactivation of PKA-regulated enzymes during carbon metabolism (39). Other changes in Ct-PKAC disruption mutants were the alteration in conidiation patterns. The three mutants randomly conidiated over a prolonged period of time. In fungi, PKA is involved during transitions from vegetative to reproductive growth. For example, PKA mutants in both budding and fission yeasts sporulated constitutively in the absence of the nutrient-starvation signal, which is normally required for wild-type strains to sporulate (72). Induced PKA activity resulted in sterility (54,101). Consistent with these data, in *N. crassa,* the crisp mutants with defective adenylate cyclase not only grew slowly, but sporulated more profusely in solid media (100). Two recent reports have provided additional evidence for PKA function during fungal reproductive growth. *U. maydis* mutants in the R subunit gene of PKA colonized maize normally, but failed to produce galls (41). Loss of sporulation was also observed in one Gi subunit disruption mutant of *C. parasitica,* and higher levels of cAMP were detected in the disruption mutants (35). In *A. nidulans,* alteration in sporulation patterns in one G subunit mutant as well as the RGS (Regulator of G protein signaling) mutant indirectly suggested that PKA regulates sporulation in this fungus, since heterotrimeric G-proteins are known PKA upstream regulators (4565; 114). In *S. cerevisiae,* PKA affects reproduction by directly regulating the expression of G1 cyclins (5,102), but the responsive molecules in *C. trifolii* and other filamentous fungi remains to be found.

The most significant change in the *C. trifolii* PKA-deficient mutants was that the mutants could not infect intact susceptible host plants. Disruptants on either detached leaves or entire plants germinated and formed appressoria like the wild type. After 48 h, the period preceeding host penetration, no obvious differences were observed between the wild type and the PKA-deficient mutants. Differences started to appear 72 hr after inoculation, the time required for wild-type strains to complete penetration (79,80). After 5 days, lesions with acervuli and abundant mycelia were found in wild type-inoculated leaves. No lesions were found in the PKA-deficient mutant-inoculated leaves, and little mycelia growth was observed on leaf surfaces. However, lesions and acervuli were observed in artificially wounded leaves inoculated either with the wild-type strain or the mutants. That these PKA-deficient mutants could form appressoria, colonize wounded leaves and produce acervuli suggests that the loss of pathogenicity is most likely by a failure in appressorial penetration. Similar results have been reported in *M. grisea* PKA mutants. Three distinct PKA C subunit mutants rarely infected intact rice leaves, but were able to grow and cause lesions after penetration through wounds (109). For *M. grisea,* high concentrations of glycerol are mainly responsible for the generation of turgor pressure during mechanical penetration by appressoria (22,49). In *C. trifolii*, molecular mechanisms for appressoria penetration are poorly understood, both mechanical force and enzymatic degradation may be involved during this process (4). Similar mechanisms of penetration are likely to operate in appressoria-forming fungi, with *C. trifolii* and *M. grisea* sharing similar biochemical bases that could generate mechanical forces via PKA. Since cAMP and PKA are important regulators in cellular responses to external stimuli as well as carbon metabolism, a number of possible reasons can be invoked to explain these data. It is clear, however, that the lack of PKA activity impaired the penetration of appressoria of *C. trifolii,* thus rendering the fungus non-pathogenic.

Lipid-Activated Protein Kinase

Members of the protein kinase C (PKC) family of phospholipid-dependent serine/threonine-specific kinases respond to extracellular signals generated by receptor-mediated hydrolysis of membrane lipids (phosphatidylinositol bisphosphate) which produces diacylglycerol (DAG) and inositol triphosphate. DAG serves as a second messenger to activate PKC. These kinases are key components in the phosphoinositol cascade, which in various animal cell types evokes a wide variety of responses such as cell proliferation, regulation of gene expression, membrane transport, and organization of the cytoskeleton (2). In addition, PKC is involved with mammalian disease, (53) is a protooncogene (88), and can drive cells to apoptosis (32).

The association of PKC with numerous mammalian developmental and disease-related phenomena promoted a search for a functional homolog in *C. trifolii.* Primers were designed for PCR based on the highly conserved Cl domain (84), which is a hallmark of all PKCs, and on domain VIII in the catalytic domain which is a more general motif for both PKA or PKC (43). By manipulating primer length and levels of degeneracy, we obtained unique fragments which were used to screen a *C. trifolii* cDNA library. We isolated a full-length clone, designated *lapk.* Sequence analysis of *lapk* identified conserved sequences found in members of the PKC family of protein kinases. Using model substrates and fungal proteins, we have not been able to demonstrate stimulation of phosphorylation by DAG or phorbol esters. A similar lack of "classic" PKC induction was also observed in *S. cerevisiae, Trichoderma reesei*, and *Metarhizium anisopliae* (67,76,93).

LAPK is a 72-kd protein. The deduced amino acid sequence of LAPK has a high degree of identity to protein kinase C-like genes from *T. reesei* (77) and *S. cerevisiae* (67) (72% and 45%, respectively). There is also a high degree of identity in the catalytic region of PKC with mammalian PKC genes (e.g., 42% with human PKC ϵ. However, as mentioned, we have not been able to induce enzyme activity of this protein when synthesized in *E. coli* or when using total fungal protein extracts, using known PKC activators such as DAG or phorbol esters. Sequence analysis is consistent with this observation, as the phorbol ester as well as the calcium-binding domains are not present. Based on these data (and those described later), we were reluctant to definitively classify this gene as an authentic PKC. Moreover, there have been recent reports of lipid-activated protein kinases, known as protein kinase N-type protein kinases, which are structurally distinct from the known PKCs, particularly in the regulatory domain (81).

Amino acid sequences of the catalytic domain in LAPK were aligned using the PILEUP command in GCG ver. 8.0 (Genetics Computer Group, Madison, WI). Alignment parameters included a gap initiation penalty of 1.0 and a gap extension penalty of 0.063. Phylogenetic analysis using parsimony as optimality criterion was carried out with PAUP 3.1.1 (96). Internal support for nodes was estimated by bootstrap analysis (100 replicates). Results reveal robust clades and strong support for *lapk* as being more closely related to (i.e., share a more recent common ancestor with) *pka* and *pkb* genes than to other *pkc* genes included in the analysis.

From Southern analysis under high and low stringencies, *lapk* is a single-copy gene in *C. trifolii* (not shown). Northern analysis indicates that the 2.4-kb mRNA for *lapk* is strongly expressed during morphogenesis, spore germination and appressoria initiation (data not shown). Mycelia and ungerminated spores exhibited barely detectable levels of mRNA. We expressed *lapk* as a fusion protein in *E. coli* and performed western blots with a panel of PKC antibodies. Only the PKC ϵ isoform strongly hybridized to LAPK. The PKC ϵ antibody also reacted with LAPK in total fungal protein extracts. Moreover, the expression pattern for the PKC ϵ antibody-reactive protein from western blots was similar to that of the *lapk* northern blots. Analyzing the sequence of the peptide from which the PKC ϵ antibody was generated, 50% identity and 83% similarity (when using conservative amino acid substitutions) to the *C. trifolii* PKC1 sequence were observed. However, the location of this sequence in *lapk* is in the C terminal end of the fungal protein, which is unusual since the peptide was made from the unique regulatory (N terminal) portion of the mammalian protein. This observation is consistent with the phylogenetic data.

To establish that LAPK is a functional protein kinase, we examined whether LAPK is capable of autophosphorylation, a characteristic of protein kinases. LAPK was expressed in *E. coli* and incubated in a kinase assay mixture. LAPK protein was immunoprecipitated with the PKC ϵ antibody. Autoradiography clearly showed that LAPK was labeled and, therefore, capable of autophosporylation.

Inhibitor studies using staurosporine-treated fungal spores showed that both conidial germination and appressoria differentiation were inhibited (112,113). Staurosporine completely inhibited appressoria formation at 100 nM, while at 1 μM, staurosporine completely prevented conidial germination. Go976, a more specific PKC inhibitor, only affected these processes at 10-fold higher concentrations. This is not altogether surprising as this inhibitor reacts primarily with the classical PKC regulatory domain which is not present in LAPK. It should be noted that, while often considered a PKC-specific inhibitor, staurosporine has reasonably strong affinities to other protein kinases (e.g., Ki PKC, 0.0007 μM; Ki PKA, 0.007μM). Thus, these studies do not confirm LAPK is in the PKC family, but do indicate that this protein is involved in morphogenesis. Following staurosporine treatment, LAPK was greatly reduced in western blots with the PKC ϵ antibody (data not shown).

Thus, these data illustrate the problem of classifying genes based on primary sequence data. In addition, even though a mammalian PKC ϵ antibody hybridized with and immunoprecipitated LAPK, based on the location of the relevant LAPK sequence (3' end), it is highly unlikely that this has any biological significance and is most likely a false positive. Despite the inability to accurately classify this gene, the inhibitor data, though far from conclusive, suggests an involvement for LAPK during appressoria formation; thus we are continuing to characterize this gene.

In a study designed to establish conditions for synchronizing large-scale *C. trifolii* populations for developmental studies (17), we found that cutin influenced spore germination. Interestingly, addition of cutin to an aqueous spore suspension, prior to placing on a contact surface (which is required for appressoria differentiation), significantly enhanced LAPK expression. Northern analysis of spores in various nutritional regimes confirmed that LAPK is rapidly induced (30 min) by plant cutin. Thus, plant cutin specifically activates *lapk* gene expression. The fact that transcription of *lapk* is rapidly and specifically induced by plant cutin, coupled with the fact that chemical inhibition of this protein prevents appressoria differentiation, encouraged further analysis. We are currently in the process of identifying the cutin constituent(s) responsible for *lapk* activation.

Conclusions

In fungi as well as other eukaryotes, cell growth and differentiation is achieved through the activation of a series of intracellular signal transduction pathways in response to external stimuli. Studies in host-pathogen interactions have led to a growing body of evidence that signals are released from both the plant and pathogen which are crucial in determining the eventual outcome of a parasitic relationship. These signal molecules are recognized and utilized by each participant to influence the compatible/incompatible relationship. Moreover, such signal elements are common in species ranging from mammals to yeast and bacteria, as well as phytopathogenic fungi. The successful, functional interchange of kinase genes among organisms (e.g., 16,58,59,83) illustrates the conservation of these genes. An increasing number of reports on the involvement of ser/thr protein kinases and phosphatases in fungal morphogenesis and pathogenicity further substantiates this situation (e.g., 27). A basic premise behind the work described in this paper is the importance and universal nature of these components and that inappropriate regulation of signal transduction pathways in fungi can have pronounced effects on infection-related morphogenesis and pathogenicity.

Aknowledgements

I thank the members of the Dickman lab including Gina Truesdell, Zhonghui Yang, Jeff Rollins, and Yange Zhang for stimulating discussion. I also thank Jeff Rollins for a review of this manuscript. This work has been supported by BARD, the U.S.-Israel Binational Agricultural Research and Development Fund and the DOE/NSF/USDA Program of Collaborative Research in Plant Biology.

Literature Cited

1. Alfano, J.R. and Collmer,A. 1996. Bacterial pathogens in plants: Life up against the wall. Plant Cell 8:1683-1698.
2. Azzi, A., Boscoboinik, D., and Hensey, D. 1992. The protein kinase C family. Eur. J. Biochem. 208:547-557.
3. Baker, B., Zambryski, P., Staskawicz, B., and Dinesh-Kumer, S.P. 1997. Signaling in plant-microbe interactions. Science 276:726-733.
4. Bailey, J.A., O'Connell, R.J., Pring, R.J., and Nash, C. 1992. Infection of *Colletotrichum* species. Pages 88-120 in: *Colletotrichum*: Biology, Pathology and Control. J.A. Bailey and M.J. Jeger, eds. CAB International, Wallingford, UK.
5. Baroni, M.D., Monti, P., and Alberghina, L. 1994. Repression of growth-regulated G1 cyclin expression by cyclic AMP in budding yeast. Nature 371:339-342.
6. Bar-Sagi, D. and Feramisco, J. R. 1985. Microinjection of the ras concogene protein into PC12 cells induces morphological differentiation. Cell 42:841-848.
7. Beckermann, J.L., Naider, F., and Ebbole, D.J. 1997. Inhibition of pathogenicity of the rice blast fungus by *Saccharomyces cerevisiae* alpha-factor. Science 276:1116-1119.

8. Berridge, M.J. and Irvine, R.F. 1989. Inositol phosphates and cell signaling. Nature 341:197-204.
9. Boguski, M.S. and McCormick, F. 1993. Proteins regulating Ras and its relatives. Nature 366:643-663.
10. Bourne, H.R., Sanders, D.A., and McCormick, F. 1990. The GTPase superfamily: A conserved switch for diverse cell functions. Nature 348:125-132.
11. Breiteneder, H., Ferreira, F., Reikerstorfer, A., Duchene, M., Valenta, R., Hoffmann-Sommergruber, K., Ebner, C., Breitenbach, M., Kraft, D., Scheiner, O. 1992. Complementary DNA cloning and expression in *Escherichia coli* of *Aln g I*, the major allergen in pollen of alder (*Alnus glutinosa*). J. Allergy Clin. Immunol. 90:909-917.
12. Breiteneder, H., Ferreira, F., Hoffmann-Sommergruber, K., Ebner, C., Breitenbach, M., Rumpold, H., Kraft, D., Scheiner, O. 1993. Four recombinant isoforms of *Cor a I*, the major allergen of hazel pollen, show different IgE-binding properties. Eur. J. Biochem. 212:355-362.
13. Breiteneder, H., Pettenburger, K., Bito, A., Valenta, R., Kraft, D., Rumpold, H., Scheiner, O., and Breitenbach, M. 1989. The gene coding for the major birch pollen allergen *BetvI*, is highly homologous to a pea disease resistance response gene. EMBO J. 8:1935-1938.
14. Bruno, K.S., Aramayo, R., Minke, P.F., Metzenberg, R.L., and Plamann, M. 1996. Loss of growth polarity and mislocalization of septa in a *Neurospora* mutant altered in the regulatory subunit of cAMP-dependent protein kinase. EMBO J. 15:5772-5782.
15. Buhr, T.L. and Dickman, M.B. 1997. Gene expression analysis during conidial germ tube and appressorium formation in *Colletotrichum trifolii*. Appl. Env. Microbiol. 63:2378-2383.
16. Buhr, T.L., Oved, S., Truesdell, G.M., Huang, C., Yarden, O., and Dickman, M.B. 1996. A kinase-encoding gene from *Colletotrichum trifolii* complements a colonial growth mutant of *Neurospora crassa*. Mol. Gen. Genet. 251:565-572.
17. Casey, P.J. 1994. Lipid modifications of G proteins. Curr. Opin. Cell Biol. 6:219-225.
18. Chiang, C.C. and Hadwiger, L.A. 1990. Cloning and characterization of a disease resistance response gene in pea inducible by *Fusarium solani*. Mol. Plant-Microbe Interact. 3:78-85.
19. Clark, G.J., Cox, A.D., Graham, S.M., and Der, C.J. 1995. *In vitro* and *in vivo* assays for Ras transformation. Meth. Enzymol. 255:395-412.
20. Clark, S. 1992. Protein isoprenylation and methylation at carboxyl-terminal cysteine residues. Annu. Rev. Biochem. 61:335-386.
21. Crowell, D.N., John, M.E., Russell, D., and Amasino, R.M. 1992. Characterization of a stress-induced, developmentally regulated gene family from soybean. Plant Mol. Biol. 18:459-466.

22. DeJong, J.C., McCormack, B., Smirnoff, N., and Talbot, N.J. 1997. Glycerol generates turgor in rice blast. Nature 389:244-245.
23. Der, C.J., Finkel, T., and Cooper, G.M. 1986. Biological and biochemical properties of human *ras* genes mutated at codon 61. Cell 44:167-176.
24. Der, C.J., Krontiris, T.G., and Cooper, G.M. 1982. Transforming genes of human bladder and and carcinoma cell lines are homologous to the *ras* genes fo Harvey and Kirsten sarcoma viruses. Proc. Natl. Acad. Sci. USA 79:3637-3640.
25. Devine, T.E., Hanson, C.H., Ostazeski, S.A., and Campbell, T.A. 1971. Selection for resistance to anthracnose (*Colletotrichum trifolii*) in four alfalfa populations. Crop Sci. 11:854-855.
26. Dickman, M.B., Buhr, T.L., Truesdell, G.M., Warwar, V., and Huang, C. 1995. Molecular signals during the early stages of alfalfa anthracnose. Can. J. Bot. 73(S):1169-1177.
27. Dickman, M.B. and Yarden, O. 1999. Serine/threonine protein kinase and phosphates in filamentous fungi. Fungal Genetics and Biology 26:99-117.
28. Elgin, J.H. and Ostazeski, S.A. 1982. Evaluation of selected alfalfa cultivars and related *Medicago* species to race 1 and race 1 anthracnose. Crop Sci. 22:39-42.
29. Ellis, R.W., DeFoe, D., Shih, T.Y., Gonda, M.A., Young, H.A., Tsuchida, N., Lowy, D.R., and Scolnick, E.M. 1981. P21 *src* genes of Harvey and Kirsten sarcoma viruses originate from divergent members of a family of normal vertebrate genes. Nature 292:506-511.
30. Emili, A., Greenblatt, J., and Ingles, C.J. 1994. Species-specific interaction of the glutamine-rich domain of Sp1 with the TATA-box binding protein. Mol. Cell Biol. 14:1582-1593.
31. Esnault, R., Buffard, D., Breda, C., Sallaud, C., El Turk, J., and Kondorosi, A. 1993. Pathological and molecular characterizations of alfalfa interactions with compatible and imcompatable bacteria, *Xanthomonas campestris* pv. *alfalfae* and *Pseudomonas syringae* pv. *pisi*. Mol. Plant-Microbe Interact. 6:655-664.
32. Freidman, A.H., Balaben, N., and Fuks, L. 1994. Protein kinase C mediates basic fibroblast growth factor protection of endothelial cells against radiation-induced apoptosis. Cancer Res. 54:2591-2597.
33. Fristensky, B., Riggleman, R.C., Wagoner, W., and Hadwiger, L.A. 1985. Gene expression in susceptible and disease resistant interactions of peas induced with *Fusarium solani* pathogens and chitosan. Physiol. Plant Pathol. 27:15-28.
34. Fristensky, B., Horovitz, D., and Hadwiger, L.A. 1988. cDNA sequences for pea disease resistance response genes. Plant Mol. Biol. 11:713-715.

35. Gao, S. and Nuss, N.L. 1996. Distinct roles for two G protein α subunits in fungal virulence morphology, and reproduction revealed by targeted gene disruption. Proc. Natl. Acad. Sci. USA 93:12122-12127.
36. Gerber, H.P., Seipel, K., Georgiev, O., Hofferer, M,. Hug, M., Rusconi, S., and Schaffner, W. 1994. Transcriptional activation modulated by homopolymeric glutamine and proline stretches. Science 263:808-811.
37. Gibbs, J.B. and Marshall, M.S. 1989. The ras-oncogene: An important regulatory element in lower eukaryotic organisms. Microbiol. Rev 53:171-185.
38. Gilman, A.G. 1987. G-proteins: Transducers of receptor generated signals. Annu. Rev. Biochem. 56:615-649.
39. Glass, D.D.B. and Krebs, E.G. 1980. Protein phosphorylation catalyzed by cyclic AMP-dependent and cyclic GMP-dependent protein kinases. Ann. Rev. Pharmacol. Toxicol. 20:363-388.
40. Gold, S., Duncan, G., Barrett, K., and Kronstad, J. 1994. cAMP regulates morphogenesis in the fungal pathogen *Ustilago maydis*. Genes Dev. 8:2805-2816.
41. Gold, S.E., Brogdon, S.M., Mayorga, M.E., and Kronstad, J.W. 1997. The *Ustilago maydis* regulatory subunint of a cAMP-dependent protein kinase is required for gall formation in maize. Plant Cell 9:1581-1594.
42. Haag, E. and Raman, V. 1994. Effects of primer choice and source of *Taq* DNA polymerase on the banding patterns of differential display RT-PCR. Biotechniques 17:226-228.
43. Hanks, S.K. and Quinn, A.M. 1991. Protein kinase catalytic domain sequence database: Identification of conserved features of primary structure and classification of family members. Methods Enzymol. 200:38-62.
44. Hardie, G. and Hanks, S. 1995. The Protein Kinase Facts Book. Academic Press, London.
45. Hicks, T.K., Yu, J.H., Kelly, N.P., and Adams, T.H. 1997. *Aspergillus* sporulation and mycotoxin production both require inactivation of the FadA Gα protein-dependent signaling pathway. EMBO J. 16:4916-4923.
46. Hirama, T. and Koeffler, H. P. 1995. Role of the cyclin-dependent kinase inhibitors in the development of cancer. Blood 86:841-854.
47. Hirt, H. 1997. Multiple roles of MAP kinases in plant signal transductions. Trends Plant Sci. 2:11-15.
48. Hollstein, M., Shomer, B., Greenblatt, M., Soussi, T., Hovig, E., Montesano, R., and Harris, C. C. 1996. Somatic point mutations in the p53 gene of human tumors and cell lines: Update compilation. Nucleic Acid Res. 24:141-146.
49. Howard, R.J., Ferrari, M.A., Roach, D.H., and Money, N.P. 1991. Penetration of hard substrates by a fungus employing enormous turgor pressures. Proc. Natl. Acad. Sci. USA 88:11281-11284.

50. Hu, Y.F., Lüscher, B., Admon, A., Mermond, N., and Tjian, R. 1990. Transcription factor AR4 contains domains that regulate diner specificity. Genes Dev. 4:1741-1752.
51. Hunter, T. 1995. Protein kinases and phosphates: The yin and yang of protein phosphorylation and signaling. Cell 80:225-236.
52. Iturriaga, E.A., Leech, M.J., Barratt, D.H.P., and Wang, T.L. 1994. Two ABA-responsive proteins from pea (*Pisum sativum* L.) are closely related to intracellular pathogenesis-related proteins. Plant Mol. Biol. 24:235-240.
53. Jarvis, W.D., Turner, A.J., Povirk, L.F., Traylor, R.S., and Grant, S. 1994. Induction of apoptotic DNA fragmentation and cell death in HL-90 human promyelocytic leukemia cells by pharmacological inhibitor of protein kinase C. Cancer Res. 54:1707-1714.
54. Jin, M., Fujita, M., Culley, B.M., Apolinario, E., Yamamoto, M., Maundrell, K., and Huffman, C.S. 1995. *Sckl*, a high copy number suppressor of defects in the cAMP-dependent protein b kinase pathway in fission yeast, encodes a protein homologous to the *Saccharomyces cerevisiae* SCH9 kinase. Genetics 140:457-467.
55. Johnson, L.N., Noble, M.E.M., and Owen, D.J. 1996. Active and inactive protein kinases: Structural basis for regulation. Cell 85:149-158.
56. Kataoka, T., Powers, S., McGill, C., Fasano, O., Strathern, J., Broach, J., and Wigler, M. 1984. Genetic analysis of yeast ras1 and ras2 genes. Cell 37:437-445.
57. Kincaid, R. 1993. Calmodulin-dependent protein phosphatases from microoganisms to man: A study in structural conservatism and biological diversity. Adv. Second Messenger Phosphoprotein Res. 27:1-23.
58. Kincaid, R.L. 1991. Signaling mechanisms in microorganisms: Common themes in the evolution of signal transduction pathways. Adv. Second Messenger Phosphoprotein Res. 23:165-183.
59. King, K., Dohlmen, H.G., Thorner, J., Caron, M.G., and Lefkowitz, R.J. 1990. Control of yeast mating signal transduction by a mammalian B_2-andronergic receptor and $G_s\alpha$ subunit. Science 250:121-123.
60. Knighton, D.R., Zheng, J., Ten Eyck, L.F., Ashford, V.A., Xuong, N-H., Taylor, S.S., and Sowadski, T.M. 1991. Crystal stucture of the catalytic subunit of cyclic adenosine monophosphate-dependent protein kinase. Science 253:407-414.
61. Knighton, D.R., Zheng, J., Ten Eyck, L.F., Ashford, V.A., Xuong, N-H., Taylor, S.S., and Sowadski, T.M. 1991. Crystal structure of the catalytic subunit of cyclic adenosine monophosphate-dependent protein kinase. Science 253-414-420.
62. Kronstad, J.W. 1997. Virulence and cAMP in smuts, blasts and blights. Trends Plant Sci. 2:193-199.

63. Landschulz, W.H., Johnson, P.F., and McKnight, S.L. 1988. The leucine zipper: A hypothetical structure common to a new class of DNA binding proteins. Science 240:1759-1764.
64. Larsen, J.N., Stroman, P., and Ipsen, H. 1992. PCR based cloning and sequencing of isogenes encoding the tree pollen major allergen *Car b I* from *Carpinus betulus*, hornbeam. Mol. Immunol. 29:703-711.
65. Lee, N.N. and Adams, T.H. 1996. FluG and flb A function interdependently to initiate conidiophore development in *Aspergillus nidulans* through brlaß activation. EMBO J. 15: 229-309.
66. Lee, Y.H. and Dean, R.A. 1993. cAMP regulates infection structure formation in the plant pathogenic fungus *Magnaporthe grisea.* Plant Cell 5:693-700.
67. Levin, D.E., Fields, F.O., Kunisawa, R., Bishop, J.M., and Thurner, J. 1990. A candidate protein kinase C gene, PKC1, is required for the *S. cerevisiae* cell cycle. Cell 62:212-224.
68. Liang, P. and Pardee, A.B. 1992. Differential display of eukaryotic messenger RNA by means of the polymerase chain reaction. Science 257:967-971.
69. Liang, P., Averboukh, L., and Pardee, A.B. 1993. Distribution and cloning of eukaryotic mRNAs by means of differential display: Refinements and optimization. Nucleic Acids Res. 21:3269-3275.
70. Ligterink, W., Kroj, T., Nieden, U., Hirt, H., and Scheel, D. 1997. Receptor mediated activation of a MAP kinase in pathogen defense of plants. Science 276:2054-2057.
71. Lowy, D. R. and Willumsen, B. M. 1993. Function and regulation of RAS. Annu. Rev. Biochem. 62:851-891.
72. Maeda, T., Watanabe, Y., Kunitomo, H., and Yamamoto, M. 1994. Cloning of the *pkal* gene encoding the catalytic subunit of the cAMP-dependent protein kinase in *Schizosaccharomyces pombe.* J. Biol. Chem. 269:9632-9637.
73. Martin, G.B., Brommonschenkel, S.H., Chunwongse, J., Frary, A., Ganal, M.W., Spirey, R., Wu, T., Earle, E.D., and Tanksley, S.D. 1993. Map-based cloning of a protein kinase gene conferring disease resistance in tomato. Science 262:1432-1436.
74. Matton, D.P. and Brisson, N. 1989. Cloning, expression, and sequence conservation of pathogenesis-related gene transcripts of potato. Mol. Plant-Microbe Interact. 2:325-331.
75. Mitchell, T.K. and Dean, R.A. 1995. The cAMP-dependent protein kinase catalytic subunit is required for appressorium formation and pathogenesis by the rice blast pathogen *Magnaporthe grisea.* Plant Cell 7:1869-1878.
76. Morawetz, R., Lendenfeld, T., Mischak, H., Goodnight, J., Mushinski, J.F., and Kubicek, C.P. 1994. A protein kinase-encoding gene, *pkt1*,

from *Trichoderma reesei* and *Aspergillus niger*. Mol. Gen. Genet. 250:17-28.

77. Morawetz, R., Lendenfeld, T., Mischak, H., Muhlbauer, M., Graber, F., Goodnight, J., deGraaf, L.H., Visser, J., Mushinski, J.F., and Kubicek, C.P. 1996. Cloning and characterization of genes (*pkc1* and *pkcA*) encoding protein kinase C homologues from *Trichoderma reesei* and *Aspergillus niger*. Mol. Gen. Genet. 250:17-28.
78. Mosch, H-V., Roberts, R.L., and Fink, G.R. 1996. Ras 2 signals via the Cdc 42/Ste 20/mitogen-activated protein kinase module to induce filamentous growth in *Saccharomyces cervisiae.* Proc. Natl. Acad. Sci. 93:5352-5356.
79. Mould, M.J., Boland, G.J., and Robb, J. 1991. Ultrastructure of the *Colletotrichum trifolii-Medicago sativa* pathosystem I. Pre-penetration events. Physiol. Mol. Plant Pathol. 38:179-194.
80. Mould, M. J., Boland, G. J., and Robb, J. 1991. Ultrastructure of the *Colletotrichum trifolii-Medicago sativa* pathosystem II. Post-penetration events. Physiol. Mol. Plant Pathol. 38:195-210.
81. Mukai, H., Miyahara, M., Surakawa, H., Shibata, H., Toshimori, M., Kitagaug, M., Shimakawa, M., Takanaga, H., and Ono, Y. 1996. Translocation of PKN from the cytosol to the nucleus induced by stresses. Proc. Natl. Acad. Sci. 93:10195-10199.
82. Mylona, P., Moerman, M., Yang, W., Gloudemans, T., van de Kerckhove, J., van Kammen, A., Bisseling, T., and Franssen, H.J. 1994. The root epidermis-specific pea gene RH2 is homologous to a pathogenesis-related gene. Plant Mol. Biol. 26:39-50.
83. Nieman, A.M. 1993. Conservation and reiteration of a kinase cascade. Trends Genet. 9:390-394.
84. Nishizuka, Y. 1988. The molecular heterogeneity of protein kinase C and implications of cellular regulation. Nature 334:661-665.
85. O'Neill, N.R., Elgin, J.H., and Baker, C.J. 1989. Characterization of induced resistance to anthracnose in alfalfa by races, isolates, and species of *Colletotrichum*. Phytopathology 79:750-756.
86. Oliveira, J.C.F., Borges, A.C.C., Marques, M.V., and Gomes, S.L. 1994. Cloning and characterization of the gene for the catalytic subunit of cAMP-dependent protein kinase in the aquatic fungus *Blastocladeilla emersonii*. Eur. J. Biochem. 219:555-562.
87. Parada, L.F., Tabin, C.J., Shih, C.O., and Weinberg, R.A. 1982. Human EJ bladder carcinoma oncogene is homologue of Harvey sarcoma virus *ras* gene. Nature 297:474-478.
88. Rahmsdorf, H.J. and Herrlich, P. 1990. Regulation of gene expression by tumor promoters. Pharmacol. Ther. 48:157-188.
89. Roberts, I.N., Oliver, R.P., Punt, P.J., and van der Hondel, C.A.M.J.J. 1989. Expression of the *E. coli* ß-glucuronidase gene in industrial and phytopathogenic filamentous fungi. Curr. Genet. 15:177-180.

90. Scheffzek, K., Ahmadian, M.R., Kabsch, W., Wiesmuller, L., Lautwein, A., Schmitz, F., and Wittinghoffer, A. 1997. The Ras-Ras GAP complex: Structural basis for GTPase activation and its loss in oncogenic Ras mutants. Science 277:333-338.
91. Seeburg, P.H., Colby, W.W., Capon, D.J., Goeddel, D.V., and Levinson, A.D. 1984. Biological properties of human c-Ha-ras1 genes mutated at codon 12. Nature 312:71-75.
92. Shoji, S., Ericsson, S., Walsh, K.A., Fisher, E.H., and Tihani, K. 1983. Amino acid sequence of the catalytic subunit of bovine type II cyclic 3',5'-phosphate dependent protein kinase. Biochemistry 22:3702-3709.
93. St. Leger, R.J., Lacetti, L.B., Staples, R.C., and Roberts, D.W. 1990. Protein kinases in the entomopathogenic fungus *Metarhizium anisopliae.* J. Gen. Microbiol. 136:1401-1411.
94. Sugiura, H. and Yamauchi, T. 1994. Changes of the expression of protein substrates of Ca^{2+}/calmodulin-dependent protein kinase II in neonate and adult rats. FEBS Lett. 341:299-302.
95. Swoboda, I., Jilek, A., Ferreira, F., Engel, E., Hoffmann-Sommergruber, K., Scheiner, O., Kraft, D., Breitenededer, H., Pittenauer, E., Schmid, E., Vicente, O., Heberle-Bors, E., Ahorn, H., and Breitenbach, M. 1995. Isoforms of *Bet v 1*, the major birch pollen allergen, analyzed by liquid chromatography, mass spectormetry, and cDNA cloning. J. Biol. Chem. 270:2607-2613.
96. Swofford, D.L. 1993. Phylogenetic analysis using parsimony (PAUP), version 3.1.1. University of Illinois, Champaign.
97. Talbot, N.J., Ebbole, D.J., and Hamer, J.E. 1993. Identification and characterization of *MPG1*, a gene involved in pathogenicity from the rice blast fungus *Magnaporthe grisea.* Plant Cell 5:1575-1590.
98. Tanaka, M., Clouston, W.M., and Herr, W. 1994. The oct-2 glutamine-rich and proline-rich activation domains can synergize with each other or duplicate of themselves to activate transcription. Mol. Cell Biol. 14:6046-6055.
99. Taylor, S.S., Buechler, J.A., and Yonemoto, W. 1990. cAMP-dependent protein kinase: Framework for a diverse family of regulatory enzymes. Annu. Rev. Biochem. 59:971-1005.
100. Terenzi, H.F., Flawia, M.M., Tellez-Inon, M.T., and Torres, H.N. 1976. Control of *Neurospora crassa* morphology by cyclic adenisine 3': 5'-monophosphate and dibutyryl cyclic adenosine 3': 5'-monophosphate. J. Bacteriol. 126:91-99.
101. Toda, T., Cameron, S., Sass, P., Zoller, M., and Wigler, M. 1987. Three different genes in *S. cerevisiae* encode the catalytic subunits of the cAMP-dependent protein kinase. Cell 50:277-287.
102. Tokiwa, G., Tyers, M., Vilpe, T., and Futcher, B. 1994. Inhibition of G1 cyclin activity by the Ras/cAMP pathway in yeast. Nature 371:342-245.

103. Tom, T. and Kolarparthi, V.S.R. 1994. Developmental decisions in *Aspergillus nidulans* are modulated by RAS activity. Mol. Cell. Biol. 14:5333-5348.
104. Truesdell, G.M. and Dickman, M.B. 1997. Isolation of pathogen/stress-inducible cDNAs from alfalfa by mRNA differential display. Plant Molec. Biol. 33:737-743.
105. Walsh, D.A., Perkins, J.P., and Krebs, E.G. 1968. An adenosine 3',5'-monophosphate-dependent protein kinase from rabbit skeletal muscle. J. Biol. Chem. 243:3763-3774.
106. Walter, M.H., Liu, J., Grand, C., Lamb, C.J., and Hess, D. 1990. Bean pathogenesis-related (PR) proteins deduced from elicitor-induced transcripts are members of a ubiquitous new class of conserved PR proteins including pollen allergens. Mol. Gen. Genet. 222:353-360.
107. Warner, S.A.J., Scott, R., and Draper, J. 1992. Characterization of a wound-induced transcript from the monocot asparagus that shares similarity with a class of intracellular pathogenesis-related (PR) proteins. Plant Mol. Biol. 19:555-561.
108. Warwar, V. and Dickman, M. B. 1996. Calcium-calmodulin mediated appressorium development in *Colletotrichum trifolii*. Appl. Env. Microbiol. 62:75-79.
109. Xu, J.R., Urban, M., Sweigard, J.A., and Hamer, J.E. 1997. The cPKA gene of *Magnaporthe grisea* is essential for appressorial penetration. Mol. Plant Microbe Inter. 10:187-197.
110. Xu, J.R. and Hamer, J.E. 1996. MAP kinase and cAMP signaling regulate infection structure formation and pathogenic growth in the rice blast fungus *Magnaporthe grisea*. Genes Dev. 10:2696-2706.
111. Yang, Z. and Dickman, M.B. 1999. *Colletotrichum trifolii* mutants disrupted in the catalytic usbunit of cAMP-dependent protein kinase are non-pathogenic. Mol. Plant Microbe Inter. 12:430-439.
112. Yang, Z. and Dickman, M.B. 1999. Molecular mapping and characterization of Ct-PKAR, a gene encoding the regulatory subunit of cAMP-dependent protein kinase in *Colletotrichum trifolii*. Arch. Microbiol. 171:249-256.
113. Yang, Z. and Dickman, M.B. 1997. Regulation of cAMP and cAMP dependent protein kinase during conidial germination and appressorium formation in *Colletotrichum trifolii*. Physiol. Mol. Plant Pathol. 50:117-127.
114. Yu, J-H., Wieser, J., and Adams, T.H. 1996. The *Aspergillus* FIBA RGS domain protein antagonizes G protein signaling to block proliferation and allow development. EMBO J. 15:5184-5194.

Chapter 14

Resistance Mechanisms of Subtropical Fruits to *Colletotrichum gloeosporioides*

Dov Prusky, Ilana Kobiler, Ruth Ardi, Dalila Beno-Moalem, Nir Yakoby, and Noel T. Keen

The Basis for Fruit Resistance to *Colletotrichum* Attack

Colletotrichum attack of fruits and vegetables can be separated into stages, including 1) the landing of spores on plant surfaces, 2) attachment of spores to those surfaces, 3) germination of spores, 4) production of appressoria, 5) penetration into the plant, 6) colonization of plant tissues, and 7i) formation of lesions followed by sporulation. Resistance has been defined as an incompatible interaction between host and pathogen. In contrast, a compatible interaction leads to disease and is the outcome, in most cases, of the ability of the pathogen to overcome various host defenses. Incompatibility, involving processes in the plant that prevent or retard pathogen growth, may be conditioned by a single gene pair-a host resistance gene and a pathogen avirulence gene (5). Unfortunately, these are qualitative terms (the disease does or does not develop) that are not generally applicable to the entire postharvest life of the fruit or vegetable. Although "gene-for-gene" interactions are extremely important and specific, they are not usually involved in the resistance of fruits and vegetables to postharvest diseases caused by *Colletotrichum.* Instead, resistance to postharvest pathogens is usually the result of several genes interacting in a way that is not well understood. Most resistance in fruit and vegetables to postharvest pathogens can be described as a "dynamic incompatibility". The response of the host's resistance genes to products of a pathogen's avirulence genes prevents or retards pathogen growth under specific host physiological conditions. The physiological state of the host changes, however, as it matures, ripens, and senesces. Storage, mechanical injury, temperature extremes, and anoxia also alter host physiology. When physiological changes in the host inhibit defense responses to pathogen activities, the interaction becomes compatible.

Pathogens such as *Colletotrichum,* which attack a broad range of hosts, usually remain quiescent during fruit development. As fruits ripen, inhibitors responsible for the quiescent state of the pathogen disappear, which allows the pathogen to resume growth. For this type of interaction, quantitative rather than qualitative reactions are the general rule in the postharvest host-pathogen interaction. Host barriers can occur at different levels of fungal pathogenicity. Dynamic incompatibility could start at the stage of inhibition of fungal penetration, inhibition of appressoria development, and inhibition of fungal colonization. The following questions apply to quantitative interactions of any stage of inhibition: 1) What are the conditions needed for triggering defense responses in fruits at different postharvest physiological stages? 2) What is the function of preformed or inducible barriers to pathogen attack? In the present chapter we will discuss mainly those interactions between the host and *Colletotrichum* that affect fungal colonization.

1. Constitutive Mechanisms of Resistance

PRESENCE OF PREFORMED BIOCIDES

Plants produce a diverse array of secondary metabolites that are toxic to fungi and can affect fungal colonization (9,21,22,25). Restriction of fungal development might be associated with host barriers and with the inhibition of fungal development by preformed compounds in the host. Constitutive or preformed resistance based on preformed compounds includes the constitutive accumulation of low molecular-weight secondary metabolites. This type of inhibition of fungal development, does not involve temporally differential gene activation, but involves the presence of biologically active compounds in healthy plants whose activity affords protection during most of the lifetime of the plant (6,24). In avocado fruits five antifungal compounds have been reported (1,16) that include 1-acetoxy-2-hydroxy-4-oxo-heneicosa 12,15, diene; 1-acetoxy-2,4-dihdroxy-n-heptadeca-16-ene; 1,2,4-trihydroxhyhepta dec-16-yne; 1,2,4-trihydroxyheptadeca-16-ene; and 1-acetoxy-2,4, dihydroxyheptadec-16-yne. These preformed antifungal compounds may directly affect the pathogen itself and delay its development until their concentrations decline to subfungitoxic levels. These compounds inhibit the penetrating hyphae of differentiated appressoria that have directly breached the host cuticle and have already overcome the first set of barriers. Some of these compounds are in a biologically active form, whereas others are inactive precursors that become activated in response to pathogen attack (8). These preformed compounds or "phytoanticipins" differ from the inducible phytoalexins that are synthesized from remote precursors in response to pathogen attack.

Of the numerous attempts to associate natural variation in levels of preformed inhibitors in plants with resistance to particular pathogens, only a few critical tests have been described (11,12). The relationship of preformed compounds to fruit resistance is unclear because of a lack of knowledge about the distribution and concentration of inhibitors in infection courts, the sensitivity of pathogens to inhibitors in the plant, inhibitor concentrations, and host resistance.

Preformed inhibitors of pathogen development tend to be concentrated in the outer layers of plant organs, providing evidence for chemical barriers against pest attack (3). More often, however, preformed compounds are compartmentalized in vacuoles or organelles in healthy plants. Kobiler et al. (7) found that 85% of an antifungal diene in avocado mesocarp (flesh) was compartmentalized in specific oil cells, whereas the peel contained uniform concentrations. The antifungal activity in the mesocarp appeared to depend on the extent of fungal damage and the amount of chemical released. Biotrophs may prevent the release of preformed inhibitors by minimizing damage to the host, while necrotrophs like *Colletotrichum* may stimulate the release of substantial amounts of these compounds. The distribution of the diene is consistent with susceptibility of wounded tissue to fungal attack, whereas the intact peel remains resistant (7). Specific signals originating from the pathogen could enhance release of preformed compounds from storage to active sites. Possible signals will be described later on in the chapter. The amount of chemical available could depend on the fruit cultivar, plant age, and environmental conditions (4,11).

The loss of natural fungicides in fruits that accompanies ripening may be related to enzymatic changes. The antifungal diene in avocado fruits is a substrate for oxidation by a lipoxygenase that is activated during fruit ripening (10,18). Lipoxygenase activity in avocado fruits is affected by an endogenous inhibitor, epicatechin, present in the avocado peel (16,19) (Fig. 1). In green fruits this flavan-3-ol competitively inhibits lipoxygenase activity. As the concentration decreases during fruit ripening, lipoxygenase activity increases until the fruits become completely susceptible.

INDUCTION OF PREFORMED BIOCIDES

Although preformed chemicals have been considered noninducible (24), significant increases in the level of constitutive fungal inhibitors have been associated with tissue exposure to different biotic and abiotic elicitors (17).

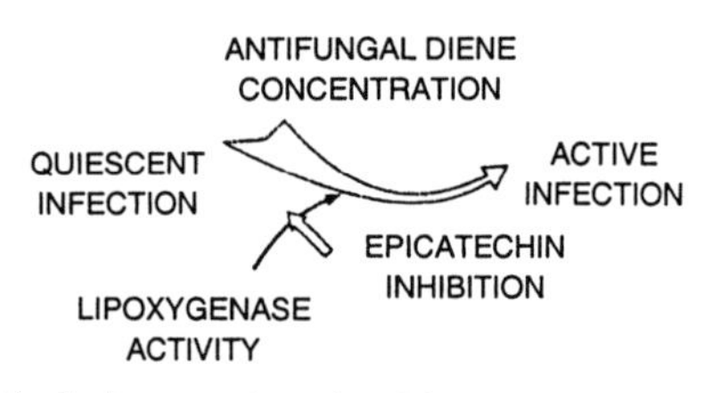

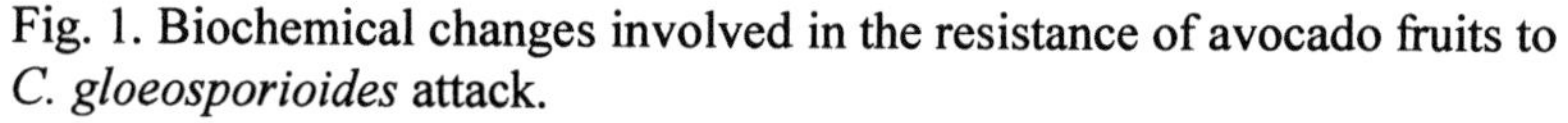
Fig. 1. Biochemical changes involved in the resistance of avocado fruits to *C. gloeosporioides* attack.

The diene can be modulated either by inducing its production and/or preventing its decrease. Most of our work has been within a few minutes in a concentration-dependent manner. Avocado cell suspension cultures-were incubated in the presence of several inhibitors to determine the signal transduction process involved in the release of H_2O_2. Incubation of cultured avocado cells in the presence of the NADPH-oxidase inhibitor diphenylene iodonium (DPI) reduced H_2O_2 release by 52% (Table 1). Inhibitors of protein kinase activity, such as sphingosine (10^{-6}M), K-252a (10^{-6}M), and staurosporine (10^{-5} M), inhibited H_2O_2 release by 70-94%, done to modulate its decrease. Challenge inoculation in unripe fruits can also induce a transient increase in the concentration of preformed antifungal compounds (15). Inoculating unripe avocado fruits with spores of the pathogenic fungus transiently induced increased levels of the antifungal diene and epicatechin (13) (Fig. 2). Inoculating unripe avocado fruits with spores of a mutant of a non-pathogenic fungus, *Colletotrichum magna,* induced a significant increase in levels of the antifungal diene. The diene remained at antifungal levels for longer than in the controls, and symptoms of disease were not observed (13). Abiotic treatments such as increased CO_2 levels also resulted in a longer induction of the diene and epicatechin (2,20). Exposure of freshly harvested fruit to a stream of 30% CO_2 for 24 hr, did not affect ripening but decay development was significantly delayed (Fig. 3). Levels of diene and epicatechin of CO_2-treated fruits were enhanced in a double-peak pattern (Fig 4). CO_2 also enhanced enzyme activity during the synthesis of epicatechin through the phenylpropanoid pathway. Activities of phenylalanine ammonia lyase (PAL), chalcone synthase and flavanon-3-hydroxylase significantly increased (Fig. 5). Transcription studies on the activation of these three enzymes suggest that elicitors affect the mechanism phosphatase inhibitor microcystin-LR (10^{-6}M) also inhibited H_2O_2 release by 75%.

Table 1. Effect of metabolic inhibitors on the elicitor-induced release of H_2O_2 from avocado cell suspension cultures.

Treatment	**H_2O_2** (Arbitrary units)[a]
Cells only	70
Elicitor (60 µg/equivalent Glu)	500
DPI (0.05 Mm) + Elicitor	276
Staurosporine (10^{-5} M) + Elicitor	153
Sphingosine (10^{-6} M) + Elicitor	134
K-252A (10^{-6} M) + Elicitor	75
Mycrocystin LR (10^{-6} M) + Elicitor	127

[a]Release of H_2O_2 from 10^6 cells within 10 min based on the oxidation of scopoletin (measured at 350/460 nm) in the presence of peroxidase

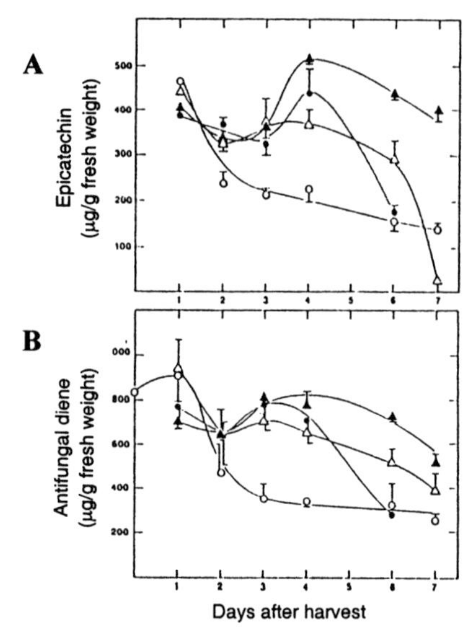

Fig. 2. Effect of inoculation with *Colletotrichum* on concentrations of epicatechin (A) and antifungal diene (B) in avocado cv. Fuerte peel. Fruit were dipped after harvesting in a conidial suspension of either *C. gloeosporioides* (●), a wild-type (Δ), or nonpathogenic mutant of *C. magna* (▲), or dipped in water (O).

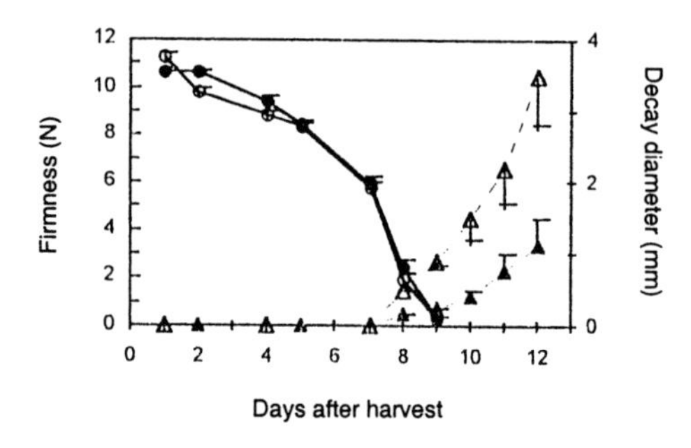

Fig. 3. Effect of CO_2 treatment on decay development and fruit ripening (firmness). Decay development by *C. gloeosporioides* (Δ▲) and the rate of ripening (O●) of cv. Fuerte avocado fruits. Freshly harvested fruits were exposed to 30% CO_2 for 24 hr (●▲) and compared to untreated fruits (OΔ). Fruits were inoculated 2 d after harvest.

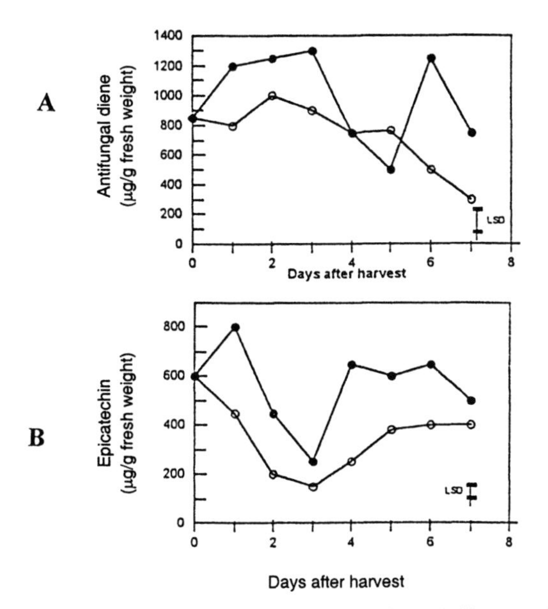

Fig. 4. Effect of CO_2 on levels of the antifungal diene and epicatechin. Levels of the antifungal diene (A) and epicatechin (B) in the peel of cv. Fuerte avocado fruits. Freshly harvested fruits were exposed to a stream of 30% CO_2 for 24 hr (●) and compared to freshly harvested fruits exposed to air (O).

To test the effect of H_2O_2 on elicitation of resistance, pericarp tissue was treated with 1 mM H_2O_2. O_2^- generation coupled with NADPH activity in of resistance by activating genes encoding the biosynthetic pathway at a transcriptional level (2).

Signal Transduction During Activation of Preformed Resistance in Avocado

One of the intriguing aspects of the elicitation of resistance is how an elicitor transfers signal(s) to activate the mechanism in fruits. Plasma membranes isolated from avocado fruit pericarp inoculated with *C. gloeosporioides* (Fig. 6) and from avocado cell suspension cultures (results not shown) incubated with cell wall elicitors from *C. gloeosporioides*, exhibited significantly increased activity of NADPH oxidase (Fig. 6). Furthermore, when cultured cell suspensions were treated with the cell wall elicitor from *C. gloeosporioides*, release of H_2O_2 was detected while the plasma membrane doubled within 30 min. PAL activity increased 1.5-fold

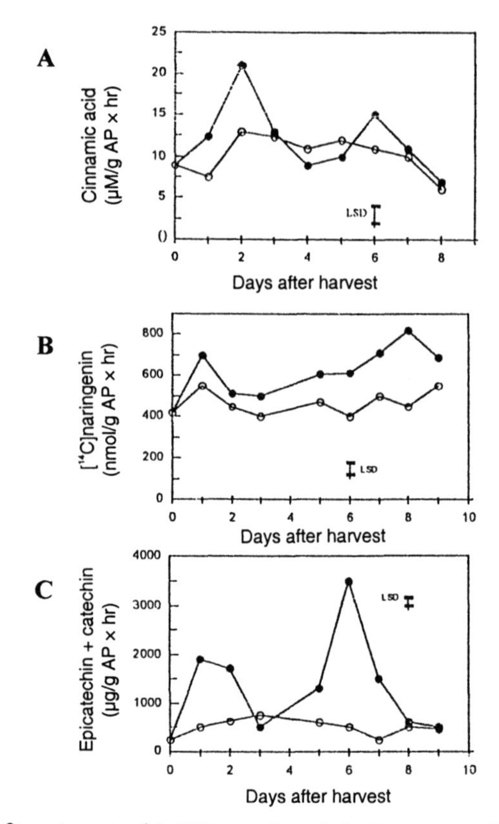

Fig. 5. Effect of treatment with CO_2 on phenylalanine ammonia lyase (PAL), chalcone synthase (CHS)and flavanon-3-hydroxylase (F3H). Activities of PAL (A), CHS (B), and F3H (C) in peel of cv. Fuerte avocado fruits. Freshly harvested fruits were exposed to a stream of 30% CO_2 for 24 hr (●) and compared to control fruits exposed to air (O).

within 60 min and 2.5-fold after 6 hr. A very similar pattern was observed for epicatechin which increased from an initial concentration of 400 μg/g fresh weight to 1200 μg/g fresh weight 6 hr after treatment. When avocado pericarp was treated with 10^{-6} M staurosporine, a protein kinase inhibitor, before exposure to 1 mM H_2O_2, PAL activity was inhibited and was similar to that in H_2O_2-untreated pericarp tissue. However, when the pericarp was exposed to H_2O_2 without a previous staurosporine treatment, a significant increase in PAL activity was detected 1-3 hr later.

This situation clearly suggests that signals are also transduced during the induction of epicatechin and indicates the importance of H_2O_2 during the initial signaling process affecting the induction of resistance. Interestingly, the O_2^- generation-coupled NADPH oxidase from the plasma membrane of microsomes extracted from pericarps of freshly harvested

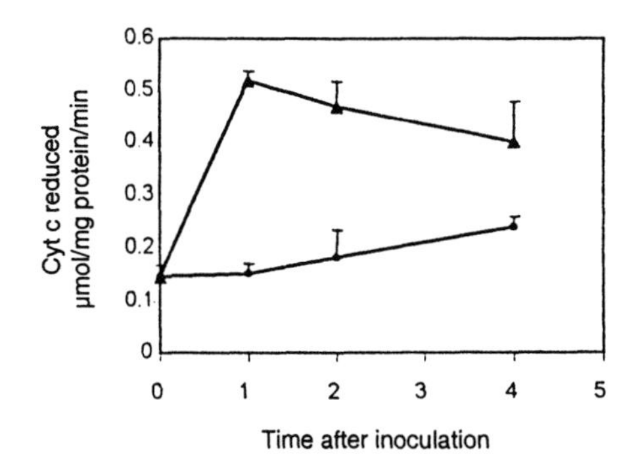

Fig. 6. Time course in NADPH-dependent cytochrome *c* reducing activity in plasma membrane isolated from pericarp tissue inoculated with *C. gloeosporioides*. Control (●), inoculated (▲).

unripe (resistant) fruits showed a higher activity (0.168 ± 0.026 µmol cyt C/mg protein/h) in NADPH oxidation, than in the same tissue extracted 10-12 days later from ripe (susceptible) fruits (0.072 ± 0.014 µmol cyt C/mg protein/h). Higher NADPH oxidase activity in unripe resistant fruits apparently is a key factor in the production of H_2O_2 and the induction of resistance during the quiescent stage of *Colletotrichum*.

2. Fungal Pathogenicity and Host Modulation

PREFORMED PHENOLS INHIBITING PATHOGENICITY FACTORS

Phenols have been reported as possible direct inhibitors of pathogenic fungi. Inactivation of pectic enzymes by inhibitors has also been hypothesized to be a mechanism of host resistance for modulating fungal pathogenicity (21). In avocado peel the epicatechin that inhibits the oxidation of antifungal compounds, as discussed previously, also inhibits pectolytic enzymes produced by *C. gloeosporioides*. Purified polygalacturonase and pectate lyase produced by *C. gloeosporioides* were inhibited *in vitro* by epicatechin (14,26,27). At 20 µg/ml, epicatechin inhibited the ability of these enzymes to macerate avocado wedges by 64%. Levels of epicatechin in unripe fruit, 350 µg/g fresh weight or ca. 270 µg/ml, greatly exceed minimum inhibitory concentrations, evidence that this flavan contributes significantly to the resistance of avocado fruits to *C. gloeosoporioides*.

Host pH Modulates the Secretion of Pectate Lyase

The possibility that host pH could affect the activation of pathogenicity factors during resistant periods was suggested recently (28). When *C. gloeosporioides* was grown in pectolytic enzyme-inducing media, the pH of the media increased from 3.5 to 6.5. Pectate lyase (PL) protein was detected on western blots when the pH of the media was over 5.8. Analysis of hyphal RNA at pH values lower than 5.8 showed increased expression of pel that did not correspond with the lack of secretion of the protein. Furthermore, PL was not secreted from mycelial mats grown in media from pH 3.5 to 6.5.

Freshly harvested fruits that cannot be attacked by *Colletotrichum* had a pH in the pericarp of 5.3. The pH values in ripe fruits increased to 6.1 when symptoms of decay appeared. *C. gloeosporioides* readily macerated (within 1 d) the mesocarp of peeled fruits of freshly harvested unripe avocado fruits having a pH of 6.2 (Fig. 7), before any physiological changes associated with ripening could be detected.
Interestingly, one avocado cultivar with resistance to *Colletotrichum* attack had an average pericarp pH of 5.4 during the entire ripening period (Fig. 8). These observations provide evidence that the pericarp pH might affect fungal secretion of pectolytic enzymes, consequently promoting quiescent infections in the peel.

The Relevance of Preformed and Inducible Barriers for Controlling Postharvest Disease

Manipulating concentrations of preformed and induced antifungal chemicals in harvested fruits and vegetables is a logical approach to

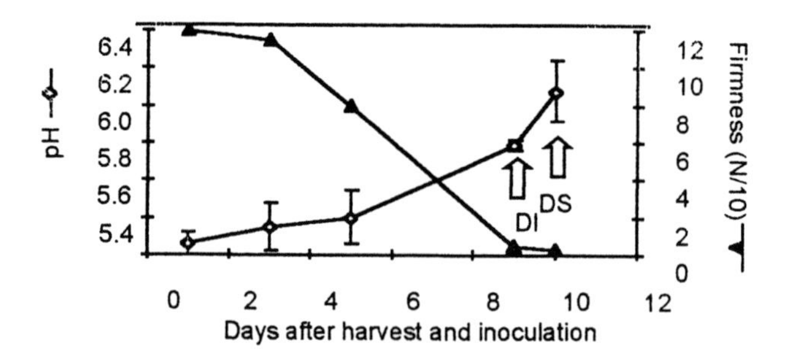

Fig. 7. Pericarp and mesocarp pH values, fruit firmness and *C. gloeosporioides* symptom development on freshly harvested avocado fruits cv. Fuerte. DI: Disease initiation, DS: Disease symptoms, N: Newtons.

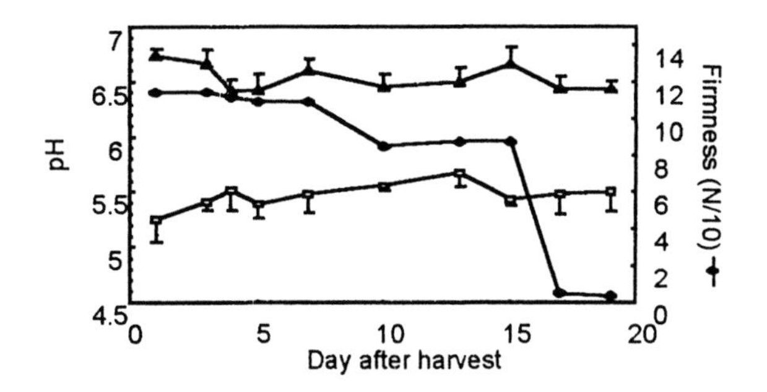

Fig. 8. Pericarp and mesocarp pH values and fruit firmness on freshly harvested avocado fruits cv. Ardit. Pericarp (□), mesocarp (▲) pH. N: Newtons.

developing new improved controls for postharvest diseases. Strategies might include preventing the loss of inhibitory compounds, enhancing the concentration or activity of inhibitory compounds, or selecting plant lines that are rich in such antifungal compounds (19). The assumptions underlying these strategies are that such compounds 1) are safe because they are natural, 2) are effective since they have evolved in nature specifically for protecting the plant against pests, and 3) have enabled existing plants to survive the selection pressure of evolution. However, naturally occurring compounds are not necessarily safe (8). Many of the world's most potent poisons are derived from plants, and some plant tissues are extremely toxic to animals because they contain protective compounds, e.g. potato, tomato, and tobacco foliage. Furthermore, the levels of preformed compounds can change during the host's lifetime. Usually they decrease as the fruit or vegetable becomes ready for market (16). Thus, a specific induction may be required to protect the plant during the marketing period. The exogenous application of naturally occurring defense compounds to protect plants would not be economical, their synthesis would be difficult, and the cost of their isolation probably high.

In spite of the possible flaws in this scenario, the protection of the host during specific periods of susceptibility by manipulating the levels of the natural compounds seems to be a safe way to preserve host resistance. For example, high level of the antifungal diene could be maintained in stored avocado fruits, but the inhibitor would be allowed to disappear before consumption of the fruit. An alternative to the external application of defense compounds is their elicitation, specific or nonspecific, within the plant. Regulation of gene expression and, hence, of the gene products, would lead to the synthesis and further accumulation of the desired compounds. Biotic and abiotic factors have been shown to stimulate plant tissues to produce higher levels of preformed antifungal compounds (6,16,

24). This type of stimulation may also increase phytoanticipin concentration. The induction of preformed compounds by inoculation of avocado with non-pathogenic strains of *C. magna*, or any other non pathogenic *C. gloeosporioides* strain, suggests that this could be done by biological means. However, the use of biological or physical elicitors (CO_2, UV-C, γ-radiation) has not always proved effective because the host, *e.g.*, avocado, is receptive to the signals only during very specific periods after harvest (17).

Enhancing resistance barriers in plants by incorporation of specific inhibitors of fungal enzymes involved in fungal colonization of host tissue has also been suggested. In the past, cultivars with higher concentrations of epicatechin, a phenol that inhibits pectolytic-degrading enzymes, are also more resistant to *Colletotrichum* attack (13). In recent work, the extracellular secretion of pectolytic enzymes was suggested to be another factor regulating fungal colonization during fruit ripening. Avocado cultivars where pH values in the pericarp were low enough not to enhance pectolytic enzyme secretion completely inhibited fungal attack. This fact could indicate that requirements for hyphae to colonize the fruit tissue could be dependent on indirect host modulation of the pathogen secretion system.

The general understanding of the plant biosynthetic pathways that lead to formation of preformed and inducible barriers, and of the pathogenic factors contributing to fungal colonization are of primary importance in manipulating these compounds to modulate resistance. Such an understanding offers the possibility that preformed barriers might be enhanced by genetic, biological, physical, and chemical means to provide novel crop protection strategies that may reduce pesticide use in the future.

Acknowledgements

BARD, GIARA and CDR funds have supported this work.

Literature Cited

1. Adikaram, N.K.B., Edwing, D.F., Karunaratne, A.M., and Wijeratne, W.M.K. 1992. Antifungal compounds from immature avocado fruit peel. Phytochemistry 31:93-96.
2. Ardi, R., Kobiler, I., Keen, N.T., and Prusky, D. 1998. Involvement of epicatechin biosynthesis in the activation of the mechanism of resistance of avocado fruits *to Colletotrichum gloeosporioides.* Physiol. Mol. Plant Pathol. 53:269-285.
3. Bennet, R.N. and Wallsgrove, R.M.. 1994. Secondary metabolites in plant defence mechanisms. New Phytol. 127:617-33.

4. Davis, R.H. 1991.Glucosinolates. Pages 202-225 in: Toxic substances in crop plants. J.P. D'Mello, C.M. Duffus, and J.H. Duffus, eds. Royal Society of Chemistry, Cambridge, UK.
5. Flor, H.H. 1971. The current status of the gene-for-gene concept. Annu. Rev. Phytopathol. 9:275-296.
6. Ingham, J.L. 1973. Disease resistance in higher plants. The concept of pre-infectional and post-infectional resistance. Phytopathol. Z. 78:314-335.
7. Kobiler. I., Prusky, D., Midland, S.L., Sims, J.J., and Keen, N.T. 1994. Compartmentation of antifungal compounds in oil cells of avocado fruit mesocarp and its effect on susceptibility to *Colletotrichum gloeosporioides*. Physiol. Mol. Plant Pathol. 43:319-328.
8. Osbourn, A. 1996a. Saponins and plant defence-A soap story. Trends Plant Sci. 1:4-9.
9. Osbourn, A. 1996b. Pre-formed antimicrobial compounds and plant defence against fungal attack. Plant Cell 8:1821-1831.
10. Prusky, D. 1988. The use of antioxidants to delay the onset of anthracnose and stem end rot in avocado fruits after harvest. Plant Dis. 72:381-384.
11. Prusky, D. 1996. Pathogen quiescence in postharvest diseases. Annu. Rev. Phytopathol. 34:413-434.
12. Prusky, D. 1997. Constitutive barriers and plant disease control. Pages 163-176 in: Environmentally safe approaches to crop disease control. N.A. Rechcigl and J.E. Rechcigl, eds. CRC Press and Lewis Publishers, Boca Raton, FL.
13. Prusky, D., Freeman, S., Rodriguez, R.J., and Keen, N.T. 1994. A non-pathogenic mutant strain of *Colletotrichum magna* induces resistance to avocado fruits to *C. gloeosporioides*. Mol. Plant-Microbe Interact. 7:326-33.
14. Prusky, D., Gold, S., and Keen, N.T. 1989. Purification and characterization of an endopolygalacturonase produced by *Colletotrichum gloeosporioides.* Physiol. Mol. Plant Pathol. 35:121-123.
15. Prusky, D., Karni, L., Kobiler, I., and Plumbley, R.A. 1990. Induction of the antifungal diene in unripe avocado fruits: effect of inoculation with *Colletotrichum gloeosporioides*. Physiol. Mol. Plant Pathol. 37:425-435.
16. Prusky, D. and Keen, N.T. 1993. Involvement of preformed antifungal compounds in the resistance of subtropical fruits to fungal decay. Plant Dis. 77:114-119.
17. Prusky, D. and Keen, N.T. 1995. Inducible preformed compounds and their involvement in the resistance of plant pathogens. Pages 139-152 in: Novel approaches to integrated pest management. R. Reuveni, ed. Lewis Publishers, Boca Raton, FL.

18. Prusky, D., Keen, N.T., and Eaks, I. 1983. Further evidence for the involvement of a preformed antifungal compound in the latency of *Colletotrichum gloeosporioides* on unripe avocado fruits. Physiol. Plant Pathol. 22:189-198.
19. Prusky, D., Kobiler, I., and Jacoby, B. 1988. Involvement of epicatechin in cultivar susceptibility of avocado fruits to *Colletotrichum gloeosporioides* after harvest. J. Phytopathol. 123:140-146.
20. Prusky, D., Plumbley, R.A., and Kobiler I. 1991. Modulation of natural resistance of avocado fruits to *Colletotrichum gloeosporioides* by CO_2 treatment. Physiol. Mol. Plant Pathol. 39:325-334.
21. Schlosser, E. 1994. Preformed phenols as resistance factors. Acta Hortic. 381:615-630.
22. Schoenbeck, F. and Schlosser, E. 1976. Preformed substances as potential protectants. Pages 653-678 in: Physiological plant pathology. R. Heitefuss and P.H.Williams, eds. Springer-Verlag, Berlin.
23. Swinburne, T.R. 1983. Quiescent infections in post-harvest diseases. Pages 1-21 in Post-harvest pathology of fruits and vegetables. C. Dennis, ed. Academic Press, London.
24. VanEtten, H.D., Mansfield, J.W., Bailey, J.A., and Farmer, E.E. 1994. Two classes of plant antibiotics: Phytoalexins versus "phytoanticipins." Plant Cell 9:1191-1192.
25. Verhoeff, K. 1974. Latent infections by fungi. Annu. Rev. Phytopathol. 12:99-110.
26. Wattad, C., Dinoor, A., and Prusky, D. 1994. Purification of pectate lyase produced by *Colletotrichum gloeosporioides* and its inhibition by epicatechin: A possible factor involved in the resistance of unripe avocado fruits to anthracnose. Mol. Plant-Microbe. Interact. 7:293-297.
27. Wattad, C., Kobiler, D., Dinoor, A., and Prusky, D. 1997. Pectate lyase of *Colletotrichum gloeosporioides* attacking avocado fruits: cDNA cloning and involvement in pathogenicity. Physiol. Mol. Plant Pathol. 50: 197-212.
28. Yakoby, N., Kobiler, I., Dinoor, A., and Prusky, D. 1998. Effect of pH on the secretion of pectolytic enzymes and pathogenicity of *Colletotrichum gloeosporioides*. 6th International Mycological Congress, Jerusalem, p.82. (Abst.).

Chapter 15

Colletotrichum Strains for Weed Control

Alan K. Watson, Jonathan Gressel, Amir Sharon, and Amos Dinoor

Native bioherbicidal pathogens of weeds are typically used as inundative inoculum to incite sufficient disease to provide weed control, typically on a seasonal basis, with a need for annual repeated applications. In some cases, introduced "alien" weeds can be controlled with a single application of an imported "classical" pathogen that classically controlled the weed in its country of origin. Classical control is of far less commercial interest and is dealt with mainly by the public sector. The use of *Colletotrichum* strains as bioherbicides was extensively reviewed by Templeton in 1992 (63). Much of the early attention in bioherbicide research was on *Colletotrichum,* based on the success of *C. gloeosporioides* (Penz.) Penz. and Sacc. in Penz. f.sp. *aeschynomene* (18,59,64,75). *Colletotrichum* was predicted to become increasingly important as a bioherbicide because of its biology and cosmopolitan distribution (63). Nineteen *Colletotrichum* strains had been studied as possible bioherbicides (63), and two strains, *C. gloeosporioides* f.sp. *aeschynomene* (Collego) and *C. gloeosporioides* f.sp. *cuscutae* (Lubao), have been used for many years. Three additional strains, *C. orbiculare* (Berk. and Mont.) von Arx f.sp. from *Malva pusilla* (previously *C. gloeosporioides* f.sp. *malvae* BioMal) (50), *C. coccodes* (Wallr.) Hughes (Velgo) (73), and *C. orbiculare* (4) were, at that time, considered to be excellent commercial prospects. BioMal was registered in Canada, but never marketed, and *C. coccodes* (Velgo) and *C. orbiculare* have not been commercially developed. The predicted commercial successes have not materialized, but *Colletotrichum* strains remain important in bioherbicide research and are being used as model systems for investigating various barriers to bioherbicide development. If such barriers can be overcome, these and other *Colletotrichum* species can meet significant agronomic needs. It is necessary to analyze the relation between bioherbicide, its production and formulation, its establishment and growth within a weed to see where these barriers to large scale agricultural use lie.

Commercial Development of Inundative Bioherbicidal Products

LUBAO

C. gloeosporioides f.sp. *cuscutae* (Lubao) has been used in China since 1966 for the control of dodder species (*Cuscuta chinensis* Lam. and *C. australis* R. Br.) parasitizing soybeans [*Glycine max* (L.) Merr.]. Inoculum had been produced and distributed locally through country factories. The present status of Lubao in China is unknown, as during a scientific mission to China, the bioherbicide was not available. A culture of the fungus was obtained and is being investigated in Israel. Production of inoculum *in vitro* has been erratic, and cultures were lost unless they produced chlamydospore-like cells (not sclerotia as is typical with other *Colletotrichum* strains). Standard inoculum production is now being developed. An intensive host-range study was conducted (E. Nof, A. Dinoor, and B. Rubin, unpublished), in which dodder-parasitized crops were inoculated. Thirty-five cultivars of 19 crop species were inoculated, including Fabaceae, Solanaceae, Cucurbitaceae, cotton, sunflowers, and corn. None were infected, while the dodder on each one was completely destroyed by the fungus. The strain was molecularly compared to strains of *C. gloeosporioides*, *C. coccodes*, and *C. acutatum* (from strawberries). The profile was different from *C. gloeosporioides* but had some similarities to *C. acutatum*. It was not pathogenic to three cultivars of strawberries, while control inoculations with *C. acutatum* were pathogenic (E. Nof, A. Dinoor, and B. Rubin, unpublished).

COLLEGO

C. gloeosporioides f.sp. *aeschynomene* (Collego), was marketed from 1982 to 1992 with approximately 2500 ha treated annually for the control of northern jointvetch (*Aeschynomene virginica* (L.) B.S.P. in rice (*Oryza sativa* L.) and soybeans in the Delta region of southeastern United States (63). After that, the EPA re-registration was not pursued, and the product was not available. Marketing of the product had passed from the initial producer to two other producers in sequence. The last decided to discontinue sale of Collego because of its low perceived market potential and the inability of their custom fermentors to produce a cost-effective product. Through the combined efforts of the University of Arkansas and Encore Technologies, Collego was re-registered in 1997, manufactured, and test marketed. The 1998 season was one of the best for Collego, with approximately 5,000 ha treated (D. Johnson, personal communication).

BIOMAL

A pathogen initially identified as *C. gloeosporioides* f.sp. *malvae*, was registered in 1992 in Canada for postemergence control of round-leaved mallow (*Malva pusilla* Sm.), lentil (*Lens esculenta* Moench.), flax (*Linum usitatissimum* L.), and wheat (*Triticum aestivum* L.), but it was never marketed. The project was terminated because of the high cost for inoculum production. Agriculture and Agri-Food Canada has recently come to an agreement with Encore Technologies to re-register BioMal as a biological herbicide for control of round-leaved mallow in field crops (K. Mortensen, personal communication). Small quantities are expected to be available in 1999.

The host specificity studies required for registration revealed, quite unexpectedly, that one cultivar of safflower (*Carthamus tinctoris* L.) was susceptible to BioMal in greenhouse and field trials (51). Recently, molecular taxonomy of *Colletotrichum* species from Malvaceae suggests that BioMal is not *C. gloeosporioides* f.sp. *malvae*, but *C. orbiculare* f.sp. from *Malva pusilla* (6). Studies of the infection process endorse placement in *C. orbiculare,* as the intracellular infection structures produced by BioMal infection (45) are characteristic of *C. orbiculare* aggregate species (6). These findings may partly explain why safflower was susceptible in the host range studies (51), since *C. orbiculare* is known to infect certain members of the Asteraceae (63). Further host specificity testing of the BioMal strain are indicated.

HAKEA

Hakea sericea Schrader (Proteaceae) (silky needlebush) is a shrub or small tree from southeastern Australia that is a major invasive weed in South Africa. An endemic strain of *C. gloeosporioides* has provided effective control of this weed in South Africa (46,47). Following the success of colonized bran pieces for treating hakae seedlings (46), a granular product was developed. Granules consisted of a gluten core coated with a soybean flour and *C. gloeosporioides*. The granules were incubated for several days to allow fungal colonization before drying. This product was granted provisional registration in 1990. Eighty hectares of seedling-infested mountainside of the Cape Nature Conservation area were treated (48) The fungus-colonized granules relied on re-wetting by rain to induce sporulation of the fungus on the granule surface and on rain-splash for spread to adjacent seedlings.

The registration lapsed due to budget cuts by the single purchaser, and production was stopped. Since then, a dried spore preparation has been supplied, on demand, by the Plant Protection Research Institute, Weeds Division, Stellenbosch, directly to users, free of charge, without re-registration. The dried spores are re-suspended in water to inoculate wounds in the lower stems of adult *Hakea sericea* plants (47). The design of a simple hand-held applicator that lightly wounds the tree and applies the fungus at the same time is also supplied to users (48).

Status of Some 'Orphaned' *Colletotrichum* Strains

Templeton (63) coined the term "orphaned" for *Colletotrichum* strains that demonstrated adequate biological potential in laboratory and field studies, but were not commercialized. Many of these strains are still being actively researched as model systems in attempts to better understand the pathosystems and to overcome constraints limiting commercial development. *C. orbiculare-Xanthium spinosum*, *C. coccodes-Abutilon theophrasti*, and *C. truncatum-Sesbania exaltata* continue to be actively studied as model systems with potential utility or which may provide usable information for other systems.

SPINY COCKLEBUR

Spiny cocklebur or Bathurst burr (*Xanthium spinosum* L.) is a widespread noxious weed in Australia infesting irrigated crops, including cotton (*Gossypium hirsuta* L.), soybean, and sheep-grazing lands. An endemic strain of *C. orbiculare* has been evaluated as a bioherbicide for spiny cocklebur control. Laboratory and field studies have quantified environmental requirements and host range (4,5). Despite an industrial partner, *C. orbiculare* failed to become a commercial product because of unreliable formulations in the absence of dew, low spore yields in submerged culture during scale-up, and small market size (B. Auld, personal communication). This pathosystem is being used a model system for continuing formulation research (38).

VELVETLEAF

Abutilon theophrasti Medik. (velvetleaf) is a major annual weed in soybeans and corn (*Zea mays* L.) in the midwestern United States, eastern regions of Canada, southern Europe, and the Mediterranean region. Velvetleaf is a very difficult weed to control because of its rapid growth rate, capacity to establish a height differential with the crop, and prolific seed production. This robust weed, naturally tolerant to many soybean and corn herbicides, has evolved resistance to some others (27). The fungal pathogen, *C. coccodes* has been shown to selectively infect *A. theophrasti* (73). Typically, this pathogen causes gray-brown foliar lesions on infected *A. theophrasti* plants. Initially, lesions appear as small flecks but later enlarge and become necrotic (73). Areas surrounding lesions become desiccated, and diseased leaves are shed prematurely. In general, velvetleaf plants are killed only when inoculated at the cotyledon stage (73). When *C. coccodes* is applied to *A. theophrasti* at later growth stages, the pathogen causes a severe anthracnose on inoculated leaves and stunted and delayed growth of infected plants; however, this vigorous annual resumes growing after shedding the diseased leaves.

Continuing research has focused on laboratory and field studies to optimize inoculum production, formulation, and application methodologies, and to examine the role of phylloplane organisms and combinations with growth regulators and chemical herbicides to enhance weed control efficacy (24,25,31,55,72,78).

HEMP SESBANIA

The annual leguminous *Sesbania exaltata* (Raf.) Rydb. ex. A.W. Hill (hemp sesbania) occurring throughout the southern regions of the United States is a major weed in soybean, rice, and cotton fields. In 1987, a host-limited anthracnose pathogen, *C. truncatum* (Schw.) Andrus and Moore, was isolated from diseased hemp sesbania (7). Its bioherbicidal efficacy and practical use has been limited by a lengthy dew period requirement (7) and production difficulties (33).

Formulation of *C. truncatum* as an invert emulsion was shown to overcome the dew period requirement (8), but utility of invert emulsions is hampered by the necessity to use specialized air-assisted nozzle technology (15). Meanwhile, other means to overcome lengthy dew requirements are being actively investigated (9).

Liquid culture fermentation is generally considered necessary to achieve a low-cost production method. *C. truncatum* is being used as a model system to examine how nutrition impacts propagule formation, yield, efficacy, and stability and to enable rational approaches to development of submerged culture production methods for microbial bioherbicides (33).

Recently Commenced Projects

Interest in *Colletotrichum* strains as bioherbicides continues. *C. truncatum* is one of several pathogens being evaluated for biological control of purple loosestrife (*Lythrum salicaria* L.) (53). *C. dematium* (Pers. Ex Fr.) Grove is a potential microbial control for fireweed (*Epilobium angustifolium* L.) (70), Strains of *Colletotrichum* are also being evaluated against *Rubus* spp. (54), marsh reed grass (*Calamagrostis canadensis* (Michaux) Beauv.) (69), and itchgrass (*Rottboellia cochinchinensis* (Lour.) W. D. Clayton) (23).

The following possible uses have apparently been abandoned: *C. truncatum* against Florida beggarweed (*Desmodium tortuosum* (Sw.) DC.) (11), *C. coccodes* against eastern black nightshade (*Solanum ptycanthum* Dun.) (1), a *Colletotrichum* strain against pitted morning glory (*Ipomoea lacunosa* L.) (12), and *C. malvarum* (A. Braum and Casp.) Southworth) against prickly sida (*Sida spinosa* L.) (36).

Colletotrichum Strains as Classical Biocontrol Agents

Colletotrichum strains have been imported and disseminated as "classical" biocontrol agents, after their "classical" hosts were imported and became alien weed species from a lack of natural enemies.

CLIDEMIA

Clidemia hirta (L.) D. Don (Melastomataceae) is a major weed of tropical American origin in Hawaiian rainforests. An isolate of *C. gloeosporioides* f.sp. *clidemiae* was obtained from diseased leaves of *Clidemia* collected in Panama and was regarded as a possible biocontrol agent of the weed (65). The fungus was released in 1986-93 in cooperation with personnel of the Division of Forestry and Wildlife, Department of Land and Natural Resources of Hawaii and was established on all islands where *Clidemia* is a pest (E. Trujillo, personal communication). The pathogen of *Clidemia* is a very effective biological control agent of the weed in areas of high humidity during the windy-rainy season of the year. The fungus sporulates readily at night on the abaxial surface of the leaves. However, for spores to become air-borne and be distributed long distances, rain with high wind velocity is required (E. Trujillo, personal communication).

MICONIA

Miconia calvescens DC. (Melastomataceae) is another alien invasive tree threatening Hawaiian tropical rainforests. Introduced to Hawaii in 1961, it is threatening the biodiversity of native forests receiving 1800-2000 mm or more annual precipitation (42). A strain of *C. gloeosporioides* was isolated in Brazil from diseased *Miconia* leaves with anthracnose leaf spots. The fungus was imported into Hawaii where tests showed that it was specific to the genus *Miconia*, causing leaf spots and premature defoliation (E. Killgore, personal communication). *C. gloeosporioides* was released on Hawaii and Maui in 1997 and 1998. The pathogen established within the release sites, but spread of the disease beyond a 10-km radius has not been detected. Natural dissemination requires wind-driven rain, which has been lacking due to extended drought (E. Trujillo, personal communication).

Constraints to Development of Bioherbicides

With the possible exception of Lubao in China, there have been no real commercial successes with *Colletotrichum* spp. Commercial development models have been based on the prevailing wisdom that bioherbicides had to be "like an herbicide" as a "stand alone" product to be used without herbicides. They should give complete kill, have a long shelf life, and compete economically with herbicides. Linkages were forged between public institutions and private industry, which may have been counter-productive because of this "like-an-herbicide" paradigm. The bioherbicides were test marketed without adequate understanding of the biology of production and infection-i.e., far from optimized. Most candidate bioherbicides, not just *Colletotrichum* spp., have failed from biological, technological, or commercial constraints impeding their commercial development (3,67). Extended dew requirements, low fecundity, low virulence, minimal shelf life, and restricted niche markets are common problems hindering commercial bioherbicide development and were inadequately studied before initial commercialization. Biological and technological constraints are being actively researched in a number of model systems, but scientists have limited opportunity to address commercial constraints other than the choice of target, and to shift away from the "industrial partner" development model.

BIOLOGICAL CONSTRAINTS

Disease development. Many pathogens with bioherbicide potential have been discovered, but most lack sufficient aggressiveness by themselves to overcome weed defenses to achieve adequate control (28). Improved understanding of mechanisms involved in determining virulence would assist in the selection of preferred isolates as biological weed control agents. Luo and TeBeest (41) have developed an infection component analysis to aid in assessing the best overall relative fitness of Collego isolates.

Weed-Crop Competition. There has been a lack of understanding of the biological parameters that must be controlled during testing biocontrol agents. The agents are typically tested on the weed alone in the absence of a crop, then lethality to the weeds is measured as in herbicide screening. This approach has proven inadequate with herbicides; many newer herbicides just stop growth, and the crop then out competes the weeds Measuring a biological interaction between a diseased weed and a crop, without the crop, is also clearly inadequate.

The *C. coccodes*-velvetleaf (*Abutilon theophrasti*) pathogen-weed system provided valuable insight into plant-pathogen interactions and, in particular, to the role of disease in interspecific competition. *C. coccodes* has a differential effect on plant competitive interactions depending on whether the weed is found in a pure stand or in a mixture with a soybean crop (21). In a pure stand, application of this selective foliar pathogen has little impact on seed yield of velvetleaf. In these plots, velvetleaf intraspecific competition stimulates vertical growth and favors the rapid replacement of diseased leaf tissue that has prematurely senesced. Nonetheless, *C. coccodes* application can accelerate height asymmetry within velvetleaf monocultures, thus resulting in a relatively few, tall individuals becoming dominant (20). In mixtures, however, *C. coccodes* differentially influences the yield of both species. In two of three field seasons in Québec, *C. coccodes* was shown to reduce velvetleaf seed yields by an average of 60% compared with yields for control (uninoculated) plants (21). The decline in velvetleaf yield was primarily attributed to the stunting effect of the pathogen, which allowed soybean plants to grow above the weed. Consequently, soybean yield losses within inoculated mixture plots were generally lower than for control plots (21). The significantly greater effect of *C. coccodes* in a competitive environment with a soybean crop has important ramifications *vis a vis* the value and accuracy of initial efficacy testing. Potential biocontrol agents should not be rated based solely on their effect in pure stands of the target weed.

Other research on this pathogen-weed system has revealed a feature that may have a substantial impact on biocontrol management strategies that depend on the use of multiple inoculations to increase disease pressure. Results from a controlled environment study showed that three sequential applications of *C. coccodes* at 10^9 spores m^{-2} were less effective in suppressing velvetleaf growth and subsequently increasing soybean performance than either one or two applications (19). As suggested, velvetleaf plants may have exhibited a compensatory response or induced systemic protection.

OVERCOMING HOST DEFENSES

Plant pathologists typically study host defenses in order to augment them in crop protection. Conversely, these same defenses must be overcome to control weeds with bioherbicides. Crucial steps in the establishment of a fungal pathogen include recognition and attachment to the plant surface, germination, formation of infection structures, penetration, and colonization of the host tissue. Fungal pathogenicity is modulated by enzymes that degrade host cell walls and intracellular polymers and detoxify phytoalexins (phytoalexin tolerance) and by toxins. Plants use various mechanisms, often simultaneously, to defend against pathogen invasion including callose deposition, hydroxproline-rich glycoprotein accumulation, production of pathogenesis-related proteins (PR-proteins) and phytoalexins, lignin and phenolic formation, and free radical generation (29). Therefore, virulence of a bioherbicidal fungus may be due to the rate of germination, rate of penetration, rate of extracellular degrading enzyme production, degradation of induced and pre-formed plant defenses, and/or toxin production. The overcoming of any one of these obstacles may synergistically change the equation leading to economic development. The simultaneous overcoming of more than one obstacle will enhance the economics even more.

Synergy. Weed defenses can be suppressed and bioherbicide efficacy improved by chemical and biological synergy (28). The chemical synergist can be a compound that specifically overcomes one of the host responses to pathogen attack (58) or a chemical that otherwise weakens or wounds the weed, facilitating pathogenicity. Biological synergy results when other organisms enhance dispersal or infection.

Chemical synergies. An example of suppressing host response can be seen with callose. A rapid increase in callose biosynthesis is typically seen in compatible responses to *C. coccodes* infection and in infection by many other *Colletotrichum* spp. (34). Lignin-like material later becomes embedded in the callose, sterically preventing callose degradation by glucanases, and thus inhibiting the progression of infective fungi (74). The enzyme callose synthase has calcium as an obligate cofactor, and the activity of this enzyme *in vitro* can be blocked by chelators (40). It was hypothesized that infecting a weed with a compatible *Colletotrichum* bioherbicide together with a calcium chelator would enhance infectivity by preventing callose production. *Colletotrichum* and many other fungi have calcium requirements that are so low that the traces in reagent grade chemicals in ostensibly calcium-free media are sufficient for fungal growth (52). *C. coccodes* could be cultivated on ethylene glycol-bis (β-aminoethyl ether) N,N,N′,N′-tetraacetic acid (EGTA) and other calcium-specific chelators without affecting growth, yet trace amounts of calcium were still routinely found in mycelia using ultra-high sensitivity equipment such as an inductively coupled plasma atomic emission spectrometer (D. Michaeli and J. Gressel, unpublished).

As most calcium chelators are too hydrophilic to penetrate plant cuticles, EGTA derivatives were synthesized with hydrophobic tails. They were applied with *C. coccodes* to *Abutilon theophrasti*. One of these compounds, "BE", as well as calcium-complexing oxalic acid, doubled the number of infection sites (D. Michaeli, A. Warshawsky, and J. Gressel, unpublished). Concomitantly, microscopic analysis showed that far less callose was present after such treatments. Immunochemical determination of *Colletotrichum* in the leaves showed greatly enhanced levels of mycelia when the calcium-complexing agents were used. Thus, calcium deprivation increased infectivity while decreasing callose content (D. Michaeli, A. Warshawsky, and J. Gressel, unpublished). This correlation fits, but does not prove the hypothesis, as calcium deprivation has many effects in plants.

Another example of weakening plants was seen when *C. coccodes* was applied simultaneously with the plant growth regulator, thidiazuron or with basagran, aciflurofen, and thifensulfuron herbicides. The two agents synergistically increased weed mortality and reduced velvetleaf height and biomass (31,68,72). The mixture of thidiazuron and *C. coccodes* increased ethylene production in velvetleaf (30) and induced changes in peroxidase activity and isoform patterns (52). The increase in guaiacol-reactive soluble peroxidase activity suggested a peroxidase-associated defense response, and the rapid stimulation of two isoforms following the combined treatment may account for the synergistic inhibition of growth (52).

When Collego was tank-mixed with an herbicide, weed control was enhanced (39). Similarly, chemical combinations with BioMal and *C. gloeosporioides* f.sp. *malvae* have increased control of round-leaved mallow (26). Many *Colletotrichum* species produce the plant hormone 3-indole-acetic acid (IAA) in culture (55), with the highest IAA amounts in cultures of *C. gloeosporioides* f.sp. *aeschynomene.* A functional analog of IAA-2,4-Dalso effectively controls round leaved mallow, (50). Both the indole-3-acetamide and the indole-pyruvic acid pathways of IAA-biosynthetic pathways contribute to IAA production in *C. gloeosporioides* f.sp. *aeschynomene*, but in culture the production of IAA was mainly through indole-3-acetamide (55). Although *in planta* production of IAA by the fungus has not been determined, IAA may be involved in fungal pathogenicity, as was determined for some plant pathogenic bacteria. Some of the symptoms caused by *C. gloeosporioides* f.sp. *aeschynomene,* e.g., epinasty and leaf deformations, are mimicked by exposing plants to IAA. The IAA levels produced by *C. gloeosporioides* f.sp. *aeschynomene in vitro* should be sufficient to evoke these responses if they are also produced *in planta.*

Interorganisal synergies. There are several interorganisal synergies that enhance *Colletotrichum* bioherbicidal performance. Phylloplane organisms positively influenced bioherbicide performance. Shoot dry weight and the number of leaves per plant of *Sesbania exaltata* were reduced when several epiphytic bacterial isolates were applied with *C. truncatum* (56). Co-inoculation with *C. coccodes* and each of three isolates of *Pseudomonas* spp. significantly increased the number of lesions and disease severity caused by *C. coccodes* on velvetleaf seedlings and also accelerated the appearance of symptoms (24). Bacterial isolates reduced conidial germination and germ-tube branching of *C. coccodes*, increased the frequency of germ-tubes with appressoria, increased the relative frequency of dark-colored appressoria without germ tubes, and decreased the total germ tube length. Certain phylloplane bacteria may, therefore, enhance the efficacy of bioherbicides by stimulating the formation of appressoria and reducing saprophytic mycelial growth of the pathogen on the phylloplane before infection. Phylloplane bacteria compete for carbon, nitrogen, and iron, limiting the saprophytic phase of the pathogen on the phylloplane and accelerating the development of the parasitic phase (25). This acceleration of infection decreases the critical period of moisture required for infection.

Synergy can also occur when two pathogens invade the same plant. The *Xanthium spinosum* pathogen *C. orbiculare* produces only hypersensitive flecks on *Xanthium occidentale* Bertol. (Noogoora burr), but a strong synergistic effect was observed when a rust *Puccinia xanthii* Schw. and *C. orbiculare* were applied sequentially to Noogoora burr plants, leading to death of the weeds (44). The *Colletotrichum* invaded through the pustules formed by the *Puccinia*, yet the rust alone did not kill the leaves.

Insects and vertebrates have also been implicated in improving bioherbicide performance. Grasshoppers and tree frogs aid in the dispersal of *C. gloeosporioides* f.sp. *aeschynomene* in rice fields (77). A leaf-feeding beetle [*Chrysolina hyperici* (Forester)] released for biological control of *Hypericum perforatum* L. (St. Johnswort) only causes up to 50% defoliation in mid-summer, but plants recover. But, more importantly, the beetles can be used as vectors for a lethal, host-specific *C. gloeosporioides* strain that causes mortality of *H. perforatum* shoots in habitats suitable for disease development, but only when spread by the beetles (49).

PRODUCTION AND FORMULATION CONSTRAINTS

Regulatory issues and market demand have hindered development of some potential bioherbicides. However, Jackson et al. (33) suggested that overall lack of commercial success stems from difficulty in producing and stabilizing the biocontrol agent, as well as the lack of consistent weed control in the field as a result of biological constraints. Commercial development requires low-cost production methods and the use of liquid culture fermentation is necessary to achieve this goal. Much of the success of Collego was due to mass production of spores and preservation by drying in commercial-scale facilities. This success has been elusive with other bioherbicides including BioMal (*C. gloeosporioides* f.sp. *malvae*) as well as for *C. orbiculare*.

Mass production. C. truncatum has been used as a model system to study optimization of liquid culture media based on defined nutrient conditions (as reviewed in 33, and references therein). Nutrition definitely impacts spore yield and spore efficacy. Carbon concentration, carbon source, and carbon-to-nitrogen ratio of the conidiation medium of *C. truncatum* influenced spore yield, morphology, and efficacy in inciting disease in *Sesbania exaltata*. A medium with a CN ratio of 15:1 to 20:1 was optimal for conidial production of *C. truncatum* (33).

Similar studies have been conducted with *C. coccodes*. A defined medium based on soy protein was developed, and the effect of carbon concentration and carbon to nitrogen ratio, as well as their interaction on *C. coccodes* growth and sporulation were studied. The highest spore yields were obtained on media with C:N ratios ranging from 5:1 to 10:1. High carbon concentrations combined with C:N ratios above 15:1 reduced both mycelial growth and sporulation and increased spore matrix production. These results indicated that C:N ratios from 10:1 to 15:1 are optimal for *C. coccodes* spore production (78).

Formulation. Formulation of bioherbicides is the blending of the active ingredient, such as fungal spores, with inert carriers to alter physical characteristics of the bioherbicide and allow drying to a form that can later be used in the dry or a hydrated form. Two key issues are addressed shelf-life and field performance. Advancements in formulation technology are being made (15), and breakthroughs in these areas should assist many of the promising "orphaned" bioherbicides to become commercial products.

Dry preparations are preferred and should have a shelf-life of at least 6 months, preferably at ambient conditions (33). Because dry conidia can be difficult to stabilize, microsclerotia were examined as a bioherbicide propagule for of *C. truncatum* (33). Nutritional conditions were identified that suppressed sporulation and promoted the production of high concentrations of *C. truncatum* microsclerotia in liquid culture.

Microsclerotia of C. *truncatum* are stable as a dry preparation and effective in controlling hemp sesbania when used as a soil amendment (33). Microsclerotial inoculum of C. *truncatum* has also been matrix-encapsulated in wheat flour-kaolin granules ('Pesta'), then dried to a favorable low water activity, which led to excellent shelf-life for up to 2 years (16,17). By reducing the water activity, conidia of the velvetleaf pathogen, C. *coccodes*, formulated in either kaolin clay or talcum powder, have remained viable and infective for at least 6 months when stored at 4 to 30°C (56).

Adjuvant formulations and delivery systems are being studied as approaches to address the lengthy dew requirements for most bioherbicide prospects (9). The simple addition of additives has had variable results. Sorbitol increased the efficacy of C. *coccodes* spray suspensions (70), whereas Tween 20 and glycerol were no better than water as a sole liquid carrier for C. *orbiculare* (37). The efficacy of C. *orbiculare* was improved by spraying wounds, as mowing significantly increased mortality in two of three field trials (38). Vegetable and mineral oil suspension emulsions have also improved bioherbicide performance of both C. *orbiculare* and C. *truncatum,* but results have been inconsistent (22,37).

The dew problem associated with many bioherbicides can be overcome with the use of invert emulsions that reduce evaporation and hold water in the spore suspension on the plant surface (9). When C. *truncatum* was applied in an invert emulsion formulation, hemp sesbania control was 100% in the absence of dew. Invert emulsion technology requires specialized air-assist nozzles, and host selectivity can be lost.

Some foliar pathogens are being successfully applied as pre-emergence soil granules to counteract lengthy dew requirements (35,66). These formulations as well as the microsclerotia formulation of C. *truncatum*, bypass the lengthy dew requirement on the plant surface by being applied to the soil surface. They take advantage of available soil moisture and attack the weeds as they emerge from the soil.

GENETIC MANIPULATIONS

Bioherbicides can potentially be modified to improve a desirable trait such as increased virulence or resistance to crop production chemicals such as the fungicide benomyl. Benomyl-resistant mutant strains of C. *gloeosporioides* f.sp. *aeschynomene* (60) and C. *gloeosporioides* f.sp. *malvae* (32) were developed to possibly facilitate use as bioherbicides within integrated crop production systems that utilize this fungicide. The use of the fungicide might have an added benefit, controlling fungi that compete with the bioherbicide. Mutant strains have been most useful in pathogenesis and epidemiology experiments. Benomyl-resistant and nitrate non-utilizing strains of C. *gloeosporioides* f.sp. *aeschynomene* are less competitive than wild-type strains because one or more infection components are reduced (41,76). Similarly, the benomyl-resistant C. *gloeosporioides* f.sp. *malvae* caused fewer lesions than the wild-type in greenhouse studies (32) suggesting that one must generate many mutants to screen for those that are unaffected by the mutagenesis process.

Bioherbicides represent excellent opportunities for the application of recombinant DNA technology to study pathogenesis and to develop strains with enhanced weed control efficacy (62). Increasing pathogen virulence through the introduction of a genetically encoded virulence factor is certainly one avenue for improving weed control efficacy. Progress in this area has been limited. Genes must be found, transformation systems improved, and traits must be expressed without reduction in fitness.

Brooker et al. (10) have transformed *C. gloeosporioides* f.sp. *aeschynomene* with the *bar* gene conferring resistance to the herbicide bialaphos. The crop host range of *C. gloeosporioides* f.sp. *aeschynomene* was not altered by the transformation. There is no benefit from co-applying the fungus with sublethal levels of bialaphos at rates of *C. gloeosporioides* f.sp. *aeschynomene* recommended for the control of northern jointvetch. Significantly more disease developed when sublethal levels of bialaphos were applied with a bialaphos-resistant isolate at reduced rates of *C. gloeosporioides* f.sp. *aeschynomene*. The co-application of the resistant isolate with sublethal levels of bialaphos extends the control range to include Indian jointvetch (*Aeschynomene indica* L.) (10). The virulence and control range of a bioherbicide could be improved if the bioherbicide could be engineered to endogenously produce a nonselective phytotoxin such as bialaphos, which in nature is produced by an actinomycete.

Potential gene exchange is an important component of risk assessment for fungal pathogens used for biological weed control, especially with the prospects of genetically engineered bioherbicide strains (62). Mating between strains of *C. gloeosporioides* is known to occur. Heterokaryosis was demonstrated with *C. gloeosporioides* f.sp. *aeschynomene*, but no putative diploids or recombinant phenotypes were detected (13). Host specialization, vegetative compatibility, and lack of a demonstrated parasexual cycle may serve to genetically isolate *C. gloeosporioides* f.sp. *aeschynomene* from other populations of *C. gloeosporioides* (13). However, isolates of *C. gloeosporioides* with different host specificities are capable of outcrossing on a parental host. They produce viable, sexually recombinant progeny (14). None of the progeny were pathogenic to northern jointvetch, but they exhibited varying degrees of pathogenicity to apple fruits (14), showing why close scrutiny is needed.

Previous attempts to improve the biocontrol properties of *C. gloeosporioides* f.sp. *aeschynomene* by genetic transformation have proven largely unsuccessful because of very low levels of stable transformants (2,10,61). Recently, a highly efficient and reproducible procedure for the transformation of *C. gloeosporioides* f.sp. *aeschynomene* by electroporation of germinating conidia has been developed (M. Robinson and A. Sharon, *submitted*). Under optimal conditions, over 80 stable transformants per cuvette were obtained. Plasmid integration was predominantly homologous, but high transformation rates were obtained, both with homologous and non-homologous vectors. A low rate of morphological changes was observed (less than 5% of colonies) suggesting that the transformation process by itself had no harmful effect on the fungus. The observed phenotype changes probably reflect mutations caused by the insertion of the plasmid. Moreover, isolates expressing a high level of the transgenic GFP (green fluorescent protein) were as pathogenic as wild-type isolates. This fact suggests that a high level of constitutive production of a transgene product should be possible without having a negative effect on the bioherbicidal capacity of the fungus (M. Robinson and A. Sharon, unpublished).

Detailed studies of the biology and pathogenic nature of the bioherbicidal fungi are essential for the development of improved bioherbicide formulations by any of the approaches mentioned, e.g., by new formulations, combination with synergistic agents, or by genetic engineering of the fungus. The new transformation procedure opens the way for heterologous gene expression, gene deletion, and generation of tagged mutants. These methods can be used to advance the use of this fungus and to obtain knowledge that will assist in genetic engineering improvement of Collego. In a broader sense, the described procedure provides a framework for developing transformation of many additional bioherbicides.

HOST RANGE AND SPECIFICITY

The specificity of pathogenicity is a key issue in developing bioherbicides in general and *Colletotrichum* bioherbicides in particular. The safe application of a bioherbicide in crops depends on specificity towards the target weed. Biocontrol agents may also persist *in situ* and affect later crops. Thorough studies of host-range among target and important crops are therefore imperative before any permit for commercial use should be considered.

An additional difficulty arises when an attempt is made to enhance or improve the aggressiveness of the pathogen. Increased efficiency, abolishment of defense mechanisms, genetic manipulation, etc., may conceivably add new hosts to the range. Therefore, any modification of the bioherbicide should be followed by a host-range study.

The taxonomy of *Colletotrichum* is not always unequivocal. For example, the strain of *C. coccodes* attacking *Abutilon theophrasti* did not attack any Canadian crops, but another strain of *C. coccodes* is a severe pathogen of potatoes. Could strains of the bioherbicide genetically engineered for increased aggressiveness pass genetic material to the potato pathogen? Even the thought of that possibility could be daunting enough to consider the use of only chemical synergies of pathogens, which are only active when the chemical is present, in cases where the pathogen has potentially introgressing relatives.

Alternative Development Models

Commercial constraints (market size, patent protection, cost of production, regulations, and profit margin) have been cited as major barriers to bioherbicide development (3). Many bioherbicide candidates target small market opportunities. Thus, alternatives to the present commercial model need to be examined. The market size for most host specific bioherbicides is just too small to be of interest to the large multinational companies dominating plant protection. Meanwhile, the manufacture of bioherbicides by smaller companies that target niche markets should be more economically viable once registration, production, and formulation hurdles have been overcome.

Government involvement in projects, for the public good where chemical herbicide use is risky or banned in areas such urban parks, rights-of-ways, nature reserves, and national parks is a possibility, and there may be government financial incentives to encourage collaboration with industrial partners. This is especially true for allergenic weeds in urban habitats.

The possible success of cottage industry production techniques for Lubao in China and various biocontrol projects in the tropics suggest that the cottage industry concept could be adaptable to many of the ongoing bioherbicide projects. Such industries are incipient in developed countries for natural enemies of insects and for biocontrol agents against pathogenic fungi.

Concluding Remarks

The anthracnose disease cycle of many *Colletotrichum* strains was suggested to make them ideal for development of bioherbicides for annual weeds in annual crops (63). Most weeds targeted for biological control with *Colletotrichum* strains are annual weeds in annual crops. Because the natural spread and persistence of inoculum is restricted, *Colletotrichum* strains are considered unsuitable for classical biological weed control. However, results with *Colletotrichum* strains against *Clidemia hirta*, *Miconia calvescens*, *Hypericum perforatum*, and *Hakea sericea* in uncultivated habitats broaden the prospects of *Colletotrichum* strains affording effective weed control.

Even the limited success of *C. gloeosporioides* f.sp. *aeschynomene* has been difficult to follow-up. Many other *Colletotrichum* strains were not blessed with high virulence nor could mass production of spores and preservation by drying in commercial-scale facilities be achieved. Research efforts are shifting away from discovering new *Colletotrichum* strains to solving production, storage, and efficacy problems, some of which will require new host range determinations.

There clearly is a need to better understand the biochemical and physiological aspects of pathogenesis by the potential *Colletotrichum* bioherbicides so that weak links in host defense can be exploited. The *Colletotrichum* bioherbicides will need the augmentation of formulants, as well as chemical and/or biological synergists to provide the lethality to weeds that the consumer desires. It is rare that this level of lethality is achieved by single organisms in nature and is too much to expect from an unassisted *Colletotrichum.*

Literature Cited

1. Andersen, R.N. and Walker, H.L. 1985. *Colletotrichum coccodes*:A pathogen of eastern black nightshade (*Solanum ptycanthum*). Weed Sci. 33:902-905.
2. Armstrong, J.L. and Harris, D.L. 1993. Biased integration in *Colletotrichum gloeosporioides* f.sp. *aeschynomene* transformants with benomyl resistance. Phytopathology 83:328-332.
3. Auld, B.A. and Morin, L. 1995. Constraints in the development of bioherbicides. Weed Technol. 9:638-652.
4. Auld, B.A., McRae, C.F., and Say, M.M. 1988. Possible control of *Xanthium spinosum* by a fungus. Agric. Ecosyst. Environ. 21:219-233.
5. Auld, B.A., Say, M.M., and Millar, G.D. 1990. Influence of potential stress factors on anthracnose development on *Xanthium spinosum*. J. Appl. Ecol. 27:513-519.
6. Bailey, J.A., Nash, C., Morgan, L.W., O'Connell, R.J., and TeBeest, D.O. 1996. Molecular taxonomy of *Colletotrichum* species causing anthracnose on the Malvaceae. Phytopathology 86:1076-1083.
7. Boyette, C.D. 1991. Host range and virulence of *Colletotrichum truncatum*, a potential mycoherbicide for hemp sesbania (*Sesbania exaltata*). Plant Dis. 75:62-64.
8. Boyette, C.D., Quimby, P.C. Jr., Bryson, C.T., Egley, G.H., and Fulgham, F.E. 1993. Biological control of hemp sesbania (*Sesbania exaltata*) under field conditions with *Colletotrichum truncatum* formulated in an invert emulsion. Weed Sci. 41:497-500.
9. Boyette, C.D., Quimby, P.C. Jr., Caesar, A.J., Birdcall, J.L., Connick, W.J. Jr., Daigle, D.J. Jackson, M.A., Egley, G.H., and Abbas, H.K. 1996. Adjuvants, formulation, and spraying for improvement of mycoherbicides. Weed Technol. 10:637-644.
10. Brooker, N.L., Mischke, C.F., Patterson, C.D., Mischke, S., Bruckart, W.L., and Lydon, J. 1996. Pathogenicity of *bar*-transformed *Colletotrichum gloeosporioides* f.sp. *aeschynomene*. Biol. Control 7:159-166.
11. Cardina, J., Littrell, R.H., and Hanlin, R.T. 1988. Anthracnose of Florida beggarweed (*Desmodium tortuosum*) caused by *Colletotrichum truncatum*. Weed Sci. 36:329-334.

12. Cartwright, D.K. and Templeton, G.E. 1996. Controlled environment assessment of a *Colletotrichum* strain as a mycoherbicide for pitted morningglory control. Page 485 in: VIII International Symposium on Biological Control of Weeds. E.S. Delfosse and R.R. Scott, eds. Canterbury, New Zealand, 2-7 February 1992. CSIRO, Melbourne, Australia.
13. Chacko, R.J., Weidemann, G.J., TeBeest, D.O., and Correll, J.C. 1994. The use of vegetative compatibility and heterokaryosis to determine potential asexual gene exchange in *Colletotrichum gloeosporioides*. Biol. Control 4:382-389.
14. Cisar, C.R., Thornton, A.B., and TeBeest, D.O. 1996. Isolates of *Colletotrichum gloeosporioides* (teleomorph: *Glomerella cingulata*) with different host specificities mate on northern jointvetch. Biol. Control 7:75-83.
15. Connick, W.J. Jr., Daigle, D.J., and Quimby, P.C. Jr. 1991. An improved invert emulsion with high water retention for mycoherbicide delivery. Weed Technol. 5:442-444.
16. Connick, W.J. Jr., Daigle, D.J., Boyette, C.D., Williams, K.S., Vinyard, B.T., and Quimby, P.C. Jr. 1996. Water activity and other factors that affect the viability of *Colletotrichum truncatum* conidia in wheat flour-kaolin granules ('Pesta'). Biocontrol Sci. Technol. 6:277-284.
17. Connick, W.J. Jr., Jackson, M.A., Williams, K.S., and Boyette, C.D. 1997. Stability of microsclerotial inoculum of *Colletotrichum truncatum* encapsulated in wheat flour-kaolin granules. World J. Micro. Biotech. 13:549-554.
18. Daniel, J.T., Templeton, G.E., Smith, R.J. Jr., and Fox, W.T. 1973. Biological control of northern jointvetch in rice with an endemic fungal disease. Weed Sci. 21:303-307.
19. DiTommaso, A. and Watson, A.K. 1995. Impact of a fungal pathogen, *Colletotrichum coccodes* on growth and competitive ability of *Abutilon theophrasti*. New Phytol. 131:51-60.
20. DiTommaso, A. and Watson, A.K. 1997. Effect of the fungal pathogen, *Colletotrichum coccodes*, on *Abutilon theophrasti* height hierarchy development. J. Appl. Ecol. 34:518-529.
21. DiTommaso, A., Watson, A.K., and Hallett, S.G. 1996. Infection of the fungal pathogen *Colletotrichum coccodes* affects velvetleaf (*Abutilon theophrasti*)-soybean competition in the field. Weed Sci. 44:924-933.
22. Egley, G.H. and Boyette, D.C. 1995. Water-corn oil emulsion enhances conidia germination and mycoherbicidal activity of *Colletotrichum truncatum*. Weed Sci. 43:312-317.
23. Ellison, C.A. and Evans, H. 1996. Present status of the biological programme for the graminaceous weed *Rottboellia cochinchinensis*. Pages 493-500 in: VIII International Symposium on Biological Control of Weeds. E.S. Delfosse and R.R. Scott, eds. Canterbury, New Zealand, 2-7 February 1992. CSIRO, Melbourne, Australia.

24. Fernando, W.G.D., Watson, A.K., and Paulitz, T.C. 1994. Phylloplane *Pseudomonas* spp. enhance disease caused by *Colletotrichum coccodes* on velvetleaf. Biol. Control 4:125-131.
25. Fernando, W.G.D., Watson, A.K., and Paulitz, T.C. 1996. The role of *Pseudomonas* spp. and competition for carbon, nitrogen and iron in the enhancement of appressorium formation by *Colletotrichum coccodes* on velvetleaf. Eur. J. Plant Path.102:1-7.
26. Grant, N.T., Prusinkiewicz, E., Makowski, R.M.D, Holmstrom-Ruddick, B., and Mortensen, K. 1990. Effect of selected pesticides on survival of *Colletotrichum gloeosporioides* f.sp. *malvae,* a bioherbicide for round-leaved mallow (*Malva pusilla*). Weed Technol. 4:701-715.
27. Gray, J.A., Balke, N.E., and Stoltenberg, D.E. 1996. Increased glutathione conjugation of atrazine confers resistance in a Wisconsin velvetleaf (*Abutilon theophrasti*) biotype. Pestic. Biochem. Physiol. 55:157-171.
28. Gressel, J., Amsellem, Z., Warshawsky, A., Kampel, V., and Michaeli, D. 1996. Biocontrol of weeds: overcoming evolution for efficacy. J. Environ. Sci. Health B31:399-405.
29. Hoagland, R.E. 1996. Chemical interactions with bioherbicides to improve efficacy. Weed Technol. 10:651-674.
30. Hodgson, R.H. and Snyder, R.H. 1989. Thidiazuron and *Colletotrichum coccodes* effects on ethylene production by velvetleaf (*Abutilon theophrasti*) and prickly sida (*Sida spinosa*). Weed Sci. 37:484-489.
31. Hodgson, R.H., Wymore, L.A., Watson, A.K., Snyder, R., and Collette, A. 1988. Efficacy of *Colletotrichum coccodes* and thidiazuron for velvetleaf (*Abutilon theophrasti*) control in soybean (*Glycine max*). Weed Technol. 2:473-480.
32. Holmstrom-Ruddick, B. and Mortensen, K. 1995. Factors affecting pathogenicity of a benomyl-resistant strain of *Colletotrichum gloeosporioides* f.sp. *malvae*. Mycol. Res. 99:1108-1112.
33. Jackson, M.A., Schisler, D.A., Slininger, P.J., Boyette, C.D., Silman, R.W., and Bothast, R. J. 1996. Fermentation strategies for improving the fitness of a bioherbicide. Weed Technol. 10:645-650.
34. Kauss, H. 1992. Callose and callose synthase. Pages 1-8 in: Practical Approaches to Molecular Plant Pathology, Vol. 2. S.J. Gurr, M.J. McPherson, and D.J. Bowles, eds. Oxford University, Oxford.
35. Kempenaar, C., Wanningen, R., and Scheepens, P.C. 1996. Control of *Chenopodium album* by soil application of *Ascochyta caulina* under greenhouse conditions. Ann. Appl. Biol. 129:343-354.
36. Kirkpatrick, T.L., Templeton, G.E., TeBeest, D.O., and Smith, R.J. Jr. 1982. Potential of *Colletotrichum malvarum* for biological control of prickly sida. Plant Dis. 66:323-325.
37. Klein, T.A. and Auld, B.A. 1995. Evaluation of tween 20 and glycerol as additives to mycoherbicide suspensions applied to Bathurst burr. Plant Protect. Quart. 10:14-16.

38. Klein, T.A. and Auld, B.A. 1996. Wounding can improve efficacy of *Colletotrichum orbiculare* as a mycoherbicide for Bathurst burr. Aust. J. Exp. Agric. 36:185-187.
39. Klerk, R.A., Smith, R.J. Jr., and TeBeest, D.O. 1985. Integration of a microbial herbicide into weed and pest control programs in rice (*Oryza sativa*). Weed Sci. 33:95-99.
40. Kohle, H., Jeblick, W., Poten, F., Blaschek, W., and Kauss, H. 1985. Chitosan-elicited callose synthase in soybean cells as a Ca^{+2}-dependent process. Plant Physiol. 77:544-551.
41. Luo, Y. and TeBeest, D.O. 1998. Behavior of a wild-type and two mutant strains of *Colletotrichum gloeosporioides* f.sp. *aeschynomene* on northern jointvetch in the field. Plant Dis. 82:374-379.
42. Medeiros, A.C. and Loope, L.L. 1997. Status, ecology and management of the invasive plant, *Miconia calvescens* DC. (Melastomataceae) in the Hawaiian Islands. Bishop Museum Occasional Pap. 0 (48):23-36.
43. Michaeli, D. 1997. Enhancing *Colletotrichum coccodes* pathogenicity on the weed using calcium chelators to inhibit callose formation. M.Sc. thesis. The Hebrew University of Jerusalem, Rehovot. 96 pp.
44. Morin, L., Auld, B.A., and Brown, J.F. 1993. Synergy between *Puccinia xanthii* and *Colletotrichum orbiculare* on *Xanthium occidentale*. Biol. Control 3:296-310.
45. Morin, L., Derby, J., and Kokko, E.G. 1996. Infection process of *Colletotrichum gloeosporioides* f.sp. *malvae*. Mycol. Res. 100:165-172.
46. Morris, M.J. 1989. A method for controlling *Hakea sericea* Schrad. seedlings using the fungus *Colletotrichum gloeosporioides* (Penz.) Sacc. Weed Res. 29:449-454.
47. Morris, M.J. 1991. The use of plant pathogens for biological weed control in South Africa. Agric. Ecosyst. Environ. 37:239-255.
48. Morris, M.J., Wood, A.R., and den Breeyen, A. 1999. Plant pathogens and biological control of weeds in South Africa: a review of projects and progress during the last decade. African Entomology Memoir no. 1:129-137.
49. Morrison, K.D., Reekie, E.G., and Jensen, K.I.N. 1998. Biocontrol of common St. Johnswort (*Hypericum perforatum*) with *Chrysolina hyperici* and host-specific *Colletotrichum gloeosporioides*. Weed Technol. 12:426-435.
50. Mortensen, K. 1988. The potential of an endemic fungus, *Colletotrichum gloeosporioides*, for biological control of round-leaved mallow (*Malva pusilla*) and velvetleaf (*Abutilon theophrasti*). Weed Sci. 36:473-478.
51. Mortensen, K. and Makowski, R.M.D. 1997. Effects of *Colletotrichum gloeosporioides* f.sp. *malvae* on plant development and biomass of non-target field crops under controlled and field conditions. Weed Res. 37:351-360.
52. Nickerson, R.G., Tworkoski, T.J., and Luster, D.G. 1993. *Colletotrichum coccodes* and thidiazuron alter specific peroxidase

activities in velvetleaf (*Abutilon theophrasti*). Physiol. Molec. Plant Pathol. 43:47-56.

53. Nyvall, R.F. and Hu, A. 1997. Laboratory evaluation of indigenous North American fungi for biological control of purple loosestrife. Biol. Control 8:37-42.
54. Oleskevich, C., Shamoun, S.F., and Punja, Z.K. 1996. Evaluation of *Fusarium avenaceum* and other candidate fungi for biological control of invasive *Rubus* spp. Phytopathology 86:S102.
55. Robinson, M., Riov, J., and Sharon, A. 1998. Indole-3-acetic acid biosynthesis in *Colletotrichum gloeosporioides* f.sp. *aeschynomene*. Appl. Environ. Microbiol. 64:5030-5032.
56. Saad, F. 1993. Formulation of *Colletotrichum coccodes* as a bioherbicide. Ph.D. Thesis. McGill University, Montreal. 150 pp.
57. Schisler, D.A., Howard, K.M., and Bothast, R.J. 1991. Enhancement of disease caused by *Colletotrichum truncatum* in *Sesbania exaltata* by co-inoculating with epiphytic bacteria. Biol. Control 1:261-268.
58. Sharon, A., Amsellem, Z., and Gressel, J. 1992. Glyphosate suppression of an elicited defense response: Increased susceptibility of *Cassia obtusifolia* to a mycoherbicide. Plant Physiol. 98:654-659.
59. TeBeest, D.O. 1982. Survival of *Colletotrichum gloeosporioides* f.sp. *aeschynomene* in rice irrigation water and soil. Plant Dis. 66:469-472.
60. TeBeest, D.O. 1984. Induction of tolerance to benomyl to *Colletotrichum gloeosporioides* f.sp. *aeschynomene* by ethyl methane sulfonate. Phytopathology 74:864.
61. TeBeest, D.O., and Dickman, M. B. 1989. Transformation of *Colletotrichum gloeosporioides* f.sp. *aeschynomene*. Phytopathology 79:1173
62. TeBeest, D.O., Yang, X.B., and Cisar, C.R. 1992. The status of biological control of weeds with plant pathogens. Annu. Rev. Phytopathol. 30:637-657.
63. Templeton, G.E. 1992. Use of *Colletotrichum* strains as mycoherbicides. Pages 358-380 in: *Colletotrichum*: Biology, Pathology and Control. J.A. Bailey and M.J. Jeger, eds. CAB International, Wallingford, UK.
64. Templeton, G.E., TeBeest, D.O., and Smith, R.J. Jr. 1984. Biological weed control in rice with a strain of *Colletotrichum gloeosporioides* (Penz.) Sacc. used as a mycoherbicide. Crop Protect. 3:409-422.
65. Trujillo, E.E., Latterell, F.M., and Rossi, A.E. 1986. *Colletotrichum gloeosporioides*, a possible biological control agent for *Clidemia hirta* in Hawaiian forests. Plant Dis. 70:974-976.
66. Vogelgsang, S., Watson, A.K., and Hurle, K. 1994. The efficacy of *Phomopsis convolvulus* against field bindweed (*Convolvulus arvensis*) applied as a preemergence bioherbicide. J. Plant Dis. Protect.14:253-260.
67. Watson, A.K. and Wymore, L.A. 1990. Identifying limiting factors in the biocontrol of weeds. Pages 305-316 in: New Directions in

Biological Control: Alternatives for Suppressing Agricultural Pests and Diseases. R.R. Baker and P.E. Dunn, eds. Alan R. Liss Inc., New York.

68. Watson, A.K., Gotlieb, A.R., and Wymore, L.A. 1986. Interactions between a mycoherbicide *Colletotrichum coccodes*, and herbicides for the control of velvetleaf (*Abutilon theophrasti* Medic.). Weed Sci. Soc. Amer. Abstracts 26:143.
69. Winder, R.S. 1997. The *in vitro* effect of allelopathy and various fungi on marsh reed grass (*Calamagrostis canadensis*). Can. J. Bot. 75:236-241.
70. Winder, R.S. and Watson, A.K. 1994. A potential microbial control for fireweed (*Epilobium angustifolium*). Phytoprotection 75:19-33.
71. Wymore, L.A. and Watson, A.K. 1986. An adjuvant increases survival and efficacy of *Colletotrichum coccodes*, a mycoherbicide for control of velvetleaf. Phytopathology 76:1115-1116.
72. Wymore, L.A., Watson, A.K., and Gotlieb, A.R. 1987. Interactions between *Colletotrichum coccodes* and thidiazuron for control of velvetleaf (*Abutilon theophrasti*). Weed Sci. 35:377-383.
73. Wymore, L.A., Poirier, C., Watson, A.K., and Gotlieb, A.R. 1988. *Colletotrichum coccodes*, a potential bioherbicide for control of velvetleaf (*Abutilon theophrasti*). Plant Dis. 72:534-538.
74. Xuei, X.L., Järlfors, U., and Kuć, J. 1988. Ultrastructural changes associated with induced systemic resistance to cucumber disease: Host response and development of *Colletotrichum lagenarium* in systemically protected leaves. Can. J. Bot. 66:1028-1038.
75. Yang, X.B. and TeBeest, D.O. 1992. Rain dispersal of *Colletotrichum gloeosporioides* in simulated rice field conditions. Phytopathology 82:1219-1222.
76. Yang, X.B. and TeBeest, D.O. 1995. Competitiveness of mutant and wild-type isolates of *Colletotrichum gloeosporioides* f.sp. *aeschynomene* on northern jointvetch. Phytopathology 85:705-710.
77. Yang, X.B., TeBeest, D.O., and Smith, R.J. Jr. 1994. Distribution and grasshopper transmission of northern jointvetch anthracnose in rice. Plant Dis. 78:130-133.
78. Yu, X., Hallett, S.G., Sheppard, J., and Watson, A.K. 1998. Effects of carbon concentration and carbon-to-nitrogen ratio on growth, conidiation, spore germination and efficacy of the potential bioherbicide *Colletotrichum coccodes*. J. Indust. Micro. Biotech. 20:333-338.

Chapter 16

Potential for Biological Control of Diseases Caused by *Colletotrichum*

Lise Korsten and Peter Jeffries

Colletotrichum is one of the most important genera of plant pathogenic fungi worldwide, particularly in subtropical and tropical regions (67). It attacks a wide range of agronomically important crops resulting in annual losses of billions of dollars in revenue (4). Warm, moist tropical climates favor the development, reproduction, and dissemination of *Colletotrichum*. These environmental conditions are also more typically found in developing countries which often lack the financial means for effective fungicide usage or the infrastructure, such as cool storage facilities, to effectively contain the disease. If we are to seriously address food shortage problems, which are almost exclusively found in developing countries, and improve the quality standards of our fresh produce on international markets, we need to develop a holistic, sustainable and safer disease control strategy. One such alternative that has been exploited with varying degrees of success is biological control. Despite the more recent glut in publications on biological control of fruit and vegetable diseases, very few success stories have been reported. Of even greater concern is that very few of these studies deal with diseases caused by *Colletotrichum* spp., despite the fact that it is one of the most important genera of plant pathogens. This chapter presents an overview of biological control of plant diseases caused by *Colletotrichum* spp. and highlights new research strategies and areas of focus for future programs.

Important Concepts and Research Strategies in Biocontrol

Definitions of biocontrol are continuously being adapted, broadened and changed to suit the needs of researchers and their perspectives of where the focus and strategies of research should be. A general broad definition of biocontrol is the reduction of inoculum density or disease-producing activities of a pathogen or parasite in its active or dormant state, by one or more organisms, accomplished naturally or through manipulation of the environment, host, or antagonist, or by mass introduction of one or more antagonists (5). Different research approaches can be followed to develop an effective biocontrol program. These might include 1) manipulation of the host through induced resistance, phytoalexin production, resistance breeding and genetic manipulation; 2) modification of the postharvest environment (temperature, humidity, atmosphere); 3) alteration of the host surface, e.g., by changing the plant nutrient status; 4) reduction of inoculum through pre- and postharvest sanitation processes, suppression of disease development and spread, or physical treatment (hot water dipping, radiation, etc.); 5) direct or indirect use of foreign or natural disease-suppressing microorganisms. Agriculture already benefits greatly from resident microbial communities that are responsible for natural biological control, but additional benefits are derived by introducing them when or where needed (13). This introduction can be achieved in three ways. The first approach involves inoculative release, which seeks to introduce the agent once or only occasionally into the environment, with the intent that it will establish as a sustained population and impose some level of control. Using a nonindigenous natural enemy/pathogen to control the nonindigenous pest species is defined as "classical biological control." Augmentative application, on the other hand, can be used to supplement the resident population of a microbial biocontrol agent by applying a microorganism already present, either naturally or because of a previous introduction. The subsequent increase of the antagonistic population to an effective density prior to economic damage caused by the target pathogen represents biocontrol. The third alternative, inundative application, could be used to instantly elevate the antagonist population to a density that will ensure maximum and rapid suppression of the target pathogen.

For the purpose of this review, we prefer to concentrate on strategies which involve the direct use of living organisms, either through manipulation of the indigenous microflora or by bioaugmentation involving inoculation with high concentrations of antagonistic microbes from native or exotic sources. The strategy may seem simple to accomplish, but in practice turns out to be extremely difficult as living organisms, by their very nature, respond differently to changing environments and can interact with other organisms in unpredictable ways. Consequently, the realization of effective biological control strategies must be supported by a firm knowledge of microbial ecology within the particular ecosystem. Microorganisms represent a largely untapped natural resource for biological control (13), that have not been fully exploited in disease management. This lack could possibly be attributed to the technical difficulties inherent in establishing a biological control program and the high costs associated with the product development, the regulatory approvals required for each strain, formulation, and use (13). The lack of success in this field is attributed to a number of factors, including 1) unrealistic research time dictated by industry, 2) financial resources decrease the longer the project takes to commercialize, 3) unrealistic expectations about product performance and efficiency, 4) discontinuation of research projects for the previous reasons, 5) a low interest in biological control by industry and economic groups, and 6) a lack of appropriate regulations and guidelines for product registration. Researchers themselves are often also to blame for the difficulties in establishing biocontrol because of inadequate selection and conservation of antagonists, lack of environmental impact assessment, unproven consistency of product performance and the absence of quality control of formulated products.

Potential for Biocontrol of Diseases Caused by *Colletotrichum*

Most *Colletotrichum* spp. are widespread and occupy specific ecological niches on plant surfaces. They can exist as saprotrophic competitors in dead and dying organic plant material, or as necrotrophic or semi-biotrophic pathogens (4), within or on living or dying plant tissues. The presence of other microorganisms on the plant surface can significantly influence the population levels and, thus, the inoculum of *Colletotrichum.* If infection potential is affected as a consequence of antagonistic interactions, then these events offer potential in biocontrol. For example, both stimulatory and inhibitory effects on conidial germination and appressorial differentiation have been reported in the presence of certain bacteria (9,36,49,50). In a broader context, the activity of the microflora associated with plant surfaces can also affect the sporulation capacity of the pathogen, either on infected tissues or in a saprotrophic mode in senescent host material. Several reports exist where application of fungicides can result in increased levels of disease by *Colletotrichum* (25), presumably by the suppression of naturally occurring antagonists. Conversely, foliar fertilizers can reduce disease by *Colletotrichum* (51), presumably by increasing population levels of natural antagonists. If these effects are truly mediated through the relative activity of the resident microflora, then these biological factors can be manipulated to control disease. Unfortunately, biologically based approaches involve complex interactions between microbial populations, the environment, and the plant itself. These interactions are often difficult to predict biological variation and adaptation inherent to living microorganisms, and *in vitro* potential is often matched by *in vivo* disappointment. Thus, the exploitation of the potential for biological control of diseases caused by *Colletotrichum*, or indeed by any other fungal pathogen, is far more complex than might at first appear, and a thorough understanding of plant surface microbiology is essential as a basis for progress. In this review, some of the appropriate stages in the infection process of *Colletotrichum* which offer targets for antagonism are highlighted, and examples illustrate where disease has been or can potentially be reduced by microbial activity.

Given the difficulty in understanding and subsequently manipulating complex natural microbial populations on plant surfaces, researchers have mostly opted for the "silver bullet approach," using one antagonist to control one pathogen. Few reports deal with the use of more than one antagonist applied in combinations or at intervals and making use of compatible *r*- and *K*-strategist combinations to control plant diseases. Controlling only one disease with the resultant increase in importance of another disease which was previously unimportant, is known as an iatrogenic disease. In addition, most researchers often evaluate the antagonist against only one disease and ignore disease complexes or economically less important diseases. To avoid this phenomenon, Korsten et al. (37,42) developed a biocontrol program targeting both pre- (*Cercospora* spot) and postharvest (anthracnose, *Dothiorella/Colletotrichum* fruit rot complex [DCC] and stem end rot [SE]) diseases of avocado.

The direct application of mostly indigenous antagonists has recently received more attention and resulted in a number of new commercial products, including Aspire™, BioSave™, Trichodex™, AQ10™, Avogreen™ and BlightBan A506™ (28,45,76). This suggests that the use of biocontrol products for aerial plant diseases is increasing. Since most of these products have been developed for postharvest applications, particular attention will be given to postharvest control of fruit anthracnose as this situation offers advantages in the search for biocontrol strategies (28). For example™, environmental conditions during fruit transport and storage are generally more uniform than in the field and can often be manipulated. The biomass of harvested produce is also much less than that of the standing crop, easier to treat in a uniform manner, and more suited to directly target the pathogen with an appropriate biocontrol formulation. Progress has already been made, and several case studies will be reviewed which have attempted to exploit this natural phenomenon to reduce levels of disease caused by *Colletotrichum*.

Targeting the Pathogen's Life Cycle for Effective Biocontrol

Inoculum of *Colletotrichum* spp. usually arrives at the infection court as rain-washed spores, and adhesive factors operate to stick them to the plant surface. During subsequent developmental stages, the fungus is vulnerable to antagonism by resident microbes and must be able to compete and resist any inhibitory effects by surface microflora. Thus, the rational design of biological control strategies which are effective *in vivo* depends on a good understanding of the behavior of the pathogen within the infection court. Although *Colletotrichum* spp. cause a wide range of plant diseases, the infection process usually follows a generalized pattern of conidial germination, appressorial formation and direct cuticle penetration (29). Spore germination offers an immediate target that may be inhibited or stimulated by metabolites of other organisms. Growth of *Colletotrichum gloeosporioides* (Penz.) Penz. & Sacc. in Penz. on guava was inhibited by extracts and volatiles from several dominant phylloplane fungi on the fruit (59). Because several *Colletotrichum* spp. also produce self-inhibitors of spore germination (71), the use of inhibitor analogues may facilitate biological control of these fungi. Alternatively, if other microorganisms could be found which produce the inhibitors, then they may have potential as antagonists. Korsten and De Jager (36) showed that naturally occurring dominant *Bacillus* spp. isolated from the phylloplane could effectively inhibit *C. gloeosporioides* spore germination *in vitro*.

After germination and prior to appressorial development, the germ tube is vulnerable to mycoparasitism or lysis. *In vitro Colletotrichum* inhibition assays with *Bacillus* spp. showed that bulb formation and subsequent lysis of the germ tube took place immediately after conidial germination (36). Appressorial formation may occur almost immediately after germination or after a period of hyphal growth over the plant surface. With *in vitro* inhibition studies, Korsten and De Jager (36) could also show an enhanced stimulation of appressorial formation. However, once formed, the mature appressorium presents a formidable barrier to antagonism as it is thick-walled and melanized; biocontrol is unlikely to be exerted during this phase of development. Microbial activity can also affect the number of appressoria that are successfully formed or may influence infection from the appressorium. Korsten and De Jager (36), for instance, showed that the presence of *Bacillus* spp. could sustain appressorial dormancy. Conversely, in a preharvest environment, biocontrol sites are targeted prior to appressorium formation. This illustrates the importance of following a pre- and postharvest strategy to control latent infections as demonstrated by the successful control of anthracnose on avocado with pre- and postharvest applications of *Bacillus subtilis* (ATCC 55466/B246) (32,35,42).

Development of the penetration peg may be influenced by external events, and the tissue can be colonized immediately, as in leaf and flower necrosis. Colonization may also be delayed during a period of quiescence, as a dormant appressorium or as limited subcuticular mycelium, similar to fruit anthracnose. The role of appressoria and quiescence in the infection process is still somewhat controversial, especially with preharvest infection of immature fruit (12). Under conditions that stimulate infection peg formation, does the fungus establish a discrete, dormant subcuticular infection or does the subcuticular infection perish as a result of lethal concentrations of phytoalexins or preformed inhibitors? This information can be very important in determining interactions with potential biocontrol agents, since the melanized appressorium is likely to resist antagonistic activities, but a subcuticular mycelium could be vulnerable to endophytic fungi or bacteria which manage to follow the fungus into the subcuticular region via the penetration channel. Both pre-appressorial and post-appressorial targets can probably be exploited as there is now sufficient evidence that both preharvest sprays that enhance microbial activity and postharvest dips of fruit in suspensions of antagonists can reduce disease levels. Whatever the target, the success of a potential biocontrol agent is obviously governed by the activities of the pathogen. Data regarding the influence of environmental conditions, such as temperature and relative humidity, the timing of spore germination, appressorial development and penetration peg formation is vital if rational approaches to biocontrol are to be implemented.

Case Histories

An earlier review of biocontrol potential for diseases caused by *Colletotrichum* (30) suggested that biological methods for control would not be more widely adopted until effective chemicals were withdrawn. Fungicides remain the first choice for agricultural workers wishing to control these diseases. Nevertheless, research on biological alternatives has continued, and new information has become available since the earlier review. Some of the recent developments are discussed below.

Diseases of Avocado

Considerable progress has been made on integrating the use of fungicides and antagonistic bacteria for control of pre- and postharvest diseases of avocado. *Cercospora* spot caused by *Pseudocercospora purpurea* (Cke Deighton) is one of the most important preharvest diseases in unsprayed orchards. *Cercospora* spot lesions also provide effective entry points for preharvest infections by preharvest infections by *C. gloeosporioides*. Postharvest diseases such as anthracnose and, to a lesser extent, DCC and SE diseases cause serious losses in export consignments exposed to maximum refrigeration conditions during long shipments. Although *C. gloeosporioides* plays an important role in all three postharvest diseases, pathogens such as *Dothiorella aromatica* (Sacc.) Petr. & Syd., *Lasiodiplodiatheobromae* (Pat.) Griffon & Maubl., *Pestalotiopsis versicolor* (Speg.) Steyart, *Phomopsis perseae* Zerova, *Thyronectria pseudotrichia* (Schw.) Seeler, and several other minor pathogens can also cause SE infections. Because the French government has restricted the use of prochloraz as a postharvest treatment, growers in South Africa (SA) face the daunting task of providing high-quality, unblemished fruit to European export markets with little chemical protection. Two to three annual preharvest copper oxychloride field sprays are currently used to control all fruit diseases, with obvious limited success.

In the mid-1980s, South African avocado growers realized that they had to develop and financially support a research project aimed at finding alternative control strategies for avocado fruit diseases; "new chemicals" were not likely to be developed by agrochemical companies for such a small market. Availability of existing chemicals was also being threatened by growing global chemophobia. According to Blakeman (8), fundamental information on the relationship between host, pathogen, and associated microflora, as well as the character of the plant surface and environment, must first be obtained before initiating a biocontrol program. Because of the complex, time-consuming nature of this approach, few researchers have opted to do so.

In 1986 a biocontrol program was initiated to select natural antagonists to control both pre- and postharvest diseases of avocado (34). Numerous bacteria, filamentous fungi, and yeasts were isolated from avocado leaf and fruit surfaces using leaf imprint and leaf and fruit washing techniques (32). From the dominant 106 representative fungal and 176 bacterial isolates, 36 fungal and 48 bacterial isolates were identified to the generic or specific level. All isolates obtained in the ecological study were initially screened *in vitro* against the most prevalent pathogen, *D. aromatica*, using both direct and indirect screening assays (32). Several *Bacillus* spp. were initially selected and screened *in vivo*, and eventually evaluated in postharvest packhouse experiments (32). The most promising of these, *B. subtilis* (B246), was further evaluated in semi-commercial pre- and postharvest experiments (32).

Preharvest *B. subtilis* sprays on their own or integrated with copper oxychloride commercial sprays effectively controlled postharvest anthracnose (32). A single early-season copper oxychloride spray followed by two sprays at monthly intervals with *B. subtilis* effectively reduced the disease. Four sprays at monthly intervals with *B. subtilis* were equally effective. These experiments were conducted at two geographically distinct farms over a 3-year period, and in both instances the biological or integrated programs were not effective the first year of spraying, but provided positive control the following 2 years. This lag indicates that the biocontrol agent first had to establish itself to a critical minimum threshold level within the tree canopy before effective control could be obtained. The establishment of a residual population on the phylloplane, which resulted in sustained protection for a prolonged period was corroborated when significantly less anthracnose occurred in the biological and integrated treatments, 1 year after the experiment was terminated.

According to Spurr and Knudsen (65), antagonists applied before harvest should be monitored in the field at several intervals starting at the time of application. This information is necessary to predict survival of the antagonist in the field and can provide information necessary to enhance biocontrol effectiveness through improved formulation, adjustment of dosages, and spray schedules. Demonstrating antagonist survival and monitoring antagonist population fluctuations in biocontrol preharvest experiments are difficult. Antagonist attachment and survival were studied by scanning electron microscopy (SEM) and leaf and fruit imprints and washings (70). *B. subtilis* could effectively attach to the avocado leaf and fruit surface within 20 seconds. Cell multiplication was evident between wax rodlets on the abaxial surface, and the antagonist effectively colonized depressions on the corrugated abaxial and, to a lesser extent, adaxial surfaces (33). A significant drop in viable cells was recorded over time with the imprint and washing techniques. This tendency was not reflected in the SEM study, which actually showed a nonsignificant increase in number of cells on the abaxial, but not the adaxial surface (70).

Monoclonal antibodies were subsequently developed to improve field detection of the antagonist. Towsen (70) optimized an enzyme-linked immunosorbent assay (ELISA) and used selected clones which belonged to the IgG2b subclass. The chemical nature of the epitope detected by these antibodies was proteinaceous (70). Both *in vitro* and greenhouse studies showed that the signal/background value obtained with the ELISA corresponded with the *B. subtilis* cell concentration used. However, similar data could not be obtained in field studies, due to either the loss of antigenicity or stage-specific expression of the antigen. Improving antagonist attachment was evaluated by comparing different stickers and spreaders to the topically applied biocontrol spray. Cell counts revealed that antagonist attachment could not be significantly increased by any of the products tested.

An important aspect of all biocontrol programs is the mode of action. Korsten and De Jager (36) postulated that *B. subtilis* had more than one possible mode of action against *C. gloeosporioides* and correlated it to growth stage of the pathogen. Microbial interactions are known to be density-dependent, and organisms can interact in more than one way. When the antagonist was applied to the avocado fruit surface 1 to 7 days prior to arrival of the pathogen, *C. gloeosporioides* and subsequent disease development were effectively inhibited. Competitive exclusion was therefore proposed as one type of microbial interaction. Nutrient competition and antibiosis were traditional modes of interactions indicated by a "checkerboard-type" assay in ELISA plates. Direct contact between *B. subtilis* cells and *C. gloeosporioides* spores in minimal liquid growth medium inhibited spore germination and bulb formation, and subsequently, germ tubes and hyphae lysed.

Most reports on biological control of postharvest fruit diseases focus on post- instead of preharvest applications because the antagonist is easier to apply to the target site (78). Environmental conditions in packhouses and cold-storage facilities, where the postharvest environment is an artificial "ecological island" separated by the buffering effect of natural microbial ecosystems (79) (79), is also easier to manipulate. Most importantly, a preharvest biocontrol strategy requires larger volumes of biocontrol product than postharvest applications. For instance, Korsten et al. (41) reported that 900 L of a 10^9 cells ml^{-1} suspension of *B. subtilis* is required to treat 1 tonne of avocado fruit at preharvest, compared to 1 L needed to treat the same number of fruit in packhouses. This is obviously a tremendous cost saving for the grower but would not be the most lucrative option for the company producing and marketing the biocontrol agent. Since *C. gloeosporioides* infects in the field and not after harvesting, a preharvest strategy of reducing inoculum and preventing infection will always remain the most feasible option. An added advantage of using a preharvest approach is that the preharvest disease, *Cercospora* spot, can be controlled simultaneously (42).

Postharvest application of *B. subtilis* proved effective in controlling anthracnose and to a lesser extent DCC and SE (32,38,40,43,45,72). Biocontrol performance, assessed over several years throughout the season in different packhouses and even in different countries (New Zealand and Australia), was variable (33), viz. equally (40), more (38) or less effective (72) than chemical control. If biocontrol cannot produce results comparable with that of commercial fungicides, it will not be a viable commercial option (69). Biological products, which can be used in combination with conventional crop protection procedures, should more readily gain acceptance. As part of our initial screening strategy, compatibility with fungicides or disinfectants were routinely tested. *B. subtilis* was compatible with copper oxychloride, benomyl, thiabendazole, and reduced concentrations of prochloraz (32). Integrated control where *B. subtilis* was combined with reduced concentrations of the fungicide prochloraz proved equally effective in controlling anthracnose as full-strength prochloraz treatments (38,40).

Different approaches to the integration of antagonist applications with fungicides or disinfectants provide different levels of control (72). An initial 10% ethanol dip followed by *B. subtilis* applied in the subsequent wax treatment proved more effective in controlling anthracnose than the two products on their own. By first "disinfecting" the fruit surface, we partially create a "biological vacuum" that can be more effectively colonized by the antagonist applied later on in the packing line. Inconsistencies nevertheless were recorded with this "synergistic" approach to plant disease control. When the previously described experiment was repeated at another packhouse and treated fruit were stored for 28 days at 5°C to simulate export conditions before ripening and evaluation, anthracnose could not be controlled.

If biocontrol can be made compatible with existing farming practices, it will add to its commercial acceptance. Current postharvest handling practices of avocado fruit in South Africa routinely include a chlorine spray at the receiving point to remove chemical residues, dust, and sooty blotch. This is usually followed by an on-line water spray rinse, hot air (52°C) drying, commercial waxing, drying once again, sorting and packing. Various modifications can be found at different packhouses, with some excluding the initial chlorine spray, or using a dip or drench application instead. Packhouses might use a prochloraz ultra-low-volume (ULV) spray before the wax application, particularly if fruit is destined for local markets. Various application points of the antagonist have been evaluated, viz. a dip, ULV spraying, and wax application (32). The latter application method was the most practical and effective in controlling anthracnose. Furthermore, the antagonist survived in the wax for several days without a drop in viability (38). This procedure enables packhouses to mix the antagonist into the wax on a daily or twice-weekly basis, without adversely affecting product performance.

Optimal antagonist concentrations required for disease control at various inoculum pressures were determined by means of a "checkerboard-type titration" assay (37). Increasing concentrations (10^5 to 10^8 cells ml^{-1}) of *B. subtilis* were effective against increasing concentrations of *C. gloeosporioides* (10^3 to 10^6 spores ml^{-1}). Using the same concentration range in packhouse tests indicated that a concentration of 10^7 cells ml^{-1} was the most effective in reducing anthracnose and SE (37). An important attribute of a successful biocontrol agent is the ability to be effective at low concentrations (81). *B. subtilis* evaluated in the previous studies conformed to this prerequisite by being generally effective against the various avocado postharvest pathogens, and diseases caused by them, at the lowest tested concentration of 10^5 cells ml^{-1} both in the laboratory and in the packhouse. In general, increased control of avocado postharvest diseases tended to be associated with increasing antagonist concentrations and decreasing pathogen spore challenge levels, a phenomenon that is common in biocontrol. Keep in mind that, under commercial conditions, fruit obtained from different growers and/or production blocks will be treated similarly in a particular packhouse. Effectiveness of any control measure will thus be affected by the difference in physiological status of the fruit, and inoculum pressure and product performance will have to be consistent against all pathogens at different inoculum levels. This eventually proved the ultimate challenge when the biocontrol product (Avogreen), originally formulated for preharvest applications, was tested commercially in six different packhouses throughout the season (45). Product performance varied from packhouse to packhouse and, within a packhouse, from week to week. Interestingly though, the agrochemicals included for comparative purposes, were also inconsistent (32,38,40,42,43,45). Establishing consistent levels of control under various conditions, with fruit obtained from various growing areas and representing various cultivars, will remain the ultimate challenge of any biocontrol agent. *B. subtilis* effectively controlled anthracnose on most of the commercially important cultivars, which includes Fuerte (32,38,43,72), Hass (32,37,40), Edranol (32), and Ryan (40).

As most avocado fruit pathogens occur in the orchard as latent skin infections (60), the significant control achieved with postharvest applications of *B. subtilis* demonstrated its effectiveness against established infections. Disease control was still evident after 21 days in the laboratory and 35 days in the packhouse, when the respective treatments were terminated, in contrast with other postharvest biocontrol studies, which mainly report temporary effects (19,53). What makes the study unique is that a holistic approach was followed from the beginning and aimed at evaluating *in vitro* and *in vivo* performance of the antagonist against the full spectrum of postharvest diseases and the various pathogens involved. Such an approach can help prevent the development of iatrogenic diseases.

DISEASES OF MANGO

Postharvest anthracnose of mango similar to avocado caused by *C. gloeosporioides* is a latent disease in which infection of immature fruit takes place in the field, but symptoms appear only after harvesting on fully ripened fruit. Control of the disease is achieved by preharvest application of fungicides (mostly copper oxychloride), usually followed by a postharvest treatment consisting of a hot water bath often with the incorporation of 500 ppm benomyl (29). Widespread use of benomyl has led to resistance to this fungicide in the pathogen population. In a series of experiments in the Philippines, both pre- and postharvest treatments with potential microbial antagonists were investigated as alternative methods of disease control. Only the postharvest approach was successful (30,31). This can possibly be ascribed to the complexities involved in establishing antagonistic microbial populations on the mango phylloplane, which is an ecologically diverse environment, as well as to the complexities involved in maintaining an ecological balance of microbes (80). For the postharvest approach, Koomen and Jeffries (31) isolated potential microbial antagonists of *C.*

gloeosporioides from blossoms, leaves, and fruit of mango and screened them using a series of assay techniques (31). In total, 648 isolates, including bacteria, yeasts and filamentous fungi, were screened for their inhibition of growth of *C. gloeosporioides* on malt extract agar. *In vitro*, 121 isolates inhibited the fungus and were tested further for their ability to affect conidia germination. Of these isolates, 45 bacteria and yeasts inhibited germination. These were inoculated onto mangoes artificially infected with *C. gloeosporioides* and assessed for their potential to reduce the development of anthracnose lesions. Seven isolates were selected for use in a trial in the Philippines using freshly harvested fruit. This final screening procedure yielded two potential candidates for further trials, isolate 204 (identified as *Bacillus cereus*) and isolate 558 (identified as *Pseudomonas fluorescens*). In postharvest trials under commercial conditions, isolates 204 and 558 were tested by different application methods and in combination with adhesive material, peptone, fruit wax, or sucrose polyester (Tables 1 and 2). Isolate 204 did not reduce disease, whereas isolate 558 significantly suppressed anthracnose development (Table 1). Incorporating the bacteria in adhesive material, peptone, fruit wax, or sucrose polyester did not add any benefit. The initial concentration of inoculum in bacterial dips was 5.5 x 10^6 colony forming units (cfu) ml^{-1} for isolate 204 and 6 x 10^7 cfu ml^{-1} for isolate 558. This resulted in bacterial numbers on the mango peel after dipping of 5 x 10^4 cfu cm^{-2} for isolate 204 and 10^5 cfu cm^{-2} for isolate 558. For isolate 204, the numbers of bacteria on the mango surface decreased to between 10^4 and 3 x 10^4 cfu cm^{-2} after 5 days. Isolate 558 could not be re-isolated from the mango surface of any of the treatments by day five. Samples were not taken before the fifth day, so the rate of decline in numbers is not known. In a further experiment, the population levels of isolate 558 were maintained for 48 hr after dipping, but declined rapidly after this period.

These experiments were continued in a second year of field trials, and results again showed that anthracnose of mango could be reduced at the postharvest stage by using both bacteria under commercial conditions. However, an alternative treatment, in which mangoes were dipped in hot water (50-55°C, 10 min) proved more effective for disease control than the biological strategy, provided fruit were dipped within 24 hr of harvest. Hot water dipping was recommended for local adoption.

In the experiments described previously, *P. fluorescens* (isolate 558) significantly inhibited germination of spores of *C. gloeosporioides*. Since spores germinate in the field before harvest, this isolate was investigated for preventing infection of mangoes by *C. gloeosporioides* (30). Sprays containing 9.4 x 10^7 cfu ml^{-1} of isolate 558 were applied at the onset of flowering and at fruit set, and compared with untreated controls or trees

Table 1. Percentage anthracnose development after 10 days in mangoes treated with bacterial isolates 204 and 558, or no bacteria applied, in combination with sticker, peptone or hot water (from ref. 31)

Treatment	Storage conditions	No bacteria	204	558	Mean	SE (d.f.)
Sticker	Dry	9.0	11.3	8.0	6.3	
	Bag	5.1	3.3	1.3		
Peptone + sticker	Dry	13.3	10.3	7.3	6.7	NS[a]
	Bag	4.7	2.4	2.0		
Hot water + sticker	Dry	18.0	14.0	8.7	7.9	
	Bag	3.3	2.2	1.3		
Mean	Total	8.9	7.3	4.8		
SE (d.f.)			1.36 (233)			
Overall mean	Dry	13.4	11.9	8.0	11.1	1.11
	Bag	4.4	2.6	1.5	2.8	(233)

[a]NS F value not significant at $P = 0.05$.

Table 2. Percentage disease development after 10 days for mangoes treated with either bacterial isolates 204 and 558 or no bacteria in combination with fruit wax or sucrose polyester (from ref. 31)

	% Anthracnose			Mean
	No bacteria	204	558	
Fruit wax	6.3	5.7	9.0	7.0
Sucrose	9.0	6.0	8.3	7.8
Mean	7.7	5.8	8.7 [a]NS	7.4

[a]NS = F value not significant at $P = 0.05$

receiving applications of benomyl. Disease levels did not differ significantly between any of the treatments. It was surprising that even the full fungicide treatment did not significantly reduce the anthracnose levels, but during the period in which the study was carried out there was very little rainfall; hence, disease levels were very low. From the results, isolate 558 apparently did not survive on the leaf surface. Samples taken immediately after leaves were sprayed with inoculum of isolate 558 showed a recovery rate of 1.6 x 10^5 cfu cm^{-2} leaf. After 1 week, however, the population of isolate 558 on treated leaves had dropped to 10^8 cfu cm^{-2} leaf, i.e., not different from the sprayed control (55 cfu pseudomonads cm^{-2} leaf), and below a reliable detectable level. At the time of harvest, no pseudomonads were isolated from the control, nutrient-, or isolate 558-treated trees. Similarly, a promising biocontrol agent failed to control disease in the field when *Pseudomonas cepacia* failed to survive on the phylloplane in trials on apple scab (48).

In other studies to isolate natural antagonists from the mango phylloplane, two *B. licheniformis* strains were recovered that were effective in controlling anthracnose in semi-commercial preharvest field sprays (44) and in postharvest packhouse dip applications (39,46). In a subsequent ecological study, De Jager et al. (17) reported the non-target effect of commercial fungicides used in the mango industry on the natural microbial population. They consequently evaluated the total microbial population for its ability to maintain natural antagonists. Of 1993 microbes isolated from the mango phylloplane, 842 represented bacteria, 623 yeasts, and 528 filamentous fungi (16). *Bacillus* spp. represented the most dominant bacterial group, while *Aureobasidium* and *Sporobolomyces* represented the dominant yeasts and *Cladosporium* and *Alternaria* the dominant fungal spp. *In vitro,* relatively few of these isolates inhibited mycelial growth of *C. gloeosporioides* at a distance within the petri dish (4% of bacteria, 4.8% yeasts and 1.8% fungi). Contact inhibition was far more evident (14.7% bacteria, 56.8% yeasts and 71.4% fungi). Only 51.1% of the epiphytic population tested represented natural antagonists of *C. gloeosporioides*. Of these, *Bacillus* spp. were the most effective. Scanning electron microscope studies of the various mango cultivars revealed that the relatively hostile phylloplane habitat may limit microbial colonization, and that the nutrient composition of the exudates may regulate *r*- and *K*-strategist colonization patterns (2). Korsten et al. (44) described the morphological features of the mango phylloplane and evaluated various stickers and spreaders to enhance antagonist attachment in preharvest field sprays. None of the tested products improved antagonist attachment and survival, and no significant difference in survival was found between ad- and abaxial surfaces. Preharvest *B. licheniformis* commercial field sprays with either one or both of two antagonistic isolates (B250 or B251), applied at monthly intervals between

October and January, effectively reduced anthracnose on the cv. Keitt. However, the biological treatments proved less effective than the commercial Cupravit (copper oxychloride) spray treatment. The same antagonists in postharvest packhouse dip treatments more effectively controlled anthracnose than did commercial benomyl or NaOCl treatments on the cultivar Sensation (46). Repeating these experiments on the cultivar Keitt resulted in effective control but only when isolate B251 was used on its own (46). The following year only B251 was used in semi-commercial experiments on Sensation cultivar fruit. Effective control was achieved when fruit was first dipped in warm water for 5 min followed by an antagonist application as a 7-min dip with Agral 90 added. This treatment was, however, not as effective as the commercial chemical treatment which included a prochloraz 5-min dip in 52 C followed by a benomyl commercial wax (39). Repeating these biocontrol experiments on the cultivar Keitt during the same year and the following year proved less effective in controlling anthracnose (18). What was apparent from these and other studies (45,52) was that the initial inoculum level greatly influenced the outcome of the treatments, particularly with regard to postharvest treatment of fruit. Consistency of product performance will remain one of the biggest challenges to address to gain commercialization and general acceptance of biocontrol.

Coffee Berry Disease

Coffee berry disease (CBD) is caused by *Colletotrichum kahawae* J.M. Waller & P.D. Bridge sp. now. After infection of immature coffee berries, anthracnose lesions develop on very young berries or on older, maturing ones. Where resistance is exhibited, the berries scab over, but scabs may become active and sporulate again later in the disease cycle (22). In common with many other diseases caused by *Colletotrichum*, the fungus also infects leaves and twigs, and an active population is found throughout the year within the canopy of mature coffee bushes. A number of saprotrophic *Colletotrichum* strains and other microorganisms, which also grow on coffee surfaces and interact with *C. kahawae,* offer potential for biocontrol. Circumstantial evidence for natural biocontrol comes from observations that fungicide applications can result in increased levels of CBD (25,26), suggesting that removal of indigenous fungal antagonists was responsible for this phenomenon. However, this effect has been difficult to demonstrate directly. For example, two saprophytic strains of *Colletotrichum,* when mixed with *C. kahawae*, exerted very little effect on infection by the pathogen (27), reflecting earlier studies (25) which also indicated that other microorganisms do not appear to interfere much in the infection process. Waller (73) found that rainwater collected from unsprayed coffee bushes reduced germination of fresh spores of *C. kahawae* by 16-30% and suggested that microbial activity in the phyllosphere was responsible. Masaba (55) studied interactions between the CBD fungus and microorganisms isolated from coffee bushes in detail and found several antagonistic species within the indigenous coffee microflora. The most marked inhibition was by *Fusarium stilboides* Wallenw., but unfortunately this fungus can also cause disease on coffee bushes. There was no marked correlation between the populations of dominant microflora and disease recorded in the field. Some yeasts appeared to inhibit spore germination and appressorium formation by *C. kahawae* on berry surfaces, but others increased appressorium frequency and infection despite reducing spore germination levels (55). Further work is now continuing in Malawi, where similar effects of increased CBD has been noted on bushes sprayed with fungicide (N. Phiri and P. Jeffries, unpublished).

Ecological Approaches to Disease Control

An increase in fungicide-tolerant saprotrophs was implicated in preventing large increases in *Colletotrichum acutatum* Simmonds populations following treatment of mango trees with benomyl (23). The population of this benomyl-tolerant pathogen was expected to rise as a result of fungicide application, but the saprotrophic fungi apparently quickly occupied the colonization sites of the mango and inhibited infection. The effects of agrochemical applications on non-target organisms are not well studied. It is clear that the epiphytic populations of microorganisms are both quantitatively and qualitatively altered when standard pesticide regimes are applied (3). If sufficient information was available regarding ecological effects of pesticides *in vivo*, biological control of plant disease could be achieved by a logical manipulation of indigenous microbial antagonists through an integrated program of pesticide use.

A similar potential for the control of anthracnose of avocado using phylloplane microorganisms has also been reported from Australia (66). Over a third of the 1050 microorganisms isolated from leaves, fruit, and flowers of avocado trees that had not been sprayed with fungicides for several years inhibited the growth of *C. gloeosporioides* on nutrient agar. Many of these organisms also significantly reduced spore germination of the pathogen on cellophane membranes, but only a few inhibited germination on avocado leaf discs. Twenty-two isolates were compared in detached fruit experiments for their ability to suppress anthracnose lesion formation. Although results were variable, consistent and significant suppression of preharvest infection by the pathogen was recorded for two antagonistic *Bacillus* spp. and a pink yeast. Lesion size was reduced only following treatment with *Bacillus* isolate 359. Examination of fruit peels showed that appressorium formation by the pathogen was variable between fruit batches, possibly explaining some of the variability of response of some antagonists in the detached fruit tests.

Yeasts were also highlighted as antagonists of *Colletotrichum musae* (Berk. & Curt.) Arx in a banana leaf disc assay devised by Postmaster et al. (61). Introducing the antagonists 48 to72 hr prior to challenge with the pathogen was essential for effective control, and some of the antagonistic yeasts did not inhibit growth of the pathogen in conventional *in vitro* assays. This phenomenon, widely reported elsewhere (28), emphasizes the need for a carefully designed screening strategy for determination of biocontrol potential, if possible avoiding a preliminary *in vitro* screen. Direct screening on fruit surfaces, albeit laborious, appears to be the most efficient method of obtaining antagonists which will remain effective in the field situation, and several rapid methods have now been developed including combinations of direct and indirect assays (36,47).

There are fewer examples of biocontrol of *Colletotrichum* diseases of non-fruit crops. A spore suspensionof *Trichoderma viride* (Pers. & Gray) has been used as a foliar spray to reduce brown blotch incidence on cowpea seedlings in the field (6) to levels comparable to those following fungicide application. This disease is caused by *C. truncatum*. In some cases, the activity of plant-growth promoting rhizobacteria has partially been explained by their inhibitory effects on plant pathogens. For example, Beauchamp et al. (7) showed that 34% of opine-utilizing rhizobacteria tested *in vitro* inhibited the growth of the potato pathogen *Colletotrichum coccodes* (Wallr.) Hughes.

Interactions with Nonpathogenic Species of *Colletotrichum*

Controlled inoculation of lower leaves of cucurbits with pathogenic strains of *Colletotrichum orbiculare* (Berk. & Mont.) Arx was one of the first examples of induced systemic resistance (ISR) in plants. This was developed in a protective role to reduce disease levels arising from further challenges with the same pathogen in the field (11). This fungus was tested successfully in similar cross-protection studies where induced resistance was evident not only after challenge inoculations with the same pathogen but also with other biotrophic and necrotrophic pathogens (15). The potential for exploiting ISR to control postharvest diseases of fruit and vegetables has been reviewed (77), and the authors concluded that the widespread occurrence of ISR augured well for its usefulness in future control strategies. This earlier work with pathogenic strains has now been followed by examples where nonpathogenic organisms are used to induce

ISR. For example, treatment of soil with a nonpathogenic strain of *Fusarium oxysporum* Schecht. emend. Snyd. & Hans. significantly reduced anthracnose symptoms on leaves of cucumber following subsequent challenge with *C. orbiculare* (54). Cross-protection using attenuated virus strains has also reportedly controlled this fungus on cucumber (1). Similarly, treatment of cucumber plants with plant growth-promoting bacteria resulted in significant protection from naturally occurring anthracnose caused by *C. orbiculare* (74). There is also evidence for induced resistance in other plant species, for example *Stylosanthes*, following inoculation with sublethal levels of pathogenic isolates of *C. gloeosporioides* (14). Salicylic acid has also been suggested to be involved in bacterially mediated ISR, but Press et al. (62) demonstrated that it was not the primary determinant of ISR when *Serratia marcescens* was used to protect cucumber plants from infection by *C. orbiculare*. A nonspecific mechanism seems to be involved; ISR to cucumber anthracnose can also be induced by amending growth substrates with composts or decomposed sphagnum peats (82). Whatever the mechanism involved, there is potential for this biologically mediated method for control of anthracnose diseases.

Disease control in cucurbits using a nonpathogenic mutant ('path-1') of *Colletotrichum magna* S.f. Jenkins & Winstead was described by Freeman and Rodriguez (24). This mutant was similar in behavior to the wild-type pathogen in levels of sporulation, spore adhesion, formation of appressoria and infection frequency, but it grew within cucurbit fruits as a nonpathogenic endophyte (see Chapter 8, Rodriguez and Redman). Pre-infection of cucurbits with 'path-1' protected plants from disease caused by the wild-type isolates of *C. magna*, with the degree of protection being proportional to the length of time between pre-treatment and subsequent pathogen challenge. Protection in this case was not systemic, but localized to regions invaded by the endophytic mutant. Protection was presumed to be mediated via 'priming' of host defenses, particularly since the protection could also be conferred against *Fusarium* pathogens. The possibility of using such nonpathogenic strains of *Colletotrichum* to induce fruit resistance to *C. gloeosporioides* was considered by Prusky et al. (63) as a novel approach for anthracnose control in avocado. In this case, the same nonpathogenic mutant of *C. magna* inhibited subsequent decay by the normal pathogen *C. gloeosporioides*. The enhanced resistance was suggested to result from the induction of epicatechin, a phenol that inhibits oxidation of an anti-fungal diene present in immature avocado peel (see also Chapter 14, Prusky et al.).

Recommendations for Future Work

From the experience gained with the numerous studies detailed above, a number of recommendations can be made:

- Alternative initial *in vitro* screening of inhibition of growth is desirable.
- Composites of antagonists are preferable to single organisms, preferably including species that utilize different nutrient sources or occupy different ecological niches.
- Organisms which tolerate fungicides will facilitate integrated control.
- Exploitation of induced systemic resistance may offer a novel alternative.
- A biocontrol program should be based on an in-depth ecological study to better understand microbial interactions and the possible nontarget effect of biocides.
- Consistency of product performance is essential and should be determined under different inoculum pressures and on fruit obtained from different geographical areas and different cultivars.
- Quality assurance of biocontrol products is essential to ensure purity, shelf-life, and viability at specified concentrations.
- Product registration and first-tier toxicological tests remain essential aspects to address before commercialization.
- Genetic stability of the biocontrol product, risk assessment, and potential build-up of pathogen resistance should be determined prior to implementing a biocontrol program.

Where biological agents have to compete in a market dominated by effective and relatively inexpensive fungicides, the challenge and demands facing biocontrol agents are great (2). Superimposed on these demands is the reality that growers are concerned primarily with the economic aspects of control, generally risk-averse, and inclined to rely on measures which act promptly (2).

Acknowledgements

P. Jeffries would like to thank John Dodd, Arnold Estrada, Irene Koomen, and Resty Bugante for their help in execution of the mango anthracnose work and the Overseas Development Administration (now DfID) for funding. L. Korsten would like to thank E. de Villiers and E. de Jager for research assistance and the South African Avocado Growers' Association for financial support.

Literature Cited

1. Ahoonmanesh, A. and Shalla, T.A. 1981. Feasibility of cross-protection for control of tomato mosaic virus in fresh market field-grown tomatoes. Plant Dis. 65:56-58.
2. Andrews, J.H. 1992. Biological control in the phyllosphere. Annu. Rev. Phytopathol. 30:603-635.
3. Andrews, J.H. and Kenerley, C.M. 1978. The effects of a pesticide program on non-target epiphytic microbial populations of apple leaves. Can. J. Microbiol. 24:1058-1072.
4. Bailey, J.A., O'Connell, R.J., Pring, R.J., and Nash, C. 1992. Infection strategies of *Colletotrichum* species. Pages 88-121 in: *Colletotrichum*: Biology, Pathology and Control. J.A. Bailey and M.J. Jeger, eds. CAB International, Wallingford.
5. Baker, K.F. and Cook, R.J. 1974. Biological control of plant pathogens. Freeman and Comp, SF. 43 pp.
6. Bankole, S.A. and Adebanjo, A. 1996. Biocontrol of brown blotch of cowpea caused by *Colletotrichum truncatum* with *Trichoderma viride*. Crop Protec. 15:633-636.
7. Beauchamp, C.J., Dion, P., Kloepper, J.W., and Antoun, H. 1991. Physiological characterization of opine-utilizing rhizobacteria for traits related to plant growth-promoting activity. Plant Soil 132:273-279.
8. Blakeman, J.P. 1985. Ecological succession of leaf surface microorganisms in relation to biological control. Pages 6-30 in: Biological Control on the Phylloplane. C.E. Windels and S.E. Lindow, eds. American Phytopathological Society, St. Paul, Minnesota. 169 pp.
9. Blakeman, J.P., and Parbery, D.G. 1977. Stimulation of appressorium formation in *Colletotrichum acutatum* by phylloplane bacteria. Physiol. Plant Pathol. 11:313-325.
10. Burkhead, K.D., Schisler, D.A., and Slininger, P.J. 1994. Pyrrolnitrin production by biological control agent *Pseudomonas cepacia* B37w in culture and in colonised wounds of potatoes. Appl. Environ. Microbiol. 60:2031-2039.
11. Caruso, F.L. and Kuć, J. 1977. Field protection of cucumber, watermelon, and muskmelon against *Colletotrichum lagenarium* by *Colletotrichum lagenarium.* Phytopathology 67:1290-1292.
12. Coates, L.M., Muirhead, I.F., Irwin, J.A.G., and Gowanlock, D.H. 1993. Initial infection processes by *Colletotrichum gloeosporioides* on avocado fruit. Mycol. Res. 97:1363-1370.
13. Cook, R.J., Bruckhart, W.L., Coulson, J.R., Goettel, S., Humber, R.A., Lumsden, R.D., Maddox, J.V., McManus, M.L., Moore, L., Meyer, S.F., Quimby, P.C., Stack, J.P., and Vaughn, J.L. 1996. Safety of

microorganisms intended for pest and plant disease control: A framework for scientific evaluation. Biol. Control 7:333-351.
14. Davis, R.D., Irwin, J.A.G., and Shepherd, R.K. 1988. Induced systemic resistance in *Stylosanthes* spp. to *Colletotrichum gloeosporioides*. Aust. J. Agric. Res. 39:399-407.
15. Dean, R.A., and Kuć, J. 1985. Induced systemic protection in plants. Trends Biotechnol. 3:125-129.
16. De Jager, E.S., Hall, A.N., and Korsten, L. 1995. Towards alternative disease control in mango. S. Afr. Mango Growers' Assoc. Yearb. 15:67-74.
17. De Jager, E.S., Sanders, G.M., and Korsten, L. 1994. Non-target effect of chemical sprays and population dynamics of the mango phylloplane microbial populations. S. Afr. Mango Growers' Assoc. Yearb. 14:43-47.
18. De Villiers, E.E. and Korsten, L. 1994. Biological treatments for the control of mango postharvest disease. S. Afr. Mango Growers' Assoc. Yearbook 14:48-51.
19. Droby, S., Chalutz, E., Wilson, C.L., and Wisniewski, M.E. 1989. Characterization of the biocontrol activity of *Debaryomyces hansenii* in the control of *Penicillium digitatum* on grapefruit. Can. J. Microbiol. 35:794-800.
20. Fernando, W.G.D., Watson, A.K., and Paulitz, T.C. 1994. Phylloplane *Pseudomonas* spp. enhance disease caused by *Colletotrichum coccodes* on velvetleaf. Biol. Control. 4:125-131.
21. Fernando, W.G.D., Watson, A.K., and Paulitz, T.C. 1996. The role of *Pseudomonas* spp. and competition for carbon, nitrogen and iron in the enhancement of appressorium formation by *Colletotrichum coccodes* on velvetleaf. Eur. J. Plant Pathol. 102:1-7.
22. Firman, I.D. and Waller, J.M. 1977. Coffee berry disease and other *Colletotrichum* diseases of coffee. Phytopathological Papers (20). Commonwealth Mycological Institute, Kew, U.K.
23. Fitzell, R.D. 1981. Effects of regular applications of benomyl on the population of *Colletotrichum* in mango leaves. Trans. Br. Mycol. Soc. 77:529-533.
24. Freeman, S. and Rodriguez, R.J. 1993. Genetic conversion of a fungal plant pathogen to a nonpathogenic, endophytic mutualist. Science 260:75-78.
25. Furtado, I. 1969. Effect of copper fungicides on the occurrence of the pathogenic form of *Colletotrichum coffeanum*. Trans. Br. Mycol. Soc. 53:325-328
26. Griffiths, E. 1972. 'Negative' effects of fungicides in coffee. Trop. Sci. 14:79-89.

27. Griffiths, E. and Furtado, I. 1972. A berry infection technique for assessment of the CBD strain of *Colletotrichum coffeanum* on coffee branchlets. Trans. Br. Mycol. Soc. 58:313-320.
28. Janisiewicz, W.J. 1998. Biocontrol of postharvest diseases of temperate fruits. Pages 171-198 in: Plant-Microbe Interactions and Biological Control. G.J. Boland and L.D. Kuykendall, eds. Marcel Dekker, New York.
29. Jeffries, P., Dodd, J.C., Jeger, M.J., and Plumbley, R.A. 1990. The biology and control of *Colletotrichum* species on tropical fruit crops. Plant Pathol. 39:343-366.
30. Jeffries, P. and Koomen, I. 1992. Strategies and prospects for biological control of diseases caused by *Colletotrichum*. Pages 337-357 in: *Colletotrichum*: Biology, Pathology and Control. J.A. Bailey and M.J. Jeger, eds. CAB International, Wallingford.
31. Koomen, I. and Jeffries, P. 1993. Effects of antagonistic microorganisms on the post-harvest development of *Colletotrichum gloeosporioides* on mango. Plant Pathol. 42:230-237.
32. Korsten, L. 1993. Biological control of avocado fruit diseases. Ph.D. thesis. University of Pretoria, Pretoria.
33. Korsten, L. 1995. Status of research on biological control of avocado pre- and postharvest diseases: an overview. S. Afr. Avocado Growers' Assoc. Yearb. 18:114-117.
34. Korsten, L., Bezuidenhout, J.J., and Kotzé, J.M. 1988. Biological control of avocado postharvest diseases. S. Afr. Avocado Growers" Assoc. Yearb. 11:75.
35. Korsten, L., Bezuidenhout, J.J., and Kotzé, J.M. 1989. Biocontrol of avocado postharvest diseases. S. Afr. Avocado Growers' Assoc. Yearb. 12:10-12.
36. Korsten, L. and De Jager, E.S. 1995. Mode of action of *Bacillus subtilis* for control of avocado post-harvest pathogens. S. Afr. Mango Growers' Assoc. Yearb. 18:124-130.
37. Korsten, L., De Jager, E.S., De Villiers, E.E., Lourens, A., Kotzé, J.M., and Wehner, F.C. 1995. Evaluation of bacterial epiphytes isolated from avocado leaf and fruit surfaces for biocontrol of avocado postharvest diseases. Plant Dis. 79: 1149-1156.
38. Korsten, L., De Villiers, E.E., De Jager, E.S., Cook, N., and Kotzé, J.M. 1991. Biological control of avocado post-harvest diseases. S. Afr. Avocado Growers' Association Yearb. 14:57-59.
39. Korsten, L., De Villiers, E.E., and Lonsdale, J.H. 1993. Biological control of mango postharvest diseases in the packhouse. S. Afr. Mango Growers' Assoc. Yearb. 13:117-121.

40. Korsten, L., De Villiers, E.E., Rowell, A., and Kotzé, J.M. 1993. Post-harvest biological control of avocado fruit diseases. S. Afr. Avocado Growers' Assoc. Yearb. 16:65-69.
41. Korsten, L., De Villiers, E.E., Wehner, F.C., and Kotzé, J.M. 1994. A review of biological control of postharvest diseases of subtropical fruits. Pages 172-185 in: Postharvest Handling of Tropical Fruits. B.R. Champ, E. Highley, and G.I. Johnson, eds. ACIAR 50, Proc. Int. Conf. Chiang Mai, Thailand.
42. Korsten, L., De Villiers, E.E., Wehner, F.C., and Kotzé, J.M. 1997. Field sprays of *Bacillus subtilis* and fungicides for control of preharvest fruit diseases of avocado in South Africa. Plant Dis. 81:455-459.
43. Korsten, L. and Kotzé, J.M. 1993. Postharvest biological control of avocado postharvest diseases. Proc. World Avocado Congr., 2nd. University of California, Riverside, and California Avocado Society 2:473-478.
44. Korsten, L., Lonsdale, J.H., De Villiers, E.E., and De Jager, E.S. 1992. Preharvest biological control of mango diseases. S. Afr. Mango Growers' Assoc. Yearb. 12:72-78.
45. Korsten, L., Towsen, E., and Claasens, V. 1998. Evaluation of Avogreen as postharvest treatment for controlling anthracnose and stem-end rot on avocado fruit. S. Afr. Avocado Growers' Assoc. Yearb. 21(in press)
46. Korsten, L., Van Harmelen, M.W.S., Heitmann, A., De Villiers, E.E., and De Jager, E.S. 1991. Biological control of post-harvest mango fruit diseases. S. Afr. Mango Growers' Assoc. Yearb. 11:65-67.
47. Krauss, U. 1996. Establishment of a bioassay for testing control measures against crown rot of banana. Crop Protec. 15:269-274.
48. Leben, C. 1985. Introductory remarks: Biological control strategies on the phylloplane. Pages 1-5 in: Biological Control on the Phylloplane. C.E. Windels and S.E. Lindow, eds. American Phytopathological Society, St. Paul, Minnesota.
49. Lenné, J.M. and Brown, A.E. 1991. Factors influencing the germination of pathogenic and weakly pathogenic isolates of *Colletotrichum gloeosporioides* on leaf surfaces of *Stylosanthes guianensis.* Mycol. Res. 95:227-232.
50. Lenné, J.M. and Parbery, D.G. 1976. Phyllosphere antagonists and appressorium formation in *Colletotrichum gloeosporioides.* Trans. Br. Mycol. Soc. 66:334-336.
51. Lim, T.K. and Khor, H.T. 1982. Effects of pesticides on mango leaf and flower microflora. Zeit. Pflanzenkrank. Pflanzensch. 89:125-131.
52. Lonsdale, J.H., and Kotzé, J.M. 1993. Chemical control of mango blossom diseases and the effect on fruit set and yield. Plant Dis. 77:558-562.

53. McLaughin, R.J., Wilson, C.L., Droby, S., Ben-Arie, R., and Chalutz, E. 1992. Biological control of postharvest diseases of grapes, peach and apple with the yeasts *Kloeckera apiculata* and *Candida guilliermondii.* Plant Dis. 76:470-473.
54. Mandeel, Q. and Baker, R. 1991. Mechanisms involved in biological control of Fusarium wilt of cucumber with strains of nonpathogenic *Fusarium oxysporum*. Phytopathology 81:462-469.
55. Masaba, D.M. 1991. The role of saprophytic surface microflora in the development of coffee berry disease (*Colletotrichum coffeanum*) in Kenya. Ph.D. thesis, University of Reading, Reading.
56. Mercer, P., Wood, R.K.S., and Greenwood, A.D. 1970. The effect of orange extract and other additives on anthracnose of French bean caused by *Colletotrichum lindemuthianum*. Ann. Bot. 34:593-604.
57. Morris, M.J. 1983. Evaluation of field trials with *Colletotrichum gloeosporioides* for the biological control of *Hakea sericea.* Phytophylactica 15:13-16.
58. Norman, D.J. and Trujillo, E.E. 1995. Development of *Colletotrichum gloeosporioides* f.sp. *clidemiae* and *Septoria passiflorae* into two mycoherbicides with extended viability. Plant Dis. 79:1029-1032.
59. Pandey, R.R., Arora, D.K., and Dubey, R.C. 1993. Antagonistic interactions between pathogens and phylloplane fungi of guava. Mycopathologia 124:31-39.
60. Peterson, R.A. 1978. Susceptibility of Fuerte avocado fruit at various stages of growth, to infection by anthracnose and stem end rot fungi. Aust. J. Exp. Agric. Anim. Husb. 18:158-160.
61. Postmaster, A., Sivasithamparam, K., and Turner, D.W. 1997. Interaction between *Colletotrichum musae* and antagonistic microorganisms on the surface of banana leaf discs. Sci. Hortic. 71:113-125.
62. Press, C.M., Wilson, M., Tuzun, S., and Kloepper, J.W. 1997. Salicylic acid produced by *Serratia marcescens* 90-166 is not the primary determinant of induced systemic resistance in cucumber or tobacco. Mol. Plant-Microbe Interact. 10:761-768.
63. Prusky, D., Freeman, S., Rodriguez, R.J., and Keen, N.T. 1994. A non-pathogenic strain of *Colletotrichum magna* induces resistance to *Colletotrichum gloeosporioides* in avocado fruits. Mol. Plant-Microbe Interact. 7:326-333.
64. Schisler, D.A., Howard, K.M., and Bothast, R.J. 1991. Enhancement of disease caused by *Colletotrichum truncatum* in *Sesbania exaltata* by coinoculating epiphytic bacteria. Biol. Control 1:261-268.
65. Spurr, H.W. and Knudsen, G.R. 1985. Biological control of leaf diseases with bacteria. Pages 45-62 in: Biological Control on the

Phylloplane. C.E. Windels, and S.E. Lindow, eds. American Phytopathological Society, St. Paul, Minnesota.
66. Stirling, A.M., Coates, L.M., Pegg, K.G., and Hayward, A.C. 1995. Isolation and selection of bacteria and yeasts antagonistic to preharvest infection of avocado by *Colletotrichum gloeosporioides.* Aust. J. Agric. Res. 46:985-995.
67. Sutton, B.C. 1992. The genus *Glomerella* and its anamorph *Colletotrichum.* Pages 1-26 in: *Colletrotrichum*: Biology, Pathology and Control. J.A. Bailey, J.A. and M.J Jeger, eds. CAB International,
68. Swadling, I.R., and Jeffries, P. 1998. Antagonistic properties of two bacterial biocontrol agents of grey mould disease. Biocontrol Sci. Technol. 8:439-448.
69. Swinburne, T.R. 1978. The potential value of bacterial antagonists for the control of apple canker. Ann. Appl. Biol. 89:94-95.
70. Towsen, E. 1996. Monitoring *Bacillus subtilis* populations in preharvest biocontrol programmes on avocado. M.Sc. thesis. University of Pretoria, Pretoria.
71. Tsurushima, T., Ueno, T., Fukami, H., Irie, H., and Inoue, M. 1995. Germination self-inhibitors from *Colletotrichum gloeosporioides* f.sp. *jussiaea.* Mol. Plant-Microbe Interact. 8:652-657.
72. Van Dyk, K., De Villiers, E.E., and Korsten, L. 1997. Alternative control of avocado post-harvest diseases. S. Afr. Avocado Growers' Assoc. Yearb. 20:109-112.
73. Waller, J.M. 1972. Water-borne spore dispersal in coffee berry disease and its relation to control. Ann. Appl. Biol. 71:1-18.
74. Wei, G., Kloepper, J.W., and Tuzun, S. 1996. Induced systemic resistance to cucumber diseases and increased plant growth by plant growth-promoting rhizobacteria under field conditions. Phytopathology 86:221-224.
75. Williamson, M.A., and Fokkema, N.J. 1985. Phyllosphere yeasts antagonize penetration from appressoria and subsequent infection of maize leaves by *Colletotrichum graminicola.* Neth. J. Pl. Path. 91:265-276.
76. Wilson, C.L. 1997. Biocontrol of aerial plant diseases in agriculture and horticulture: Current approaches and future prospects. J. Industrial Microbiol. Biotechnol. 19:188-191.
77. Wilson, C.L., El Ghaouth, A., Chalutz, E., Droby, S., Stevens, C., Lu, J.Y., Khan, V., and Arul, J. 1994. Potential of induced resistance to control postharvest diseases of fruits and vegetables. Plant Dis. 78:837-844.
78. Wilson, C.L. and Wisniewski, M.E. 1989. Biological control of postharvest diseases of fruits and vegetables: An emerging technology. Annu. Rev. Phytopathol. 27:425-441.

79. Wilson, C.L., Wisniewski, M.E., Biles, C.L., McLaughlin, R., Chalutz, E., and Droby, S. 1991. Biological control of postharvest diseases of fruits and vegetables: Alternatives to synthetic fungicides. Crop. Protec. 10:172-177.
80. Windels, C.E. and Lindow, S.E. 1985. Biological control on the phylloplane. American Phytopathological Society, St. Paul, Minnesota.
81. Wisniewski, M.E. and Wilson, C.L. 1992. Biological control of postharvest diseases of fruits and vegetables:Recent advances. Hort. Sci. 27:94-98.

Chapter 17

Colletotrichum Diseases of Strawberry in Florida

Daniel E. Legard

Colletotrichum is one of the most important genera of strawberry pathogens. This group of pathogens causes a variety of diseases that are particularly important on strawberries grown in annual production systems. These diseases include fruit rot, crown rot, root rot, anthracnose of the stolon and petiole, flower blight, bud rot, black leaf spot, and irregular leaf spot (23,26). Crown rot and fruit rot are the most important of these diseases in Florida. The three primary species of *Colletotrichum* known to cause these diseases are *Colletotrichum fragariae*, *C. gloeosporioides*, and *C. acutatum*. *C. dematium* can also occasionally cause a fruit rot (2). Traditionally, these diseases have been called anthracnose (23). However, several of these diseases do not produce the dark sunken lesions characteristic of anthracnose, and it is more appropriate to refer to them as *Colletotrichum* diseases of strawberry.

The earliest report of a *Colletotrichum* disease on strawberry was by Brooks (3), who found sunken lesions on strawberry runners (runner spot) and petioles in Florida and identified the new pathogen as *Colletotrichum fragariae*. Subsequently, Brooks found the pathogen also caused a wilt and crown rot of strawberry (4). *C. fragariae* also causes a fruit rot (19) and black leaf spot (21). Anthracnose fruit rot (black spot) was first reported by Sturgess (33,34) in Australia and by Wright et al. (38) in the USA. Although originally described as being caused by *Gloeosporium* spp., Simmonds (30) subsequently studied the Australian isolates and renamed them *C. acutatum*. Simmonds (30) also described *C. acutatum* isolated from leaf spots. Crown rot and wilt caused by *C. acutatum* was first reported by Smith and Black (31). Freeman recently described a root necrosis, stunting, and chlorosis caused by *C. acutatum* in Israel (15). A black leaf spot, crown rot, and fruit rot caused by *C. gloeosporioides* were first reported by Howard and Albregts (21,22).

Colletotrichum diseases have historically been a serious problem on strawberries in Florida and the southeastern United States. In the 1930s, strawberry anthracnose caused by *C. fragariae* was reported in Florida (3). By the 1980s, both *C. gloeosporioides* (22) and *C. acutatum* (23,31) were also found to cause disease on strawberry in Florida. When susceptible cultivars are grown in Florida, up to 80% of plants in commercial fruiting fields can die from crown rot caused by *C. fragariae* or *C. gloeosporioides* (23). Severe fruit rot epidemics periodically cause growers to abandon production fields or to resort to stripping fruit in an attempt to control the disease (23). In Florida, annual yield reductions due to these diseases average 5-10%. During 1997-1998, a severe epidemic of anthracnose fruit rot caused industry-wide yield losses of 15-20%.

Conidia of the three major *Colletotrichum* species that cause disease on strawberry are primarily produced in sticky masses in acervuli and, therefore, are dispersed by water or by adhering to harvesters, farm equipment, and insects (26). No teleomorph stage has been reported for strawberry isolates of either *C. acutatum* or *C. fragariae*. For *C. gloeosporioides,* the teleomorph (*Glomerella cingulata*) has not been reported on strawberry in the field. However, preliminary RAPD marker analysis of *C. gloeosporioides* isolated from crown rot diseased plants within fields in Florida has revealed populations that are highly polymorphic, suggesting that the teleomorph may play an important role in the epidemiology of crown rot (A. Urena and D.E. Legard, unpublished). Extensive research on splash dispersal of *C. acutatum* has shown that rain is only effective in moving the conidia up to 30 cm (27,39). Since conidia of these pathogens are not wind dispersed, their intermediate and long-distance movement is primarily restricted to activities associated with transplanting (i.e., movement on diseased or infested plants) and harvesting. In annual strawberry production systems, secondary spread of the pathogen is probably due to harvesting operations. During harvesting, fruit pickers brush their hands through the canopy of each strawberry plant searching for and removing ripe fruit. They undoubtedly pickup and transfer spores of *Colletotrichum* and other pathogens that are present within the canopy.

An important question about the epidemiology of *Colletotrichum* diseases of strawberry is the source of primary inoculum. Transplants are probably the most important source of primary inoculum for *Colletotrichum* diseases of strawberry. Since both nursery and fruiting fields are started with live plants, these plants are a likely source of inoculum. *C. acutatum* has been recovered from nursery transplants grown in California and Israel (13,15) and *C. fragariae* from transplants grown in Louisiana (28). Howard, et. al. (23) reported finding *C. fragariae* or *C. gloeosporioides* on transplants from most of the nursery production areas supplying Florida. Observations of recent anthracnose fruit rot *(C. acutatum*) and *Colletotrichum* crown rot (*C. fragariae*) epidemics in Florida revealed a strong correlation between nursery source and outbreaks of these diseases (D.E. Legard, unpublished). These findings highlight the inadequacy of current certification procedures and standards used in North America and elsewhere for the production of disease-free strawberry plant stock for nursery and fruiting fields.

Current certification procedures were primarily developed to eliminate viruses from strawberry (5,8). The lack of virus disease problems in Florida suggests that certification has been successful in this effort (25). However, because of the relatively wet environmental growth conditions for strawberries in Florida, improved certification procedures targeted for fungal and bacterial diseases are needed to eliminate or reduce the severity of *Colletotrichum* diseases and other important diseases, including powdery mildew (*Mycosphaerella fragariae*) and angular leaf spot *(Xanthomonas fragariae*).

Overwintering (or oversummering in winter production regions) of inoculum in soil or on plant debris is another potentially important source of inoculum. In California, *C. acutatum* was recovered from soil and buried strawberry tissue over a 9-month period from commercial fields with a recent history of anthracnose, but not from fields that had been fumigated (13). In Ohio, *C. acutatum* was found to overwinter on mummified fruit buried in soil (37). In Florida, because of the hot and wet summer conditions, *C. acutatum* is not believed to oversummer between seasons in soil or plant residue (23), and recent results suggest that *C. gloeosporioides* does not oversummer in strawberry plant debris (A. Alvaro and D. Legard, unpublished). In addition, because most annual production strawberry fields are preplant fumigated with methyl bromide and chloropicrin, these pathogens are unlikely to survive in the soil or debris.

Alternate hosts are another likely source of primary inoculum for *Colletotrichum* diseases. Many commercial nursery and fruit production fields are located in regions with wild strawberry species (9). The alpine strawberry (*Fragariae vesca*) has a near global distribution, but no wild strawberry species are known to survive in Florida. However, a majority of transplants used in Florida come from the northeastern USA and southeastern Canada where the meadow strawberry (*F. virginiana*) is common. In the eastern USA and Canada, hedgerows, waste areas, and ditches are typically found adjacent to commercial strawberry fields. A large number of different weed species often colonizes these areas. Other cultivated crops are also commonly grown within and adjacent to strawberry production fields. If any of these plant species are alternative hosts of *Colletotrichum* isolates pathogenic on strawberry they may be important sources of primary inoculum. Both *C. gloeosporioides* and *C. acutatum* are known to have extensive host ranges (12,29) within the species, and many of these hosts can be found growing adjacent to strawberry production fields. *C. fragariae* has an apparently very limited host range, including strawberry and a few other species (19,35). From observation of epidemics in Florida, some *Colletotrichum* crown rot epidemics in specific fields may be associated with nearby wild plant species. *C. gloeosporioides* has been readily isolated from these wild plant species (D.E. Legard, unpublished), and the relationship between these isolates and those from strawberry are being determined. In North Carolina, a study designed to develop screening methods for developing resistance to

C. acutatum found that most of the isolates of *C. acutatum* from strawberry were more virulent on strawberry than isolates from other hosts (1).

Currently, control of *Colletotrichum* diseases of strawberry involves the combination of cultural, genetic, and chemical measures. In Florida and other production regions, numerous cultural practices are used to reduce the severity of *Colletotrichum* and other diseases. These include the use of drip irrigation, raised bed plasticulture, increased plant spacing, resistant cultivars, field sanitation, and certified transplants. Unfortunately, fungicides currently labeled for control of *Colletotrichum* on strawberry are either ineffective or cannot be used in fruiting fields. In Florida, captan, the most effective labeled product for control of *Colletotrichum* diseases of strawberry, fails to provide complete control even when applied up to twice a week throughout the season (23). This fungicide is also ineffective in Israel (16). In the case of benomyl, *C. acutatum* is insensitive to benzimidazole fungicides, whereas isolates of *C. gloeosporioides* are variable in their sensitivity (24,32). Therefore, accurate characterization of benzimidazole resistance in *C. gloeosporioides* can be important in making control recommendations. Other fungicides not labeled for strawberry in the USA are known to have activity against *Colletotrichum* species. Freeman et al. (16) demonstrated that prochloraz-Mn and prochloraz-Zn provided better control of *C. acutatum* on strawberry than azole fungicides and captan. In Israel, prochloraz dips are routinely used to treat strawberry to effectively improve control of *C. acutatum*.

Further modification of cultural practices should improve management of *Colletotrichum* diseases, especially anthracnose fruit rot. Because of the dependence of conidia of *Colletotrichum* on water for infection and dispersal (11,14,27,34,36,39), reducing the amount of overhead irrigation (i.e., for freeze protection and plant establishment) and minimizing field activities (i.e., harvesting, cultivating, sanitation, spraying, etc.) when the foliage is wet should reduce the spread and severity of these diseases. In Florida, green-top transplants are typically overhead irrigated for 10 to 14 days to allow the plants to establish new root systems. Cut-top transplants (i.e., plants with their leaves removed) that do not require as much overhead irrigation are not used in Florida due to the commercial need for early fruit production in Florida. However, plug plants could be used to eliminate the need for overhead irrigation during plant establishment. Small and large tunnels are used in many production regions of the world to protect the plants from light freezes and rain. The warmer temperatures inside tunnels can also promote earlier fruit production. Preliminary work in Florida has shown that strawberries cultured in large tunnels have greatly reduced leaf wetness periods (D.E. Legard and C.K. Chandler, unpublished), which corresponded with a reduced anthracnose fruit rot on plants under tunnels compared with plants in the open.

Resistant varieties are potentially an excellent method for controlling *Colletotrichum* diseases, and in situations where production of disease-free plants is not practical or possible, they may be the only option. Several sources of resistance to *Colletotrichum* have been incorporated into breeding programs in the USA, although most commercial varieties grown in the USA and the rest of the world are susceptible to *Colletotrichum* diseases (6,7,17,20). The University of Florida has released two strawberry varieties (7,20) that are adapted for production in Florida and are resistant to these diseases. The highly resistant variety "Sweet Charlie" has been a commercial success and has been grown in approximately 40% of the Florida production acreage for the past three seasons. This variety is essentially field immune to anthracnose fruit rot but is only moderately resistant to *Colletotrichum* crown rot. Despite the success of this resistant variety, marketing concerns dictate that the industry grow different varieties to spread out the timing of fruit harvest and to cater to consumer and marketing demands. The USDA has also produced several lines with anthracnose resistance (17). Research has been conducted on the genetics of resistance to *Colletotrichum* diseases. Studies of segregation ratios of *Colletotrichum* resistance have shown a bimodal distribution, suggesting that the resistance trait is dominant and possibly monogenic (10,18). Plants resistant to one *Colletotrichum* species are apparently resistant to other strains or species of *Colletotrichum* although the level of resistance may vary (18,23).

Colletotrichum species cause many important yield-limiting diseases on strawberry, with anthracnose fruit rot and *Colletotrichum* crown rot being the most important in Florida. Many important etiological and epidemiological questions need to be resolved about these diseases before more effective control methods can be developed. Questions about the source of primary inoculum are critical to this process. Questions about the role of contaminated transplants, alternate hosts, and other potential inoculum sources remain. Controlling *Colletotrichum* diseases in Florida and elsewhere will require the improvement and implementation of effective certification standards for disease-free plant stock. The development of molecular markers and techniques for using them with symptomless plant material would facilitate the improvement of clean plant production methods and standards. The wide host range within the species *C. acutatum* and *C. gloeosporioides* and evidence that some isolates have restricted host ranges suggest that effective use of molecular techniques will require the development of markers that can distinguish between strawberry pathogens and nonpathogens. An improvement in the ability to detect *Colletotrichum* in strawberry plant stocks will allow nursery producers to develop cleaner plant propagation methods and fruit producers to obtain higher quality transplants.

Literature Cited

1. Ballington, J.R. and Milholland, R.D. 1993. Screening strawberries for resistance to *Colletotrichum acutatum* in North Carolina. Acta Hortic. 348:442-448.
2. Beraha, L. and Wright, W.R. 1973. A new anthracnose of strawberry caused by *Colletotrichum dematium*. Plant Dis. Reptr. 57:445-448.
3. Brooks, A.N. 1931. Anthracnose of strawberry caused by *Colletotrichum fragariae*, n. sp. Phytopathology 21:739-744.
4. Brooks, A.N. 1935. Anthracnose and wilt of strawberry caused by *Colletotrichum fragariae*. Phytopathology 25:973-974 (Abstr.).
5. Broome, O.C. and Goff, L.M. 1987. The production, testing and certification of virus-tested strawberry stock. Adv. Strawberry Production 6:3-5.
6. Chandler, C.K., Legard, D.E., and Simms, C.A. 1997. 'Rosa Linda' strawberry. HortScience 32:1134-1135.
7. Chandler, C.K., Albregts, E.E., Howard, C.M., and Brect, J.K. 1997. 'Sweet Charlie' strawberry. HortScience 32:1132-1133.
8. Converse, R.H. 1979. Recommended virus-indexing procedures for new USDA small fruit and grape cultivars. Plant Dis. Reptr. 63:848-851.
9. Darrow, G.M. 1966. The strawberry; history, breeding, and physiology. Holt, Rinehart and Winston, New York. 447 pp.
10. Denoyes-Rothan, B. 1997. Inheritance of resistance to *Colletotrichum acutatum* in strawberry (*Fragaria x ananassa*). Proc. Third Int. Strawberry Symp. H.A.Th. van der Scheer, F. Lieten, and J. Dijkstra, eds. Acta Hortic. 439(2):809-812.
11. Dodd, J.C., Estrada, A.B., Matcham, J., Jeffries, P., and Jerger, M.J. 1991. The effect of climatic factors on *Colletotrichum gloeosporioides*, casual agent of mango anthracnose, in the Philippines. Plant Pathol. 40:568-575.
12. Dyko, B.J. and Mordue, J.E.M. 1979. *Colletotrichum acutatum*. CMI descriptions of pathogenic fungi and bacteria. No. 630. Commonwealth Mycological Institute, Kew, United Kingdom. 2 pp.
13. Eastburn, D. M. and Gubler, W.D. 1990. Strawberry anthracnose: Detection and survival of *Colletotrichum acutatum* in soil. Plant Dis. 74:161-163.
14. Fitzell, R.D. and Peak, C.M. 1984. The epidemiology of anthracnose disease of mango: Inoculum sources, spore production and dispersal. Ann. Appl. Biol. 104:53-59.
15. Freeman, S. and Katan, T. 1997. Identification of *Colletotrichum* species responsible for anthracnose and root necrosis of strawberry in Israel. Phytopathology 87:516-521.

16. Freeman, S., Nizani, Y., Dotan, S., Even, S., and Sando, T. 1997. Control of *Colletotrichum acutatum* in strawberry under laboratory, greenhouse, and field conditions. Plant Dis. 81:749-752.
17. Galletta, G.J., Smith, B.J., and Gupton, C.L. 1993. US70, US159, US292 and US438 anthracnose crown rot resistant strawberry parent clones. Hortscience 28:1055-1056.
18. Gupton, C.L. and Smith, B.J. 1991. Inheritance of resistance to *Colletotrichum* species in strawberry. J. Amer. Soc. Hort. Sci. 116:724-727.
19. Howard, C.M. 1972. A strawberry fruit rot caused by *Colletotrichum fragariae*. Phytopathology 62:600-602.
20. Howard, C.M. and Albregts, E.E. 1979. Dover—a firm-fruited strawberry with resistance to anthracnose *Colletotrichum fragariae*, cultivars. Circular S. Florida Agric. Exp. Stations. Aug 1979. (267) 5 pp.
21. Howard, C.M. and Albregts, E.E. 1983. Black leaf spot phase of strawberry anthracnose caused by *Colletotrichum gloeosporioides* (= *C. fragariae*). Plant Dis. 67:1144-1146.
22. Howard, C.M. and Albregts, E.E. 1984. Anthracnose of strawberry fruit caused by *Glomerella cingulata* in Florida. Plant Dis. 68:824-825.
23. Howard, C.M., Maas, J.L., Chandler, C.K., and Albregts, E.E. 1992. Anthracnose of strawberry caused by the *Colletotrichum* complex in Florida. Plant Dis. 76:976-981.
24. LaMondia, J.A. 1995. Inhibition with benomyl to growth *in vitro* of *Colletotrichum acutatum* and *C. fragariae* and strawberry fruit infection by benomyl-resistant isolates of *Colletotrichum acutatum*. Adv. Strawberry Res. 14:25-30.
25. Legard, D.E., Whidden, A.J., and Chandler, C.K. 1998. Incidence and occurrence of strawberry diseases in Florida from 1991-1996. Adv. Straw. Res. 16:35-47.
26. Maas, J.L., ed. 1998. Compendium of strawberry diseases. 2nd ed. APS Press, St. Paul, MN. 98 pp.
27. Madden, L.V. 1992. Rainfall and the dispersal of fungal spores. Pages 39-79 in: Advances in Plant Pathology. Vol. 8. J.H. Andrews and I. Tommerup, eds. Academic Press, London.
28. McInnes, T.B., Black, L.L., and Gatti, J.M. Jr. 1992. Disease-free plants for management of strawberry anthracnose crown rot. Plant Dis. 76:260-264.
29. Mordue, J.E.M. 1971. *Glomerella cingulata*. CMI descriptions of pathogenic fungi and bacteria. No. 315. Commonwealth Mycological Institute, Kew, United Kingdom. 2 pp.
30. Simmonds, J.D. 1965. A study of the species of *Colletotrichum* causing ripe fruit rots in Queensland. Queensland J. Agric. Anim. Sci. 22:437-459.

31. Smith, B.J. 1986. First report of *Colletotrichum acutatum* on strawberry in the United States. Plant Dis. 70:1074.
32. Smith, B.J. and Black, L.L. 1990. Morphological, cultural, and pathogenic variation among *Colletotrichum* species isolated from strawberry. Plant Dis. 74:69-76.
33. Sturgess, O.W. 1954. A strawberry ripe fruit rot. Queensland Agr. J. 78:269-270.
34. Sturgess, O.W. 1957. A ripe fruit rot of the strawberry caused by a species of *Gloeosporium*. Queensland J. Agr. Sci. 14:241-251.
35. Welty, R.E. 1984. Blue lupine as a host for *Colletotrichum trifolii* from alfalfa and for *C. fragariae* from strawberry. Plant Dis. 68:142-144.
36. Wilson, L.L., Madden, L.V., and Ellis, M.A. 1990. Influence of temperature and wetness duration on infection of immature and mature strawberry fruit by *Colletotrichum acutatum*. Phytopathology 80:111-116.
37. Wilson, L.L., Madden, L.V., and Ellis, M.A. 1992. Overwinter fruit. survival of *Colletotrichum acutatum* in infected strawberry fruit in Ohio. Plant Dis. 76:948-950.
38. Wright, W.R., Smith, M.A. Ramsey, G.B., and Beraha, L. 1960. *Gloeosporium* rot of strawberry fruit. Plant Dis. Reptr. 44:212-213.
39. Yang, X., Wilson, L.L., Madden, L.V., and Ellis, M.A. 1990. Rain splash dispersal of *Colletotrichum acutatum* from infected strawberry Phytopathology 80:590-595.

Chapter 18

Biology and Control of Anthracnose Diseases of Citrus

L.W. Timmer and G.E. Brown

Colletotrichum spp. cause three diseases of citrus: postbloom fruit drop (PFD), lime anthracnose and postharvest anthracnose. PFD produces serious yield losses of sweet oranges and other citrus in the humid Americas. It is caused by *C. acutatum* which attacks petals and induces abscission of fruitlets and production of calyces. During bloom, conidia are splash-dispersed from acervuli on petals. *C. acutatum* persists between blooms as appressoria on vegetative tissues. Upon initiation of the next bloom, the appressoria germinate to produce a few conidia which are splashed to flowers to reinitiate the cycle. A predictive model based on the current number of infected flowers and the rainfall total for the last 5 days is used to forecast the percentage of the flowers infected 3-4 days in advance and to time fungicide applications. Lime anthracnose, caused by *C. acutatum*, attacks flowers, young leaves and fruit of Mexican lime only and is a limiting factor in production of that fruit in humid areas. Lime anthracnose isolates cause PFD symptoms, but PFD isolates do not affect leaves and fruit of Mexican lime. *C. gloeosporioides* is a common saprophyte and weak parasite in citrus orchards in humid areas. Conidia produced from acervuli on dead tissue form appressoria on fruit that may develop after harvest to produce postharvest anthracnose. The disease is enhanced by ethylene used early in the season for degreening of fruit and is reduced by proper harvesting and handling and by benzimidazole fungicides.

Biology and Control of Anthracnose Diseases of Citrus

Currently, three anthracnose diseases of citrus are recognized: postbloom fruit drop (PFD) and lime anthracnose, both caused by *Colletotrichum acutatum* J. H. Simmonds, and postharvest anthracnose caused by *C. gloeosporioides* (Penz.) Penz. & Sacc. Historically, anthracnose diseases have been of limited importance on citrus. In the field, anthracnose symptoms only developed on tissues damaged by freezes, hail storms or some other agent. Postharvest anthracnose on fruit was a relatively minor problem until the introduction of susceptible cultivars (37) and increased production of citrus in more southerly areas of Florida that required more extensive ethylene degreening (10). Lime anthracnose was a serious problem in humid production areas, but was limited to a single variety. PFD was not described until 1979 and initially was a rather local problem. Since that time, it has developed into a significant production problem throughout the humid American citrus-growing areas (42).

Historically, *C. gloeosporioides* was considered to be a saprophyte or weak parasite of damaged tissue in the field and a postharvest pathogen of fruit. The causal agent of lime anthracnose was described originally as *Gloesporium limetticola* R. E. Clausen (14). Subsequently, *Gloeosporium* was no longer considered a valid genus and *G. limetticola* was placed in the collective species, *C. gloeosporioides* (41,48). The causal agent of PFD was described as a form of *C. gloeosporioides* based on work in Belize (19). In Florida, the causal agent of all anthracnose diseases was considered initially to be *C. gloeosporioides*, but differences between strains were recognized. Sonoda et al. (40) recovered two strains, a fast-growing gray (FGG) and a slow-growing orange (SGO) strain, from citrus trees affected by PFD. Agostini et al. (4) differentiated those two strains as well as the lime anthracnose (KLA) strain. The FGG strain grew rapidly, forming large, gray colonies and producing conidia with mostly rounded ends, large, lobulate appressoria (8.3 x 6.0 µm), and setae in culture and on host tissue. In contrast to the FGG strain, the SGO strain grew more slowly, producing orange colonies, more fusiform conidia, smaller, clavate appressoria (6.1 x 4.7 µm) and no setae. The KLA strain was identical to the SGO strain except that appressoria were round rather than clavate. The FGG strain is highly sensitive to benomyl whereas the SGO strain is moderately tolerant (40).

The FGG (referred to as type 1) and the SGO (type 2) strains could be differentiated on the basis of forms of ribosomal DNA (30). A cloned DNA fragment from the non-transcribed spacer region hybridized with the FGG, but not with the SGO isolates. Variation in isozymes of cutinase, restriction fragment length polymorphisms, and chromosome sizes have also been used to differentiate the FGG and SGO strains (25,29,30). In pathogenicity tests, the FGG strain did not produce PFD on sweet orange [*Citrus sinensis* (L.) Osb.] flowers or anthracnose symptoms on limes [*C. aurantifolia* (Christin.) Swing.] (4). The SGO strain produced all of the symptoms of PFD on sweet orange flowers, but was only weakly pathogenic on limes. In contrast, the KLA strain caused all the symptoms of PFD on sweet orange flowers as well as anthracnose on limes. The FGG strain causes postharvest anthracnose on citrus fruit, but the SGO strain does not affect fruit (43). Molecular characterization demonstrated conclusively that the FGG strain is *C. gloeosporioides* and that the SGO and KLA strains are *C. acutatum* (13). In essence, the taxonomy of anthracnose pathogens of citrus has come full circle and the lime anthracnose pathogen again is separated from the common field saprophyte and postharvest pathogen.

Postbloom Fruit Drop

OCCURRENCE AND DISTRIBUTION

The first description of the disease and its causal organism was published by Fagan in 1979 in Belize, but the disease was noticed as early as 1956 (19). Shortly thereafter, the disease was reported throughout Central America and the Caribbean and as far south as Argentina and Brazil 28,33,36,42). It is uncertain whether the reports reflected a new problem or whether the disease was present previously. However, the presence of the persistent calyces, which provide a year-round diagnostic symptom, and the devastating nature of the disease make it likely that the problem did not go unnoticed for long. In Florida, the disease was first noted on Tahiti limes in southwestern production areas in 1983 (31). By 1988, PFD was widespread in the state and caused significant yield losses.

PFD is a common problem in the humid tropical areas of southern Mexico and Central America. In these high rainfall areas, PFD has consistent effects on yields and profitability of citrus groves. Elsewhere, the disease is sporadic and can be nearly absent in some years and devastating in others. Major epidemics have occurred in Florida in 1988, 1993, 1994, and 1998. A serious outbreak in Brazil in 1993 caused major concern among growers (28). Losses were also severe when the disease first appeared in the late 1980s in Jamaica and the Dominican Republic.

SYMPTOMS

Lesions produced by *C. acutatum* on petals are initially water-soaked and become peach to orange-brown with age and sporulation of the fungus (18,19,42). The fungus can attack unopened and even pinhead (?) flowers if inoculum is high and conditions are favorable, but open flowers are much more susceptible (19). Following severe attacks, the orange-to-brown petals remain attached to the inflorescence, whereas petals from healthy blossoms abscise and fall quickly. On affected flowers, the fruitlet usually abscises leaving persistent calyces, commonly called buttons, attached to the peduncle. Healthy flowers that do not set fruit always abscise at the base of the peduncle leaving no evidence of their existence. On affected inflorescences, some fruitlets may not abscise but remain small and never develop. Leaves surrounding affected inflorescences are usually small and twisted with enlarged veins.

The fungus apparently infects only petal tissue. The fruit drop, production of persistent buttons, and the associated leaf distortion may be caused by a volatile hormone produced ~~on~~ by diseased petals. Occasionally, a young fruit from an earlier bloom which has already set will abscise due to the presence of a nearby infected flower cluster. The nature of this hormone effect has never been investigated.

THE DISEASE CYCLE

C. acutatum infects flower petals and produces acervuli on these tissues in 4-5 days under optimal conditions (55). Conidia are splash-dispersed from diseased to healthy flowers to continue the cycle during bloom (16,19,42). Conidia that fall on vegetative tissues germinate to form appressoria and quiescent infections, and the fungus persists in this manner between blooms (3,55). Unlike most *Colletotrichum* spp., *C. acutatum* does not invade citrus tissues from quiescent infections even though it is capable of doing so under artificial conditions (55). Even fruit inoculated with a high concentration of conidia and exposed to the most favorable conditions for anthracnose development, do not develop anthracnose symptoms immediately 43). Rather, appressoria are stimulated to germinate directly, probably by exudates of the first flowers in a bloom (Fig. 1A). These hyphae produce a few conidia that are then splashed to healthy flowers to complete the cycle 42,55).

C. acutatum persists as appressoria on leaves, twigs, and buttons of affected trees (55). These appressoria gradually lose viability, and propagule densities decline with time in the absence of a bloom (3). Inoculum also declines through normal leaf drop and mortality of buttons and twigs. The fungus is capable of persisting for one year and probably longer in the absence of reproduction on flower petals (3). Despite the ability of *C. acutatum* to persist in the absence of a bloom, PFD is continually a serious problem, mostly in areas with multiple blooms per year. Frequent blooms assure that inoculum is present at high levels almost continually. Whenever rainfall occurs during bloom, some infection is a virtual certainty. Even in areas where multiple blooms are uncommon, the disease tends to occur most commonly in groves with declining trees or young trees which tend to bloom off-cycle and help maintain the inoculum levels.

EPIDEMIOLOGY AND ENVIRONMENTAL EFFECTS

Development of PFD epidemics is dependent on the availability of susceptible flowers, rainfall for splash dispersal of conidia, adequate periods of wetness for infection and warm temperatures for fungal growth (18,45). Since the pathogen reproduces only on flower petals, the disease is strongly affected by environmental effects on the flowering pattern of the host, as well as by -direct effects on the pathogen.

As already noted, the disease is prevalent in locations and groves with multiple or off-season bloom which allows buildup and maintenance of inoculum. At the same time, PFD is a minor or sporadic problem in more temperate regions even though they may have high rainfall. Cool winter temperatures keep trees dormant with no off-season bloom, and when

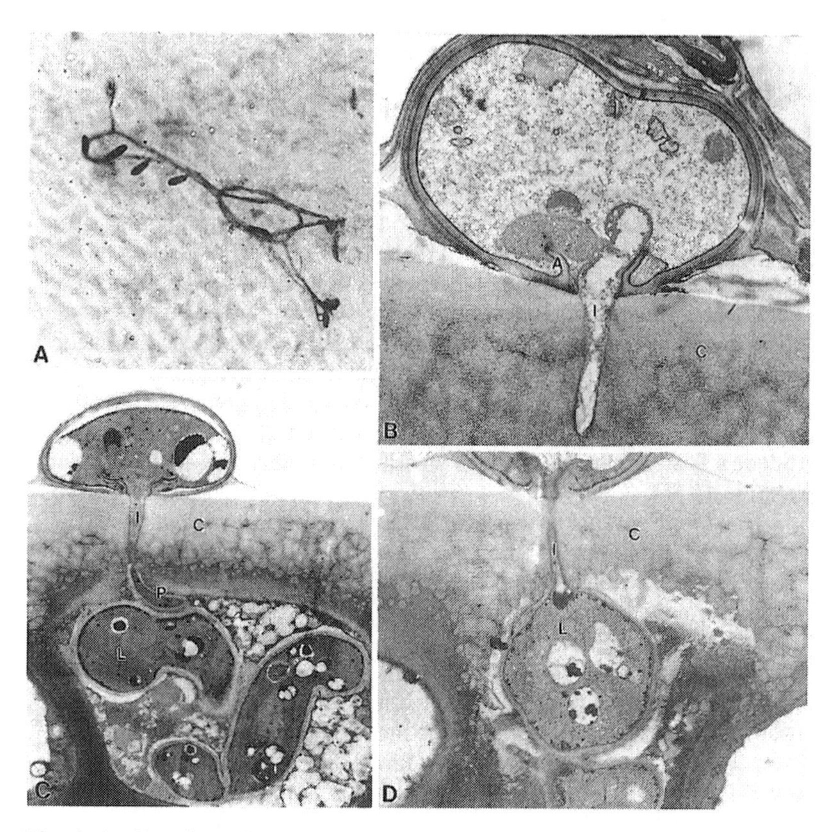

Fig. 1. A, Germinated appressorium of *Colletotrichum acutatum* on the leaf surface with hyphae and a few conidia produced after treatment with a petal extract; B, Appressorium of *C. gloeosporioides* with infection hypha growing in fruit cuticle; C, Primordial and large hyphae of *C. gloeosporioides* formed from the infection hypha after entry into an epidermal cell; D, Large hypha of *C. gloeosporioides* formed directly from the infection hypha after penetration through the cuticle into the epidermis. Legend: C = cuticle; A = collar; I = infection hypha; L = large hypha; P = primordial hypha.

temperatures increase in the spring, trees bloom for too short a period to allow inoculum buildup before flowering ends.

C. acutatum has an optimum temperature for mycelial growth of 24-27 C, but still grows at 15-20 C (2,19). In studies of environmental effects on disease development, temperature was not a significant factor (45). Low temperatures slow disease development but also slow flower development and extend the bloom period, thus increasing the likelihood that rain will occur during bloom. Outbreaks of PFD in Florida are most severe in years when bloom duration is extended by intermittent cold periods.

Other than inoculum availability, rainfall is the most critical factor in disease development. Total rainfall was strongly correlated with disease development, but duration of leaf wetness played a minor role (45). Rainfall amount and intensity are highly related to each other. Proliferation of *C. acutatum* is highly dependent on splash dispersal of conidia from initial infections (17,19,42). Dews usually provide adequate periods of moisture for infection in areas where PFD is prevalent. Leaf wetness duration is correlated to some extent with disease incidence because rain events of long duration cause more disease development than rains of short duration. However, extended periods of dew and fog cause little disease development in the absence of rainfall to disperse the conidia. Symptoms are typically more severe on flowers in the lower canopy because rain washes conidia downward and the area remains wet longer (17).

In our studies, most of the dispersal in inoculated plots was attributable to rain splash and windblown rain (1). Isolated foci developed perhaps from spread by bees or other insects. Any insect which visits affected flowers can acquire the pathogen and presumably carry it at least short distances to healthy trees (34). The pathogen may also be spread on equipment and grove workers, especially if operations are conducted during moist periods. Long distance spread may result when infected flower petals fall onto equipment or vehicles or into picking sacks and are carried to other locations. Harvesting operations for Valencia oranges are of particular concern because the previous year's fruit crop is often harvested during the current year's bloom. Long distance movement with bee colonies may also occur, but has never been documented.

VARIETAL SUSCEPTIBILITY AND YIELD LOSSES

Under field conditions, considerable differences have been observed in losses caused by PFD on different species and cultivars of citrus, but no direct comparisons have been made under controlled conditions. In Florida, navel oranges suffer the most severe damage from PFD, and in Brazil, Natal sweet oranges are the most affected. Both of these varieties produce large bouquet blooms that produce large amounts of inoculum when infected. These varieties also tend to bloom off-cycle more frequently than others. Valencia oranges often suffer serious damage in Florida, and spread may be more common in Valencia groves because crops are often harvested at bloom time. Early oranges, such as Hamlins, were considered tolerant to the disease, but some Hamlin groves suffered serious damage in Florida in 1998. In Central America where the disease is endemic, early oranges have fewer problems with PFD than late oranges. Tangerines (*C. reticulata* Blanco) and their hybrids are apparently highly susceptible, but are usually grown in cooler areas where PFD is less of a problem. PFD has not caused economic losses on grapefruit (*C. paradisi* Macf.) in Florida. Grapefruit is also less affected than other citrus species in Belize (19). Flowers of all citrus species are susceptible when inoculated, but the ability of the pathogen to reproduce on petals of different species may differ (1).

Yield loss to PFD can be severe (20,28,42,44). Citrus trees produce large numbers of flowers and typically set fruit on only about 0.5-2.0% of them. Infection of flowers by *C. acutatum* results in persistent buttons, regardless of whether the flower would have set a fruit or not. Our calculations indicate that on navel and Valencia oranges in Florida about six fruit are lost to PFD for every 100 buttons formed (46). PFD-affected trees tend to shed less fruit during the normal May-June drop period than do healthy trees. Also, trees with lower fruit numbers produce larger fruit. We found that plots with up to 20% blossom blight caused by PFD suffer no yield loss (44). Thus, low levels of PFD can be tolerated. However, as yield decreases, harvesting costsincrease because of the low increase and production costs per ton of fruit increase sharply. In years with severe PFD outbreaks in Florida, some groves have not been harvested because too few fruit remained. Those plantings are usually put on minimal care programs to reduce costs.

DISEASE CONTROL

Alterations in cultural practices can be helpful in reducing the incidence of PFD. Changing from overhead irrigation to under-the-tree sprinklers reduces wetness periods and dispersal of the pathogen. Prompt removal of decline trees from groves decreases the amount of inoculum carried over between blooms. In Brazil, flowering usually begins with the spring rainy season. Use of irrigation systems allows initiation of flowering during the dry season and helps avoid PFD problems. Any measures which avoid off-season flowering help reduce PFD, but often bloom is temperature-controlled although manipulating nitrogen fertilization and irrigation may be useful.

In most cases, fungicide applications are necessary for control of PFD outbreaks. Benomyl has proven quite effective despite the fact that *C. acutatum* is moderately tolerant to this fungicide (20,40,44). In Florida, it is widely used alone or in combination with ferbam. Captafol is very effective but no longer registered for use on citrus in most areas (16,20). Protectants such as maneb, ferbam, and captan provide some control but are not highly effective when used alone (20). Fungicides which inhibit sterol biosynthesis are generally effective against PFD (54), but are not yet registered in many areas.

Timing of fungicide applications is critical for disease control. In some years in Florida, no applications are needed throughout the entire state. In other years, many groves need to be sprayed three times during bloom, the maximum number permitted. Because spraying all plantings on a preventive basis is economically prohibitive, groves in need of treatment must be identified. Identifying those groves or blocks that have high numbers of persistent buttons from the previous year helps focus needs for scouting on those most likely to have PFD problems (46). Noting the occurrence of scattered late winter or early spring bloom and the amount of infection on these flowers also helps target those plantings where inoculum is building and PFD is likely to be a problem (46).

A predictive model has been developed and validated to determine the need for fungicide applications and their timing. The original model (47) was based on the inoculum level, the current number of diseased flowers counted on 20 trees, and the total rainfall for the last 5 days. It predicted the percentage of the open flowers that would be affected 3-4 days hence. The model has been widely used by growers in principle, although not always followed in detail. In 1996 and 1997, the model predicted that no fungicide applications were needed in virtually any area of the state. Some growers applied preventive sprays to the navel orange groves at the beginning of the bloom which could have been avoided had they followed the model. This model is conservative in that it occasionally triggers a spray which would not be needed, especially when high amounts of rainfall occur in a short period of time. Florida citrus growers are not accustomed to using thorough scouting programs. The biggest problem in the use of the model is that small numbers of infected flowers must be detected, and superficial scouting efforts are often insufficient to locate them.

When the original model was developed, only inoculum and rainfall data were included because they were the most important factors and were most readily available to growers. Leaf wetness was a significant, if minor, factor, but was not readily available to growers. Currently, growers can purchase that information directly or by contract from automated weather stations. Thus, leaf wetness has been incorporated into the model and the modified equation is:

$$y = -13.63 + 1.16\sqrt{TD} + 0.48\sqrt{100\,R} + 1.77\sqrt{LW \times 5}$$

Where y =percentage of flowers infected in 4 d; however, if $y < 0$, then $y = 0$.

TD = the total number of infected flowers on 20 trees; however, if $TD < 75$, then $TD = 0$.
R = total rainfall for the last 5 d in mm.
LW = average number of daily leaf wetness for the last 5d – 10 h

This equation reduces the forecasted incidence of PFD when rainfalls are of short durations and increases the predicted incidence with rains of long durations.

As previously, we recommend that the model be applied initially when the first flowers open during the major bloom, and large numbers of pinhead and unopened blossoms are present. Fungicide application is indicated when the model predicts that 20% of the flowers will be infected. A fungicide application should protect the bloom for 10-14 days and scouting should resume again 10 days after an application. Applications are discontinued once all flowers are open and no future bloom needs protecting.

The model functions in Florida because inoculum levels are low at the beginning of bloom and can be maintained at low levels by properly timed fungicide applications. Buildup and dispersal of the pathogen are heavily dependent on rainfall. The model may be useful in areas with climates similar to Florida. However, in the hot, humid tropics, inoculum is constantly present at high levels and rainfall is frequent during bloom. Under these conditions, the model functions, but nearly always predicts the need for a fungicide application. Fungicidess need to be applied when large numbers of unopened and pinhead blossoms are present on the trees to protect as much of the crop as possible.

Lime Anthracnose

Lime anthracnose occurs throughout the humid areas of the Americas, in Zanzibar, and probably in other areas as well (14,24,50,51). Key lime, otherwise known as Mexican or West Indian lime, is the only known host affected by this disease. However, as noted previously, isolates of the causal agent of this disease also produce PFD on other citrus. The disease is devastating in high rainfall areas and precludes commercial production of this species in such areas. Tahiti limes (*C. latifolia* Tan.) are unaffected and can be grown in humid areas.

SYMPTOMS

This disease affects the flowers, producing the same symptoms on flowers and persistent buttons as does PFD. However, on Key limes, young leaves, shoots, and fruits are also affected. In severe cases, leaves become totally blighted and drop, shoot tips die back producing the withertip symptoms, and fruitlets abscise. If conditions are less favorable or tissues more mature, localized necrotic lesions of variable size develop.

CAUSAL AGENT, DISEASE CYCLE, AND EPIDEMIOLOGY

Lime anthracnose is caused by *C. acutatum* which was originally described as *G. limetticola* (14). In contrast to the *C. acutatum* strain which produces PFD on citrus flowers, the lime anthracnose strain produces acervuli on all affected tissues-leaves, twigs, flowers, and fruit. Morphologically, it is nearly identical to the PFD pathogen (4), and, indeed, the PFD strain may have arisen from lime anthracnose isolates. The epidemiology of this disease has not been studied but is presumed to be similar to PFD. The only difference is that the pathogen reproduces readily on all tissues and inoculum is probably readily available on infected tissues year-round.

CONTROL

The most effective fungicides for control of lime anthracnose are mancozeb and captafol (32). Copper fungicides are somewhat effective and benomyl, triadimenol, and propiconazole appear promising for disease control. In western Mexico, applications need to be made frequently during the rainy season from July to October (32). In areas with high rainfall, the disease is nearly impossible to control even with fungicides which are effective against the pathogen.

Postharvest Anthracnose

OCCURRENCE AND DISTRIBUTION

Even though *C. gloeosporioides* is ubiquitous in citrus-growing regions, significant postharvest anthracnose has been reported mostly from Florida (7,10,37) and Australia (5,53), usually in association with degreening of early-season, green-colored, ripe fruit. Degreening is the commercial practice of treating fruit with ethylene (5-10 $\mu l\ l^{-1}$) at high relative humidities to remove chlorophyll from the fruit rind (49). Exposure of the natural red, yellow, and orange pigments of the fruit rind is required to achieve fruit colors acceptable to the consumer. In other citrus-producing regions, anthracnose sporadically develops on injured surfaces of mature fruit (21,27,52), irrespective of ethylene treatment.

SYMPTOMS

Anthracnose caused by *C. gloeosporioides* can be manifested in several forms. Rind blemishes in the form of tear-staining patterns down the side of the fruit develop after germination of large numbers of conidia in water on the fruit surface (21,27). Many of the appressoria produced by conidia form infection pegs that only penetrate epidermal cells where they cause some necrosis and darkening of the rind surface. The blemish does not develop further after harvest nor is it stimulated by ethylene degreening, but the blemish is often severe enough to preclude using the fruit for fresh consumption.

Infections associated with fruit damage caused by punctures, bruising, rind breakdown of the stem end, or sunburn (21,27,44,52) cause brown to black spots 1 to 3 cm in diameter. As these lesions progress, the fruit softens, particularly under high humidity. Masses of conidia produced on the lesion surface appear pink or salmon-colored. Under dry conditions, lesions remain brown to black.

Other symptoms of anthracnose are associated with ethylene treatment of fruit. Ethylene causes appressoria on the fruit surface to produce penetration pegs, invade the tissue and form lesions on any portion of the fruit surface where appressoria are present at high densities (7). At an early stage of development, the affected rind is silvery gray and leathery (52). The surface of the lesion remains at the same degree of firmness and elevation as the adjacent healthy rind. As the decay develops, tissue turns brown, then black, and eventually softens and becomes covered with pink or salmon conidia. Lesions may also develop around the fruit button. Quiescent infections of the pathogen established on the button are able to colonize it quite rapidly since ethylene induces rapid senescence of the button tissue. After colonizing the dead button, the fungus spreads into the surrounding healthy rind to form brown to black lesions 1 to 2 cm in diameter at the stem end of the fruit. This disorder has been observed primarily on mandarins (10). Similar lesions can also be formed on early-season degreened fruit at very minor injuries involving only a few oil glands that result in oleocellosis. The fungus attacks the tissue damaged by the peel oil then spreads into surrounding healthy tissue causing small brown to black lesions. This problem is common on early-season Florida grapefruit (10).

DISEASE DEVELOPMENT

Many of the appressoria on the surface of citrus fruit remain quiescent at fruit maturity (7). The appressoria are stimulated to resume growth when these green-colored fruit are exposed to ethylene during commercial degreening. Washing before degreening removes much of the fungus (7,38,39), but if washing is delayed for 3 days at 5μl l^{-1} of ethylene or 1 day at 50 μl l^{-1} during degreening for a total of 4 days, washing does not effectively remove the inoculum (7). As expected, if fruit are degreened with ethylene at 50 μl l^{-1} for 4 days, washing does not effectively at an ethylene concentration of 50 μl l^{-1} for 4 days, washing after 2 days of degreening has no effect on disease severityremove the inoculum (7). Similarly, disease incidence is increased greatly by degreening tangerines with 50 rather than 10 μl l^{-1} of ethylene (12).

The concentration of appressoria on fruit surfaces can also have a significant effect on the incidence of anthracnose. A high percentage of tangerines develop anthracnose after 5 days of degreening with a concentration of 50 μl l^{-1} of ethylene when 300 appressoria are present on 1 mm^2 of fruit surface, but no symptoms develop after the same amount of time on similar surface areas when less than 100 appressoria are present (7).

Ultrastructural studies of germinating appressoria on the rind of tangerines show that the appressorium forms a pore in the wall adjacent to the fruit cuticle (Fig. 1B). The pore opening is surrounded by a funnel-shaped collar that is formed from the inward extension of the appressorial wall (8). The wall of the infection hypha is formed on the surface of the collar and extends from within the appressorium to the pore where it continues to develop as the infection hypha emerges. The infection hypha maintains its original diameter of less than 0.5 µm during penetration of the cuticle and causes no physical depression of the cuticle during penetration. Upon approaching the epidermal cell wall, the infection hypha can develop in three distinct patterns to penetrate the host (8). The most frequent manner of penetration occurs by extension and enlargement of the infection hypha to form a primordial hypha ca. 1-2 µm in diameter (Fig. 1C). These hyphae form in the anticlinal and periclinal walls of the epidermal cells and the cell lumen. This structure, separated by a septum, leads to the formation of larger hyphae 3-5 µm in diameter that are responsible for most of the infection process. Less frequently, the primordial hyphae forms in the lower part of the cuticle and proceeds to grow subcuticularly and then grows between epidermal cells before finally developing into the larger hyphae. Finally, in the third manner of penetration, the infection hypha penetrates the cuticle and immediately enlarges into the large hyphae either between or within the lumen of the epidermal cells (Fig. 1D) (8). In susceptible rind, the intracellular large hyphae of *C. gloeosporioides* do not cause major cellular destruction in advance of penetration, and are constricted during penetration of the host wall. Chloroplasts of adjacent non-invaded cells appear most sensitive and responses are noted one to two cells in advance of the hyphae. In these chloroplasts, the lamellae are swollen and they eventually disintegrate, and the lipid bodies are weakly stained. The cuticle remains intact during early stages of invasion of the epidermal and adjacent two or three cell layers of the rind. A similar mode of penetration for *C. gloeosporioides* was described by Coates et al. (14) in studies with avocado fruit.

With advancing maturity, chlorophyll disappears from the rind of citrus fruit as it develops natural orange and yellow colors. Anthracnose does not normally develop on this fruit when it is subjected to degreening conditions similar to those of the early-season, green-colored fruit. This resistance was first noted when anthracnose was most prevalent on the greener grapefruit or on yellower oranges and mandarins in the same lot of fruit following degreening. Fruit that develop more yellow and orange color apparently experience certain physiological changes associated with maturity that cause fruit to resist infection. The orange-colored mandarins exhibit a hypersensitive response to infection hyphae (9). Ultrastructural studies show morphological changes in cells well in advance of hyphal penetration. Electron-dense products occlude cells invaded by fungal hyphae. Staining and light microscopy show the presence of phenolic and lignin materials in association with resistance of the rind to fungal invasion. Interestingly, resistance is induced in green fruit if they are washed to remove natural appressoria and treated with ethylene to induce color-break before inoculation and exposure to ethylene a second time (12). Preharvest sprays of ethephon 5-7 days before harvest significantly reduce the natural development of anthracnose in tangerines (6). Control is attributed to the accumulation of ethylene in the internal portion of the fruit which induces the physiological changes required for development of fruit resistance without stimulating germination of the appressoria on the fruit surface. This internal ethylene contrasts with the postharvest scenario where appressoria are exposed to ethylene when fruits are placed in a degreening room.

Ethylene causes germination and formation of appressoria by conidia of *C. gloeosporioides* (23). Ethylene, a natural plant ripening hormone which evolves from climacteric fruit, also signals conidia to differentiate into multiple infection structures and begin the infection process. However, under natural conditions, conidia dispersed in surface water on immature citrus fruit are perfectly capable of forming appressoria in the apparent absence of ethylene. Constituents of the fruit epicuticular wax reportedly also play a role in stimulating spore germination and appressorium formation (35), and these may play a more important role in the infection of citrus. The C_{24} and longer-chained fatty alcohols induce appressorium formation. Protein phosphorylation may be involved in the induction of appressorium formation by either surface wax or ethylene (22).

The role of ethylene in the germination of appressoria has not been elucidated. Ethylene could have a direct effect upon the fungus and stimulate germination as it does with spores, and/or provoke changes in the fruit that trigger germination. Germination of appressoria formed on nitrocellulose membranes in the absence of the host was increased upon exposure to a high (100 $\mu l\ l^{-1}$) concentration of ethylene (10). These results suggest one possible role may be that of a direct effect of ethylene on appressorium germination. Because incidence of disease is dependent on ethylene concentration (12), ethylene may also alter the host physiology and subsequent susceptibility.

CONTROL

Several cultural and handling practices can be used to minimize anthracnose. Good production practices will create healthier trees with less dead wood and subsequent inoculum. Harvest can be delayed or early fruit with better natural color can be spot-picked to reduce degreening time to less than one day. Ethylene concentrations should be accurately controlled during degreening to maintain no more than 5 $\mu l\ l^{-1}$, and packing conditions that allow good air exchange should be selected (11,26). Spraying with benomyl preharvest or predegreening drenches of benomyl, prochloraz, or thiabendazole (11,37,38,53) combined with short durations of degreening will provide even better control.

Literature Cited

1. Agostini, J.P., Gottwald, T.R., and Timmer, L.W. 1993. Temporal and spatial dynamics of postbloom fruit drop of citrus. Phytopathology 83:485-490.
2. Agostini, J.P., and Timmer, L.W. 1992. Selective isolation procedures for differentiation of two strains of *Colletotrichum gloeosporioides* from citrus. Plant Dis. 76:1176-1178.
3. Agostini, J.P. and Timmer, L.W. 1994. Population dynamics and survival of strains of *Colletotrichum gloeosporioides* on citrus in Florida. Phytopathology 84:420-425.
4. Agostini, J.P., Timmer, L.W., and Mitchell, D.J. 1992. Morphological and pathological characteristics of strains of *Colletotrichum gloeosporioides* from citrus. Phytopathology 82:1377-1382.
5. Anonymous. 1990. Degreening room burn or anthracnose. Queensland Citrus Bulletin. Spring. p. 7-8.
6. Barmore, C.R. and Brown, G.E. 1978. Preharvest ethephon application reduces anthracnose on Robinson tangerines. Plant Dis. Rep. 62:541-544.
7. Brown, G.E. 1975. Factors affecting postharvest development of *Colletotrichum gloeosporioides* in citrus fruits. Phytopathology 65:120-123.
8. Brown, G.E. 1977. Ultrastructure of penetration of ethylene degreened Robinson tangerines by *Colletotrichum gloeosporioides.* Phytopathology 67:315-320.
9. Brown, G.E. 1978. Hypersensitive response of orange-colored Robinson tangerines to *Colletotrichum gloeosporioides* after ethylene treatment. Phytopathology 68:700-706.
10. Brown, G.E. 1992. Factors affecting the occurrence of anthracnose on Florida citrus fruit. Proc. Int. Soc. Citriculture 3:1044-1048.
11. Brown, G.E. and Barmore, C.R. 1976. The effect of ethylene, fruit color, and fungicides on susceptibility of 'Robinson' tangerines to anthracnose. Proc. Fla. State Hortic. Soc. 89:198-200.
12. Brown, G.E. and Barmore, C.R. 1977. The effect of ethylene on susceptibility of Robinson tangerines to anthracnose. Phytopathology 67:120-123.
13. Brown, A.E., Sreenivasaprad, S., and Timmer, L.W. 1996. Molecular characterization of slow-growing orange and Key lime anthracnose strains of *Colletotrichum* from citrus as *C. acutatum.* Phytopathology 86:523-527.
14. Clausen, R.E. 1912. A new fungus concerned in wither tip of varieties of *Citrus medica.* Phytopathology 2:217-236.
15. Coates, L. M., Muirhead, I. F., Irwin, J. A. G., and Gowanlock, D. H. 1993. Initial infection processes by *Colletotrichum gloeosporioides* on avocado fruit. Mycol. Res. 97:1363-1370.

16. Denham, T.G. 1979. Citrus production and premature fruit drop disease in Belize. ANS 25:30-36.Proc. Natl. Acad. Sci. USA 25:30-36.
17. Denham, T.G. and Waller, J.M. 1981. Some epidemiological aspects of postbloom fruit drop (*Colletotrichum gloeosporioides*) in citrus. Ann. Appl. Biol. 98:65-77.
18. Fagan, H.J. 1971. Pathology and nematology in British Honduras. Ann. Rep. Citrus Research Unit. University of West Indies, p. 10-21.
19. Fagan. H.J. 1979. Postbloom fruit drop, a new disease of citrus associated with a form of *Colletotrichum gloeosporioides.* Ann. Appl. Biol. 91:13-20.
20. Fagan, H.J. 1984. Postbloom fruit drop of citrus in Belize: II. Disease control by aerial/ground spraying. Turrialba 34:179-186.
21. Fawcett, H.S. 1936. Citrus diseases and their control.Diseases and Their Control. 2nd ed. McGraw-Hill Book Co., New York.
22. Flaishman, M.A., Hwang, C.S., and Kolattukudy, P.E. 1995. Involvement of protein phosphorylation in the induction of appressorium formation in *Colletotrichum gloeosporioides* by its host surface wax and ethylene. Physiol. Molec. Plant Pathol. 47:103-117.
23. Flaishman, M.A. and Kolattukudy, P.E. 1994. Timing of fungal invasion using host's ripening hormone as a signal. Proc. Natl. Acad. Sci. 91:6579-6583.
24. Fulton, H.R. 1925. Relative susceptibility of citrus varieties to attack by *Gloeosporium limetticolum* (Clausen). J. Agric. Res. 30:629-635.
25. Gantotti, B.V. and Davis, M.J. 1991. Detection of pectinase polymorphism in *Colletotrichum gloeosporioides.* Phytopathology 81:1176 (abstr.).
26. Hagenmaier, R.D., and Shaw, P.E. 1992. Gas permeability of fruit coating waxes. J. Amer. Soc. Hortic. Sci. 117:105-109.
27. Klotz, L.J. 1973. Color Handbook of Citrus Diseases. 4th ed. Univ. Calif. Press, Berkeley, California. 122 pp.
28. Lima, J.E.D. 1994Atualizaçâo sobre a podridã floral de *Colletotrichum* dos citros (PFC) em Sao Paulo. Laranja & CIA 34(6):5.
29. Liyanage, H.D., Koller, W., McMillan, R.T., and Kistler, H.C. 1993. Variation in cutinase from two populations of *Colletotrichum gloeosporioides* from citrus. Phytopathology-83:113-116.
30. Liyanage, H.D., McMillan, R.T., Jr., and Kistler, H.C. 1992. Two genetically distinct populations of *Colletotrichum gloeosporioides* from citrus. Phytopathology 82:1371-1376.
31. McMillan, R.T. and Timmer, L.W. 1989. Outbreak of citrus postbloom fruit drop caused by *Colletotrichum gloeosporioides* in Florida. Plant Dis. 73:81.
32. Orozco-Santos, M. 1995. Enfermedades presentes y potenciales de los cítricos en Mexico. Universidad Autónomo, Chapingo, Mexico.
33. Orozco Santos, M. and Gonzalez Garza, R. 1986. Caída de fruto pequeño y su control en naranja 'Valencia' en Veracruz. Agric. Tec. Mex. 12(2):259-269.

34. Peña, J.E. and Duncan, R. 1990. Role of arthropods in the transmission of post bloom fruit drop. Citrus Ind. 71(4):64-69.
35. Podila, G.K., Rogers, L.M., and Kolattukudy, P.E. 1993. Chemical signals from avocado surface wax trigger germination and appressorium formation in *Colletotrichum gloeosporioides.* Plant Physiol. 103:267-272.
36. Schwarz, R.E., Klein, E.H.J., and Monsted, P. 1978. Fungal infection of citrus flowers: probable cause of abnormal fruit drop in the Parana mist zone of Misiones, Argentina. Third Int. Congr. Plant Pathol., Munich, Germany. p. 130 (abstr.).
37. Smoot, J.J. and Melvin, C.F. 1967. Postharvest decay of specialty hybrid citrus fruits in relation to degreening time. Proc. Fla. State Hortic. Soc. 80:246-250.
38. Smoot, J.J. and Melvin, C.. 1972. Decay of degreened citrus fruit as affected by time of washing and TBZ application. Proc. Fla. State Hortic. Soc. 85:235-238.
39. Smoot, J.J., Melvin, C.F., and Jahn, O.L. 1971. Decay of degreened oranges and tangerines as affected by time of washing and fungicide application. Plant Dis. Rep. 55:149-152.
40. Sonoda, R.M. and Pelosi, R.R. 1988. Outbreak of citrus postbloom fruit drop caused by *Colletotrichum gloeosporioides* from lesions on citrus blossoms in the Indian River of Florida. Proc. Fla. State Hortic. Soc. 101:36-38.
41. Sutton, B.C. 1980. The Coelomycetes.KEW: Commonwealth Mycological Institute, Kew, UK. 696 pp.
42. Timmer, L.W., Agostini, J.P., Zitko, S.E., and Zulfiqar, M. 1994. Postbloom fruit drop of citrus, an increasingly prevalent disease of citrus in the Americas. Plant Dis. 78:329-334.
43. Timmer, L.W., Brown, G.E., and Zitko, S.E. 1998. The role of *Colletotrichum* spp. in postharvest anthracnose of citrus and survival of *C. acutatum* of fruit. Plant Dis.82:415-418.
44. Timmer, L.W. and Zitko, S.E. 1992. Timing of fungicide applications for control of postbloom fruit drop of citrus in Florida. Plant Dis. 76:820-823.
45. Timmer, L.W. and Zitko, S.E. 1993. Relationships of environmental factors and inoculum levels to the incidence of postbloom fruit drop of citrus. Plant Dis. 77:501-504.
46. Timmer, L.W. and Zitko, S.E. 1995. Early season indicators of postbloom fruit drop of citrus and the relationship of disease incidence and fruit production. Plant Dis. 79:1017-1020.
47. Timmer, L.W. and Zitko, S.E. 1996. Evaluation of a model for prediction of postbloom fruit drop of citrus. Plant Dis. 80:38-383.
48. Von Arx, J.A. 1970. A revision of the fungi classified as *Gloeosporium.* Bib. Mycol. 24:1-203.

49. Wardowski, W.F., and McCornack, A.A. 1973. Recommendations for degreening Florida fresh citrus fruits. Florida Coop. Ext. Serv. Circ. 1170, IFAS, Univ. Fla., Gainesville.
50. Wheeler, B.E.J. 1963. Withertip disease of limes (*Citrus aurantifolia*) in Zanzibar. Ann. Appl. Biol. 51:237-251.
51. Wheeler, B.E.J. 1963. Studies on *Gloeosporium limetticola* causing withertip disease of limes in Zanzibar. Trans. Br. Mycol. Soc. 46:193-200.
52. Whiteside, J.O. 1988. Symptomless and quiescent infections by fungi. Page 30 in: Compendium of Citrus Diseases. J.O. Whiteside, S.M. Garnsey, and L.W. Timmer, eds. APS Press. St. Paul, MN, 80 pp.
53. Wild, B.L. 1990. Ethylene gas burn of Washington navel oranges-a form of anthracnose induced by degreening and controlled by brushing or applying fungicides. Australian J. Exp. Agric. 30:565-568.
54. Zitko, S.E. and Timmer, L.W. 1992. Evaluation of fungicides in vitro for control of *Colletotrichum gloeosporioides* from citrus, 1991. Fungicide and Nematicide Tests 47:335.
55. Zulfiqar, M., Brlansky, R.H., and Timmer, L.W. 1996. Infection of flower and vegetative tissues of citrus by *Collectotrichm acutatum* and *C. gloeosporioides.* Mycologia 88:121-128.

Chapter 19

Occurrence and Management of Anthracnose Epidemics Caused by *Colletotrichum* Species on Tree Fruit Crops in California

J.E. Adaskaveg and H. Förster

In recent years, widespread epidemics of anthracnose on almond and citrus in California have resulted in considerable crop losses. Correlated with increased rainfall and warm temperatures, these epidemics have been attributed either to a new or to a chronic occurrence of the pathogen. On almonds, anthracnose was a new outbreak where the causal organism, *C. acutatum*, is a virulent pathogen infecting blossoms, leaves and fruit (1). On citrus, the outbreak developed from endemic populations of the causal organism, *C. gloeosporioides*. This fungus is a weak pathogen generally surviving on senescent or injured tissues (29). Incidence of anthracnose has also increased on peach in California, but without significant crop losses. This review will focus, therefore, on anthracnose of almond and citrus. We will present recent research data from laboratory and field studies which help us to understand the nature of these diseases and will facilitate their management in the future.

Almond Anthracnose

Historic Background and Geographic Distribution of the Disease

Almond anthracnose was first reported from Italy in 1896, and the pathogen was identified as *Gloeosporium amygdalinum* Brizi (6). In France, Israel, and South Africa, almond anthracnose causes serious crop losses and is commonly known as kernel rot, almond gummosis, and gumming of almond (9,15,22). Cultures of the fungus isolated from almond in Israel and submitted to von Arx at the Centraalbureau voor Schimmelcultures by Shabi and Katan (22) were identified as *C. gloeosporioides*. In 1957 von Arx (3) placed *G. amygdalinum* in synonymy with *C. gloeosporioides* Penz. and Sacc. in Penz. In California, the disease was reported as a leaf spot and fruit rot of almond in Napa and Alameda counties in 1916 (8) and 1925 (27). Since these first reports, almond anthracnose was only observed sporadically in California; in 1941 Smith (24) did not even include the disease in "Diseases of Fruit and Nuts." Extension specialists and farm advisors observed anthracnose again in California in the late 1980s. In the early 1990s, the disease caused crop losses in localized areas and was reported to the Almond Board of California by the senior author. In 1995 almond anthracnose was widespread in the northern Sacramento and central San Joaquin valleys of California (Fig. 1), and the disease caused substantial crop losses in five primary counties—Butte, Glenn, Merced, San Joaquin, and Stanislaus. Anthracnose was also found in almond orchards of the San Joaquin valley. Here, however, no crop losses occurred because of the drier climate. Economic losses were most severe in 1998 when the warm, unusually wet spring and early summer weather caused by El Niño provided highly conducive conditions for disease development.

Fig. 1. Distribution map of almond anthracnose in California in 1996-98. Two epidemic centers in northern (Butte Co.) and central (Stanislaus Co.) California affected thousands of acres. The disease occurs throughout almond-growing regions of central California; however, crop losses have only been reported in counties adjacent to the two epidemic centers. Symbols indicate counties where the disease has been observed.

Disease Symptoms and Environmental Conditions for Disease Development

Symptoms of almond anthracnose may occur on blossoms, fruit, leaves, and branches at all stages of development. On blossoms, the disease looks similar to brown rot blossom blight caused by *Monilinia laxa*. Infected anthers and stigmata become brown, shriveled, and necrotic. Entire blossoms may collapse as the infection develops from the hypanthium to the peduncle. Orange spore masses are produced under wet conditions from acervuli and appear as droplets on the outside and inside surface of the hypanthium. The principal damage, however, is to developing fruitlets (Fig. 2A) or mature fruit. In years of extended rainfall from spring to early summer, new anthracnose fruit infections may develop on immature fruit, as well as on the hull of mature fully developed fruit through the hull split stage of fruit development. On mature fruit, infections develop as circular, tan to orange, necrotic, sunken lesions 0.5 cm or more in diameter (Fig. 2B). Acervuli often occur in the center of lesions. Internally, infections develop through the hull and into the kernel. Infections of immature fruit result in shriveled fruit that appear similar to aborted, non-pollinated fruit ("blanks"). On half-grown and full-grown mature fruit, infections result in profuse gumming from developing lesions. Many of the infected fruit become mummified and either drop or remain on the tree throughout tree dormancy. Young lesions on almond fruit may be confused with green fruit rot caused by *Monilinia, Botrytis,* or *Sclerotinia*

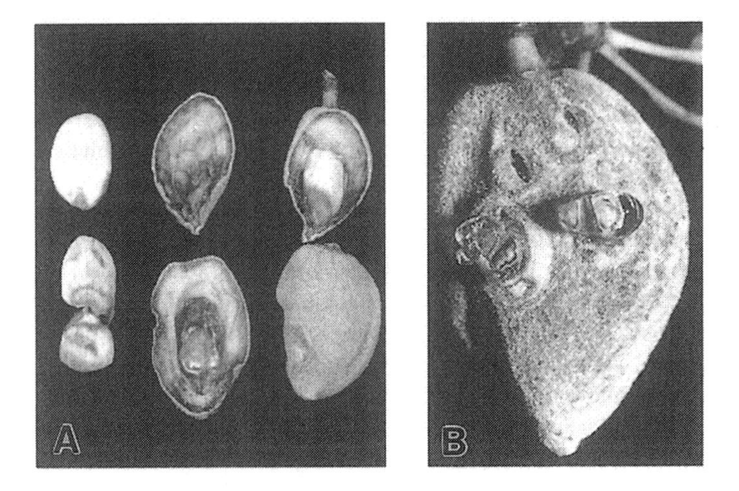

Fig. 2. Almond anthracnose fruit symptoms. A. Infected kernels (left), cross sections of fruitlets showing hull and kernel infections (center and upper right), and sunken lesion on hull of fruitlet (lower right). B. Multiple lesions with gumming exudates on hull of mature fruit. (Figures are actual size).

species. Leaf infections may occur at the tips and along the margins of leaves forming irregular lesions. These infections initially develop as water-soaked lesions that die and become tannish-brown or bleached. Infected leaves may drop or remain attached to the tree. As blossom and fruit infections progress into almond spurs, leaves on branches distal to the infection often wilt, collapse, turn yellowish-tan, and die, but often remain attached for a short time. On branches bearing diseased fruit, dieback may occur and continue into larger branches up to 2-3 cm in diameter in the second year. Disease progress is favored by warm (20-25°C), wet weather. The exact environmental conditions for optimum development, however, still need to be defined.

HOST SUSCEPTIBILITY

All currently cultivated almond varieties can be affected by the disease. Field observations suggest that NePlus Ultra, Peerless, Harvey, Thompson, Price, Wood Colony, and Merced are among the most susceptible cultivars. Moderate susceptibility has been observed with Carmel, Mission, Butte, Fritz, Sonora, and Monterey. Fruit infections of Padre and Nonpareil are less common; however, the latter cultivar is susceptible to leaf infections.

CAUSAL ORGANISM OF ALMOND ANTHRACNOSE IN CALIFORNIA

The causal organism of almond anthracnose in California, characterized using conidial morphology, temperature relationships, molecular techniques, and fungicide sensitivity, was identified as *Colletotrichum acutatum* J.H. Simmons (1). Consequently, on a worldwide scale almond anthracnose appears to be caused by two species of *Colletotrichum*, *C. acutatum* as identified in California and *C. gloeosporioides* as reported from Israel (13,22). Unfortunately, reports of the disease from other countries do not include a detailed description of the causal pathogen.

C. acutatum and *C. gloeosporioides* can be separated by a number of characters (23,26) (Table 1). *C. gloeosporioides* has gray mycelium with orange spore masses on potato-dextrose agar, grows more rapidly than *C. acutatum* at optimal temperatures, and produces elongated ellipsoid conidia (16.4 x 5.2 µm) with rounded ends. Cultures of *C. acutatum* are characterized by pink to gray mycelium, orange spore masses on potato dextrose agar, slower growth rate, and lower optimal temperature than *C. gloeosporioides*. Elongated ellipsoidal conidia average 14.8 x 5.2 µm and are usually pointed at one or both ends. The teleomorph of *C. gloeosporioides* is *Glomerella cingulata* (Ston.) Spauld. & Schrenk, whereas the teleomorph of *C. acutatum* has been unknown with only one recent report (16, see Correll et al. Chapter 10 in this book). Wild-type isolates of *C. gloeosporioides* are very sensitive to the fungicide benomyl, whereas those of *C. acutatum* are quite insensitive. Generally, the conidial morphology of California almond isolates was more

Table 1. Characteristics separating *C. acutatum* (collected from almond) from *C. gloeosporioides* (collected from citrus)[1]

Characteristic	*C. gloeosporioides*	*C. acutatum*
Growth rate	Fast (>10 mm/day at 30°C)	Slower (<10 mm/day at 25°C)
Optimum growth temperature	30°C	25°C
Conidium shape	Rounded on both ends	Pointed on at least one end
Sensitivity to benomyl	Sensitive	Quite insensitive
Molecular detection using *C. gloeosporioides*-specific primers	+	-
Molecular detection using *C. acutatum*-specific primers	-	+

[1]Based on California isolates of each species.

similar to reference cultures of *C. acutatum* from strawberry and other stone fruit crops than to cultures of *C. gloeosporioides* from citrus and papaya that were grown on the same media. However, in comparisons between *Colletotrichum* species grown on pea straw agar, conidial size was overlapping. Therefore, size and shape of conidia were not reliable criteria for identifying species of *Colletotrichum*. Based on temperature relationships of isolates grown at 10 to 35°C for 5 days, *C. gloeosporioides* had an optimum growth rate at 30°C and grew faster than *C. acutatum* isolates at their optimum of 25°C (1). Using species-specific primers developed from the ITS I region of ribosomal DNA (19) in PCR amplification reactions, DNA from the California almond isolates yielded amplification products of the expected size with the *C. acutatum* primers CaInt-1 and CaInt-2 but not with the *C. gloeosporioides* primers CgInt-1 and CgInt-2 (Fig. 3). In addition, restriction fragment patterns of ribosomal DNA from the almond isolates were more similar to those of a reference culture of *C. acutatum* from strawberry than to those of *C. gloeosporioides* from citrus. All these results indicate that the fungus isolated from almond in California is *C. acutatum* and not *C. gloeosporioides*.

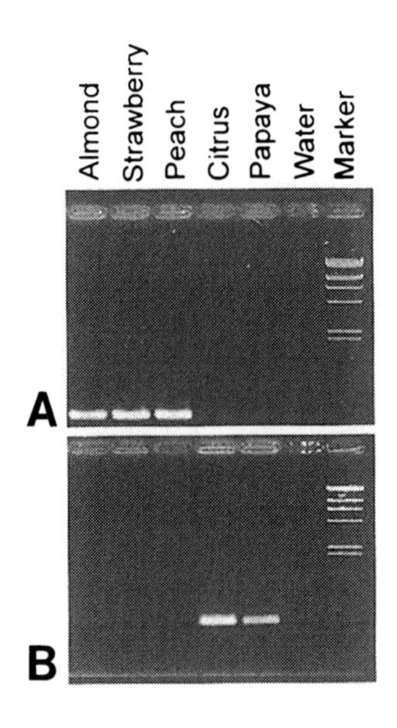

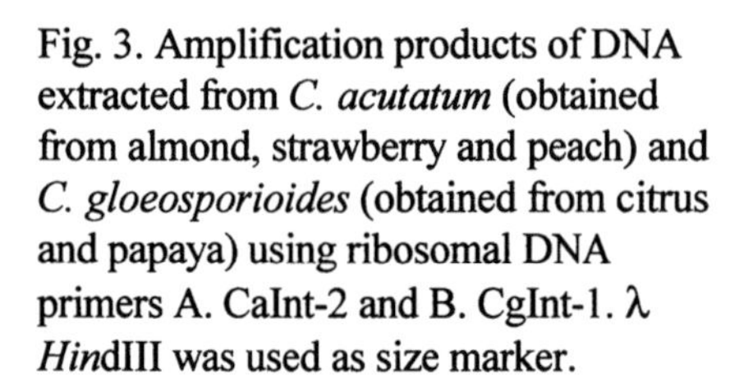

Fig. 3. Amplification products of DNA extracted from *C. acutatum* (obtained from almond, strawberry and peach) and *C. gloeosporioides* (obtained from citrus and papaya) using ribosomal DNA primers A. CaInt-2 and B. CgInt-1. λ *Hin*dIII was used as size marker.

POPULATION STRUCTURE OF *C. ACUTATUM* ON ALMOND IN CALIFORNIA

Two distinct subpopulations of *C. acutatum* were recognized during field samplings in 1996 and 1997 (11). The majority of cultures (62%) had a pink pigmentation on the colony surface and the reverse of the culture when freshly isolated. The remaining cultures were white to gray with the reverse of the colony cream to gray, and the cultures did not sporulate as profusely on agar. The conidia of the cultures with pink pigmentation were clearly pointed at each end. Those of the gray cultures had less pointed, broadly rounded ends (Fig. 4). The growth rates of both colony types were slower than that of *C. gloeosporioides*. There was no difference in disease symptoms of fruit that yielded either colony type of *Colletotrichum*. Both types sometimes were isolated from the same orchard or even the same fruit, without any evident cultivar or geographic relationships. The population structure is currently being analyzed in more detail. The two *C. acutatum* subpopulations also differed in their sensitivity to the fungicide benomyl at low conidial concentrations ($<3 \times 10^5$ conidia/ml) in a filter-paper disk assay, and in the mycelial growth study on amended media, the pink colony type was more sensitive. Interestingly, the two cultural types reacted differently with the two *C. acutatum*-specific primers CaInt-1 and CaInt-2. Whereas with primer CaInt-1, amplification products were more abundant using DNA from the pigmented colonies, generally more product was present with primer CaInt-2 using DNA from the gray colonies. Molecular differences between the pigmented and gray almond isolates were much more pronounced using simple repeat primers (13) or short random primers during PCR amplifications (11). Compared to the species-specific primers, these latter primers are much more sensitive in detecting DNA differences. After electrophoretic separation of the amplification products, very distinct banding patterns were evident for the two colony types (Fig. 5), and all representatives of each type had identical patterns with all primers suggesting the presence of two clonal subpopulations or genotypes (P and G).

RELATIONSHIPS BETWEEN ISOLATES OF *C. ACUTATUM* FROM ALMOND AND OTHER CROPS

RAPD banding patterns of the almond isolates were distinct from those of *C. acutatum* isolated from strawberry in California and North Carolina and apple and peach in North Carolina (11). All four California peach isolates that were available for our study were identical, and their banding patterns were the same as those of the California almond subpopulations producing pigmented colonies (genotype P). Because anthracnose on peach is not a significant problem in California, peach is an unlikely source of inoculum for almond (11). More isolates from different hosts will have to be collected in the future to gain insight into the origin and the population dynamics of almond anthracnose.

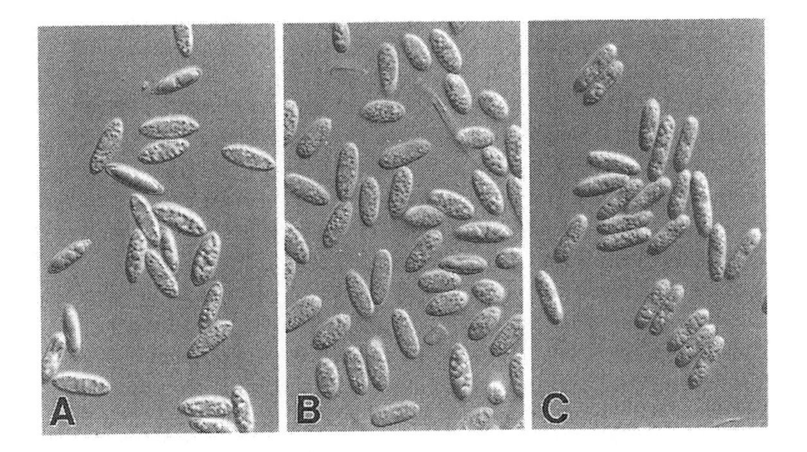

Fig. 4. Conidial morphology of California *Colletotrichum* isolates from almond and citrus grown on straw agar (x550). A. *C. acutatum* genotype P from almond. B. *C. acutatum* genotype G. C. *C. gloeosporioides* from citrus.

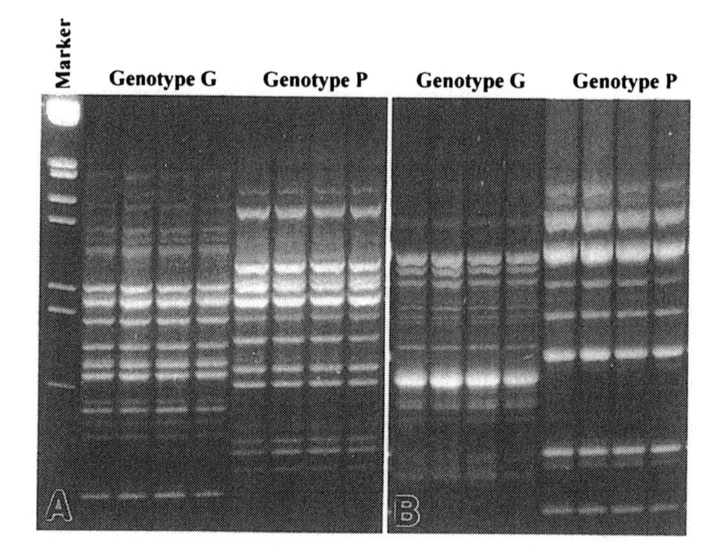

Fig. 5. DNA banding patterns of *C. acutatum* genotypes P and G from almond after PCR amplification with (A) a random 10mer primer (OPX-12) or (B) a simple repeat primer $(CAG)_5$. λ DNA/*Eco*RI+*Hin*dIII was used as a size marker.

CULTURING FROM INFECTED ALMOND TISSUE

Isolates were obtained from blossoms, fruit, peduncles, spurs, leaves, and wood tissue (11). Generally, isolations from blossoms and from fruit showing sunken, orangish lesions were the most successful. High recovery rates were also obtained from peduncles and spurs, whereas isolations from leaf and wood tissue were rarely successful in spite of extensive branch dieback and leaf yellowing (0 to 10% recovery rates). Therefore, the pathogen may not commonly invade these tissues. Isolations from branches, however, may have been obscured by extensive growth of secondary organisms. Isolations from leaf tissue generally were not obscured by secondary organisms, except by the omnipresent species of *Aureobasidium*. Leaf isolations were more successful early (spring) and late (fall) in the season, perhaps because favorable weather conditions at these times enable the fungus to infect leaf tissue. During the summer, however, host responses to secondary metabolites could be responsible for the symptoms (see next section).

In February of 1997, (late bud dormancy) isolations were carried out from fruit mummies collected in four orchards. *Colletotrichum* could be recovered from hull, shell, and endosperm tissue of mummified fruit (11). The recovery rate in two orchards (based on the number of mummies investigated) was 100%. This high recovery rate from overwintering fruit mummies indicated that fruit mummies are inoculum sources for early spring infections caused by rain-splash dispersal of spores. In the winter of 1998, although disease levels were low in the 1997 season, the fungus could still be isolated from mummified fruit. Therefore, removal of anthracnose mummified fruit is likely to reduce the initial disease level. Future studies are required to determine if peduncles and spurs may also serve as inoculum sources in early spring.

In Vitro Production of Phytotoxic Compounds by *C. acutatum* and Their Possible Involvement in Symptom Expression

Culture filtrates of *C. acutatum* (genotype P) from almond were obtained after growing the fungus for 2 weeks in modified Richard's solution (14) on a shaker. When rooted tomato seedlings or peach or almond cuttings were incubated in diluted (1:4) crude *C. acutatum* culture filtrate or filtrate that was passed through a 0.45 μm filter, severe phytotoxic effects were evident. No such symptoms were evident after incubation in diluted growth medium. Peach and almond cuttings lost their turgor after less than 24 h, and after 2 to 3 days the cuttings were completely wilted. Lesions on leaves first appeared watery, later necrotic. These symptoms were quite similar to those often found on field-infected material. The tomato seedlings started wilting after 2 days incubation and were completely wilted or dead after 3 or 4 days. These observations led to the hypothesis that leaf and dieback symptoms of almond branches in the field may be associated with the production of phytotoxic compounds by the fungus that are transported from infected blossoms and fruit into peripheral parts of the tree. Other species of *Colletotrichum* produce phytotoxic compounds that may be involved in symptom development (2,12,14). Our research on the phytotoxic compounds of *C. acutatum* from almond is ongoing; we hope to chemically characterize possible phytotoxic metabolites in the future. We have already demonstrated that the toxic substances in the culture filtrate are highly water-soluble and not extractable into the organic phase.

Postharvest Damage of Almond Kernels in Storage

The pathogen of anthracnose also has been isolated from processed almond kernels in storage (11). Symptoms included brown and purplish discolorations of the kernel tissue (Fig. 6). Infections commonly originated at almond sutures and developed hemispherically into the kernel. Quality assurance tests with positive samples of kernels with anthracnose were 0, 2.1, 9.8, and 10.0% for Nonpareil, Butte, Carmel, and Mission, respectively (Table

Fig. 6. Postharvest kernel symptoms of almond that resulted from preharvest anthracnose infections (actual size). Fully developed sliced kernels (left) showing brownish to purplish discoloration, as well as shriveled and discolored whole kernels (right).

Table 2. Postharvest anthracnose damage in processed almond kernels caused by *Colletotrichum acutatum*

Variety	Total tests	No. of tests with anthracnose affected kernels[a]	Tests with anthracnose affected kernels (%)[a]
Nonpareil	38	0	0
Butte	241	5	2.1
Carmel	133	13	9.8
Mission	420	42	10.0

[a]Tests with one or more anthracnose-discolored kernels in a 0.72-kg sample.

2). Isolations from blanched fruit were negative; however, isolations from non-blanched, halved fruit were all positive for *C. acutatum*. Both processed kernels had similar symptoms, but blanched fruit were steam-treated to remove the pellicle of the kernel. This heat treatment during the blanching process probably killed the fungus. Additional isolations were carried out from five almond lots. *Colletotrichum* was mainly isolated from kernels that showed discolorations when cut open (50% of the isolations were successful). Isolations from shriveled kernels or kernels with gumming were much less successful.

To define conditions that lead to postharvest kernel damage, laboratory-inoculated almond kernels were incubated for 3 months at selected relative humidities using salt solutions (17). The fungus did not grow on the kernels with a moisture content below 10.1% (11). The fungus, however, was viable at kernel moistures of 4.3% or above after the same time period. Since the moisture content of almond kernels in storage is kept between 5% and 10%, our results indicate that new infections of almond kernels are unlikely to occur in storage. Postharvest damage, therefore, is caused by preharvest infections in the field (11). We cannot rule out, however, that the fungus may grow from pre-existing infections over long periods (up to 1 year) in storage.

PATHOGENICITY OF *C. ACUTATUM* ISOLATES

Koch's postulates were followed to demonstrate the pathogenicity of *C. acutatum* and causality of almond anthracnose in California (1). The fungus was re-isolated from inoculated fruit showing symptoms similar to those in the field. In additional laboratory studies using detached fruit, both subpopulations of *C. acutatum* were pathogenic on wound-inoculated fruit, both causing sunken orange to tan lesions. Genotype P, however, was more virulent than genotype G in these assays. Average lesion diameters on the almond variety Wood Colony after 2 weeks of incubation at 20°C were 17.3 mm and 12.9 mm for genotypes P and G, respectively. Isolations from naturally field-infected tissues demonstrated, however, that both subpopulations are able to cause severe damage in the field.

IN VITRO SENSITIVITY OF FUNGAL ISOLATES TO SELECTED FUNGICIDES

Chemical control strategies are dependent on the species of *Colletotrichum* (5). Thus, we evaluated almond isolates for their sensitivity to fungicides according to spore germination and colony formation in a filter-paper disk assay, mycelial growth on fungicide-amended media, and spore germination on fungicide-amended media. Isolates of *C. acutatum* were less sensitive to benomyl than isolates of *C. gloeosporioides*, whereas isolates of both species were sensitive to captan. Using the filter-paper disk assay, propiconazole, tebuconazole, and captan were the most effective against *C. acutatum*, whereas propiconazole, tebuconazole, and benomyl were the most effective against *C. gloeosporioides*. The two subpopulations of *C. acutatum* just described responded very similarly to most of the fungicides evaluated; however, they differed in their sensitivity to benomyl. This difference was dependent on spore concentration; isolates of both populations were completely insensitive to benomyl at spore concentrations greater than 3×10^5. Mycelial growth from agar plugs and from germinated spores was inhibited similarly in both fungicide-amended media and disk assays. The results for benomyl, captan, myclobutanil, propiconazole, and tebuconazole are displayed in Table 3. Propiconazole and tebuconazole had very similar inhibitory effects against the two *C. acutatum* genotypes from almond when compared with *C. gloeosporioides*. However, captan and benomyl were more effective against *C. gloeosporioides* than against *C. acutatum* from almond. Conidial germination was not inhibited at concentrations less than 50 μg/ml of myclobutanil, propiconazole, tebuconazole, and benomyl for the almond isolates. Captan completely inhibited germination at 50 μg/ml, as did concentrations of 0.5 μg/ml of fluazinam or chlorothalonil. Unfortunately, fluazinam is not registered

Table 3. Relative inhibition of radial growth of *Colletotrichum acutatum* and *C. gloeosporioides* on fungicide-amended media[a]

			Inhibition of growth (%)[b]	
Fungicide	Formu-lation	Conc. (μg/ml)	*C. acutatum*	*C. gloeosporioides*
Benomyl	Benlate 50WDG	0.1	44.8	66.1
		5	55.8	99.2
		50	61.5	100
Captan	Captan 50WP	50	30.1	68.4
		500	70	97
		1000	74.1	93.7
Myclobutanil	Rally 40WP	0.5	43.2	14.5
		5	75.4	58.6
		10	79.5	70.2
Propiconazole	Break 45WP	0.1	25.3	23.9
		1	56.4	68.6
		10	79.5	97
Tebuconazole	Elite 45DF	0.1	33.8	20.6
		1	75	64.7
		10	95.8	93.9

[a]Cultures were grown on media amended with each fungicide at selected concentrations.
[b]Percent inhibition is relative to growth of each fungus on nonfungicide-amended potato dextrose agar.

in the United States, and chlorothalonil with its persistent residues on almond hulls and potential environmental hazards will probably not be registered on almonds in California.

MANAGEMENT OF ALMOND ANTHRACNOSE IN THE FIELD

In the spring of 1996, fungicide field trials were conducted to determine efficacious compounds, rates, and rotation programs. Two plots were established, one in northern and the other one in central California representing the two major regions where epidemics have occurred. Trials were conducted using an air-blast sprayer. Tebuconazole (Elite 45DF), propiconazole (Break 3.6EC), fenbuconazole (Indar 75WP), myclobutanil (Rally 40WP), strobilurin compounds (Abound 80WDG/2F; Flint 50DF), thiophanate-methyl (Topsin 75WP), captan, and maneb were evaluated as blossom, petal-fall, and developing-fruit treatments. Additionally, programs with fungicide rotations or mixtures of compounds were studied (e.g., myclobutanil-captan, myclobutanil-maneb, azoxystrobin-captan, azoxystrobin-thiophanate-methyl, and iprodione-captan). All treatments significantly reduced the disease; however, tebuconazole and propiconazole had the lowest incidence in 1996. In 1998 these two compounds and the strobilurins most effectively reduced the incidence of fruit and spur infections from an average of 95.5 infections per tree in the control treatment to less than 12 infections per tree for the propiconazole and strobilurin treatments that were applied six times every 10-14 days in the spring. None of the treatments, however, eradicated the disease and all performed as protectants. When treatments stopped and conducive conditions continued, the disease subsequently increased. Thus, fungicide programs are dependent on continued applications prior to conducive wet conditions throughout the spring. Rotation programs between fungicides of different classes were also effective and, thus, these programs reduced the total number of any one fungicide from six or seven to three or four applications per season. Management strategies to reduce the potential for developing populations of the pathogen resistant to any one class of fungicide are needed because of the extended period of susceptibility of almond fruit and continued spring rains that can occur during fruit maturation. Epidemiological studies to improve timing of fungicide applications are currently being conducted and models for predicting potential increases in disease severity will be developed as was done with postbloom drop of citrus (28). In 1997-1999, emergency registrations for propiconazole applications on almond were approved by the United States Environmental Protection Agency (EPA). In 1999and 2000, azoxystrobin was also labeled as a special local need (Section 24c) for management of almond anthracnose in California.

Peach Anthracnose

Anthracnose or bitter rot of peach is characterized by sunken, circular, tan to brown necrotic, firm lesions that develop concentric rings as the fungus grows and sporulates. Lesions may separate cleanly from healthy tissue when pressure is applied. The disease is of moderate importance in the peach-growing areas of the southern and southeastern United States (20). Based on spore morphology, colony color, and growth rate, two species of *Colletotrichum, C. gloeosporioides* and *C. acutatum,* have been identified infecting peach fruit in South Carolina (4). In California, anthracnose on harvested peaches and nectarines was observed by the senior author during the summer of 1990 in Fresno County; however, the pathogen was not characterized. A very limited outbreak that occurred in 1995 in Sutter/Yuba County was caused by *C. acutatum* (1). Cultural characteristics were very similar to genotype P of the almond anthracnose population. Random amplified polymorphic DNA confirmed the identity of the peach and the almond genotype populations. In addition, DNA banding patterns had similarities with those of four *C. acutatum* peach isolates from South Carolina (obtained from E. Zehr).

In California, development of management practices has not been needed. In the southeastern United States, captan and benzimidazoles (e.g., benomyl) have been used; however, the latter class of fungicide is only effective against the disease when *C. gloeosporioides* is the pathogen. Cultural practices that remove alternate hosts or dead branches in trees have been suggested as methods to reduce inoculum levels and potential disease severity in orchards.

Citrus Anthracnose

History, Geographic Distribution, and Causal Organism

Citrus tear stain was first described in Florida by Rolf (21) and since then has been reported from most citrus-growing regions of the world including China, the Philippines, Indonesia, Japan, islands of the south Pacific, and California and Florida of the United States (10). The causal organism of tear stain and wither tip damage of leaves in California has been attributed to *C. gloeosporioides*. Other species of *Colletotrichum* cause Key lime anthracnose (*Gloeosporium limetticolum*) and anthracnose of Satsumas (*G. follicolum),* but these two species have been placed in synonomy with *C. gloeosporioides* by von Arx (3), or the former was identified as *C. acutatum* by Timmer et al. (29). Recently, postbloom fruit drop (PFD) of citrus in Florida has been attributed to *C. acutatum* (7). PFD has not been identified in California to date. For additional information on these diseases in Florida and other subtropical areas, see Chapter 18 by Timmer.

Epidemics of anthracnose on citrus fruit have occurred as tear staining and pitting in California in the spring seasons of 1997 and 1998. Navel and Valencia oranges in the Central Valley (north of Porterville) up into the Sacramento Valley and grapefruit in southern California were affected. In some cases, defoliation of grapefruit trees by *C. gloeosporioides* has also occurred in addition to fruit infections.

DISEASE SYMPTOMS AND EPIDEMIOLOGY

In Florida and California, the fungus is ubiquitous on dead twigs and leaves (18) and normally does not cause severe damage. The fungus can cause quiescent infections of leaf, stem, and fruit tissue generally without causing damage to these tissues. Under conducive environments (wet and warm weather) for fungal growth or conditions (biotic or abiotic) that may be injurious to host tissue; however, quiescent infections may become active or new leaf and fruit infections may occur. Leaf symptoms have been described as irregular, tan to brown, necrotic lesions. Acervuli or lenticular conidioma of the fungus form in concentric rings around the edge of the lesions and appear as characteristic black spots to the unaided eye. Leaves with lesions usually drop, defoliating the trees. The fungus also causes "wither tip" damage on the leaves. Symptoms include branch dieback and withering of young leaves.

In California, three types of fruit symptoms have been associated with anthracnose of citrus: staining, pitting, and rotting. The most common symptom is tear staining of the outer rind (flavedo) or pericarp (Fig. 7A). Conidia that have formed on dead branches and leaves are splash-dispersed in rain or dew drops onto fruit. Discoloration of the pericarp surface (Fig. 7B) results in part from fungal growth on the fruit surface. Under prolonged periods of conducive environments such as warm temperatures and wetness from dew, fog, or rain, conidia of the fungus commonly germinate and produce darkly pigmented appressoria (Fig. 7C) on the surface of the fruit that contribute to the discoloration symptoms. The dull reddish to reddish-green staining develops as streaks or bands from the stem to the blossom end of the fruit where wetness and water run-off have occurred. The staining, which may occupy a large portion of the fruit surface, is superficial and can develop on apparently uninjured tissue. Later in the season, symptoms on ripe fruit develop into brownish stains that resemble rust-mite

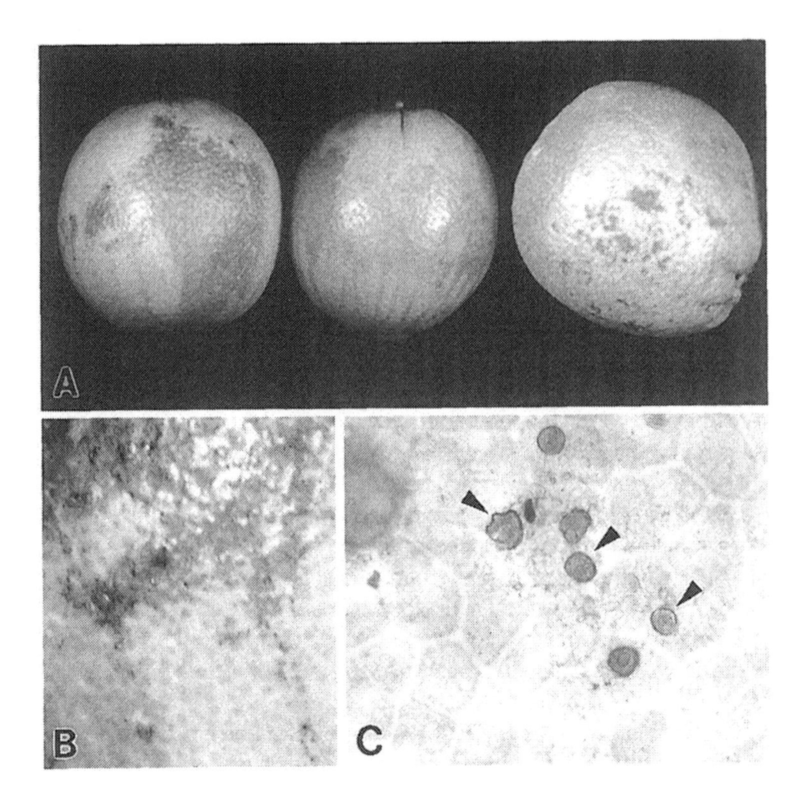

Fig. 7. Symptoms and signs of citrus anthracnose on navel orange. A. Tear staining (left and center) and pitting of outer pericarp or rind (right) (x0.5). B. Tear staining effecting collenchyma and oil gland cells as seen under a dissecting scope (x40). C. Appressoria of *C. gloeosporioides* (arrowheads) on the outer surface of the pericarp (x280).

damage. The staining cannot be washed off and the disorder generally results in off-graded fruit.

Occasionally, anthracnose will cause pitting (Fig. 7A) or spotting of fruit which is probably preceded in most cases by slight mechanical injuries or cold damage of the fruit. Pitting symptoms include small, irregular, brownish depressions of the pericarp that may merge into larger lesions, resulting in active decay that rots the fruit. Soft-rotting, the third symptom on fruit, is uncommon in uninjured, fresh, vigorous fruit, but may occur on very mature, overripe fruit. The rot develops through the stem end of the fruit or from advanced stages of tear stain or pitting. The decay slowly advances along the inner rind and pulp.

In the past, anthracnose has been controlled with copper, especially in the San Joaquin valley. Copper is only a protectant and has been reported to inhibit germination of spores of *C. gloeosporioides* (25). Bordeaux mixtures and fixed copper formulations have been applied in late November and early December for anthracnose control in California. In years with high rainfall in the fall, growers are unable to re-apply these treatments until later in the spring. Thus, the disease has not been consistently and effectively managed with copper fungicides. Preliminary evaluations of copper toxicity to *C. gloeosporioides* in laboratory assays have indicated that copper is relatively ineffective against spore germination and mycelial growth of California isolates of the fungus.

In the first year of field trials, two applications of benomyl (Benlate 50WP) or azoxystrobin (Abound 2F) in late December and later in January significantly reduced tear stain of fruit from 77.3% in the untreated trees to 30% and 38% in the benomyl and azoxystrobin treated trees, respectively. Future studies are planned to determine the effectiveness and optimal timing of these treatments for managing the disease. In 1998 a special local need registration was approved for benomyl application on citrus in California.

Concluding Remarks

The species identification of the causal organism of almond anthracnose in California has been one focus of our research. Considering the controversial taxonomy of the genus *Colletotrichum* with its poorly defined species concepts and the phenotypic plasticity of individual isolates (26), we aimed to compare the causal organism in California with the one in Israel that has been described as a clonal population of *C. gloeosporioides*. Conidial shape, slower growth rate, lower optimum temperature for growth, and insensitivity to the fungicide benomyl are traditional characters that differentiate *C. acutatum* from *C. gloeosporioides*. Genotype P of the California anthracnose population with its pink colony appearance and distinctly pointed conidia clearly fits the species description of *C. acutatum*. Genotype G, although less typical, was much more similar to *C. acutatum* in all characters than it was to *C. gloeosporiodes*. Additional support for the species identification came from DNA molecular studies. In recent years, molecular methods have become essential tools for the characterization of organisms, especially when a limited number of morphological and physiological characters are available. Restriction fragment patterns of ribosomal DNA, and DNA amplification using two *C. acutatum*-specific primers both confirmed that *C. acutatum* is the causal organism of almond anthracnose in California. The two genotypes may represent distinct, and possibly not very closely related, subgroups within the species in addition to subgroups identified from other hosts such as strawberry. Consequently, colony pigmentation may be an unreliable character in *Colletotrichum* systematics, and assigning gray cultures to *C. gloeosporioides* and pink cultures to *C. acutatum* without considering additional species characteristics may be misleading. The finding that both pink and gray colonies may be found within *C. acutatum* together with the accumulation of more molecular data will require an amended species definition in the future.

Based on preharvest losses in 1996 and 1998 and the postharvest damage to the crop, almond anthracnose potentially represents one of the most serious threats to the almond industry in California. Due to an unusually dry spring, the disease was not a problem in 1997; however, the fungus persisted and caused problems in orchards with high-angle sprinkler irrigation. In 1998 the disease also occurred in orchards that were symptomless in 1997. Postharvest damage to almond kernels was the result of a preharvest disease that went unmanaged and possibly represents a new perennial problem for the almond industry in years with conducive preharvest environments. Incidence of the disease is dependent on wetness and the splash dispersal of conidia of the fungus. Considering that blossoms, foliage, fruit, and spurs are susceptible to infection and that fruit and leaves are susceptible throughout the spring, management of the disease solely with the use of fungicides will be difficult. Thus, integrated approaches are currently being evaluated. These approaches include removal of mummified fruit during the dormant season with mechanical shakers to reduce inoculum levels, pruning programs to remove dead branches and open canopies to improve air-movement and shorten wetness periods, and rotation of fungicides with single- and multiple-site modes of action to prevent the development of resistant populations of the pathogen.

Literature Cited

1. Adaskaveg, J.E. and Hartin, R.J. 1997. Characterization of *Colletotrichum acutatum* isolates causing anthracnose of almond and peach in California. Phytopathology 87:979-987.
2. Amusa, N.A. 1994. Production, partial purification and bioassay of toxic metabolites of three plant pathogenic species of *Colletotrichum* in Nigeria. Mycopathologia 128:161-166.
3. Arx, J.A. von. 1957. Die Arten der Gattung *Colletotrichum* Cda. Phytopathol. Z. 29:413-468.
4. Bernstein, B., Zehr, E.I., Dean, R.A., and Shabi, E. 1995. Characteristics of *Colletotrichum* from peach, apple, pecan, and other hosts. Plant Dis. 79:478-482.

5. Bernstein, B. and Miller, R.W. 1995. Anthracnose. Pages 18-19 in: Compendium of Stone Fruit Diseases. J.M. Ogawa, E.I. Zehr, G.W. Bird, D.F. Ritchie, K. Uriu, and J.K. Uyemoto, eds. American Phytopthological Society, St. Paul, MN.
6. Brizi, U. 1896. Eine neue Krankheit (Anthracnosis) des Mandelbaumes. Z. PflKrankh. 6:66-72.
7. Brown, A.E., Sreenivasaprasad, S., and Timmer, L.W. 1997. Molecular characterization of slow-growing orange and Key lime anthracnose strains of *Colletotrichum* from citrus as *C. acutatum*. Phytopathology 86:523-527.
8. Czarnecki, H. 1916. A *Gloeosporium* disease of the almond, probably new to America. Phytopathology 6:310.
9. Dippenaar, B.J. 1931. Anthracnose disease of almonds. Farm. S. Africa 6:133-134.
10. Fawcett, H.S. 1936. Citrus Diseases and their Control. McGraw-Hill, New York and London.
11. Förster, H. and Adaskaveg, J. E. Identification of subpopulations of *Colletotrichum acutatum* and epidemiology of almond anthracnose in California Phytopathology 89:1056-1065.
12. Frantzen, K.A., Johnson, L.B., and Stuteville, D.L. 1982. Partial characterization of phytotoxic polysaccharides produced in vitro by *Colletotrichum trifolii.* Phytopathology 72:568-573.
13. Freeman, S., Katan, T., and Shabi, E. 1996. Characterization of *Colletotrichum gloeosporioides* isolates from avocado and almond fruit with molecular and pathogenicity tests. Appl. Environ. Microbiol. 62:1014-1020.
14. Goodman, R.N. 1959. Observations on the production, physiological activity and chemical nature of colletotin, a toxin from *Colletotrichum fuscum* Laub. Phytopath. Z. 37:187-194.
15. Grosclaude, C. 1972. L'anthracnose de l'amandier en France. C.R. Acad. Agric. Fr. 58:1392-1395.
16. Guerber, J.C. and Correll, J.C. 1997. The first report of the teleomorph of *Colletotrichum acutatum* in the United States. Plant Dis. 81:1334.
17. Hopp, H. 1936. Control of atmospheric humidity in culture studies. Bot. Gaz. 98:25-44.
18. Klotz, L.J. 1978. Fungal, bacterial, and nonparasitic diseases and injuries originating in the seedbed, nursery, and orchard. Pages 1-66 in: The Citrus Industry, Vol. IV. W. Reuther, E.C. Calavan, and G.E. Carman, eds. University of California, Division of Agricultural Sciences.
19. Mills, P.R., Sreenivasaprasad, S., and Brown, A.E. 1994. Detection of the anthracnose pathogen *Colletotrichum*. Pages 183-189 in: Modern Assays for Plant Pathogenic Fungi. A. Schots, F.M. Dewey, and R. Oliver, eds. CAB International, University Press, Cambridge.

20. Ogawa, J.M. and English, H. 1991. Diseases of Temperate Zone Tree Fruit and Nut Crops. University of California, Division of Agriculture and Natural Resources, Oakland, CA. Publication 3345. 461 pp.
21. Rolfs, P.H. 1904. Withertip and other diseases of citrus trees and fruits. U.S. Department Agr. Bur. Plant Indus. Bull. 52. 20 pp.
22. Shabi, E. and Katan, T. 1983. Occurrence and control of anthracnose in Israel. Plant Dis. 67:1364-1366.
23. Smith, B.J. and Black, L.L. 1990. Morphological, cultural, and pathogenic variation among *Colletotrichum* species isolated from strawberry. Plant Dis. 74: 69-76.
24. Smith, R.E. 1941. Diseases of fruits and nuts. Calif. Agric. Exp. Serv. Circ. 120. 168 pp.
25. Solel, Z. and Oren, Y. 1978. Laboratory and field evaluation of fungicides to control anthracnose stain of citrus fruit. Phytoparasitica 6:59-64.
26. Sutton, B.C. 1992. The genus *Glomerella* and its anamorph *Colletotrichum*. Pages 1-26 in: *Colletotrichum*: Biology, Pathology and Control. J.A. Bailey and M.J. Jeger, eds. CAB International, Wallingford, UK.
27. Taylor, R.M. and Philp, G.L. 1925. The almond in California. Univ. Calif. Agri. Exp. Stn. Circ. 284.
28. Timmer, L.W. and Zitko, S.E. 1996. Evaluation of a model for prediction of postbloom fruit drop of citrus. Plant Dis. 80:380-383.
29. Timmer, L.W., Brown, G.E., and Zitko, S.E. 1998. The role of *Colletotrichum* spp. in postharvest anthracnose of citrus and survival of *C. acutatum* on fruit. Plant Dis. 82:415-418.

Chapter 20

Recent Advances in Understanding *Colletotrichum* Diseases of Some Tropical Perennial Crops

J.M. Waller and P.D. Bridge

Colletotrichum or its *Glomerella* teleomorph is well represented in the warm, moist environments of the humid and subhumid tropics where it can often be isolated from both healthy and diseased tissues of stems, leaves, flowers, and fruit. *Colletotrichum* species are significant components of the microflora of tropical perennial crops, occurring as saprobes or secondary invaders of moribund tissue, as endophytes and as pathogens. They are opportunistic pathogens, causing particularly troublesome diseases on tropical perennial crops. The characteristics of the main diseases caused by *Colletotrichum* spp. on perennial and other cash crops in the tropics has been previously reviewed (19), together with the main epidemiological features relevant to control strategies.

Colletotrichum diseases often directly reduce the quantity or quality of the harvested yield, most noticeably when the pathogen directly damages the harvested portion of the crop, often the fruit. However, damage to shoot tissues and flowers can also reduce vigor and productivity and may be the major cause of loss in some crops. In economic terms, losses also include the cost of labor and material for control measures (which are seldom completely successful), and lost opportunity costs through the disease acting as a developmental constraint. Fruit crops (mango, avocado, papaya, banana, citrus) suffer particularly damaging effects from *Colletotrichum* worldwide, especially from postharvest depredations on the primary yield component of these crops.

Many *Colletotrichum* species and forms have been recorded from tropical perennial crops. Isolates vary widely in both their morphological characters in culture and their pathogenicity and host range. The *C. gloeosporioides* group, which embraces a wide diversity of forms, is the most predominant *Colletotrichum* pathogen on the crops under consideration being widely associated with anthracnoses of tropical perennial crops, particularly of young, expanding, ripe or senescent tissues. However, *C. acutatum* is now becoming more apparent probably through improved recognition. Although often referred to under the teleomorph name *Glomerella cingulata*, many isolates of *C. gloeosporioides* do not produce the *Glomerella* state; in the field *Glomerella* ascocarps are most frequently encountered in dead tissue, but infection by ascospores is known to occur widely. Recent advances have included improved knowledge of the relationships, both taxonomic and ecological, of the various species or forms associated with several crops.

Colletotrichum Variability on Tropical Perennial Crops

Although *Colletotrichum* spp. can be readily isolated from both diseased and apparently healthy tissue of many crops, the biological characteristics of these isolates in terms of their host specificity and interactions are often poorly understood. The pathogen may cause distinct diseases of separate organs of the same plant but similar diseases on different plant species. The ease with which *Colletotrichum* can be isolated has often led to confusion as to the true etiology of diseases, particularly as different forms of the fungus and other pathogens can co-inhabit lesions. Consequently, problems associated with this complex often relate to pathogenic variability and ways of identifying or confirming special pathogenic forms. Is pathogenic variability merely one aspect of the phenotypic plasticity typical of this genus or are pathogenic variants genetically discrete and biologically separate populations? What role does the general *Colletotrichum* gene pool in the ecosystem play in facilitating or moderating pathogenicity to crop plants? Knowledge of these aspects should enable a better understanding of the epidemiology of anthracnose diseases leading to improved, more sustainable control methods.

Variability Between Crops

The wide host range of the *C. gloeosporioides* group and the morphological variability of isolates in culture unrelated to host origin has prompted the view that the species is an unspecialized generalist pathogen. However, other considerations such as the hemibiotrophic nature of lesions and the apparent endophytic behavior of the fungus suggest that the variability within the species may extend to pathogenic specialization. The mixed perennial crop gardens of Sri Lanka provide a unique habitat to investigate the specificity of *Colletotrichum gloeosporioides* isolated from different perennial crops grown in juxtaposition. Crops such as mango, durian, rambutan, mangosteen, avocado, coffee, and a range of spice crops, all hosts of *C. gloeosporioides*, are grown as individual trees interplanted randomly, allowing ready transfer of inoculum. Anthracnose lesions associated with *C. gloeosporioides* are common on the foliage and fruits of these plants, but anthracnose diseases are only of economic significance on some crops (e.g., mango and papaya fruits).

An investigation (1) of the host range of isolates from perennial crops in these gardens clearly demonstrated quantitative differential pathogenicity in detached leaf assays. Although there was some cross-infectivity at high inoculum levels, there was none below 10^3 spores/ml so that at inoculum levels generally encountered in the field there would be very little cross-infection between the different host species despite the ready transfer of inoculum between species. Isolates from mango, durian, and avocado showed the greatest degree of host specialization, but these crops were also the most susceptible to isolates from other crops. Isolates from the relatively resistant pini-jambu and mangosteen were the most pathogenic to other crops suggesting that isolates from these crops had developed a higher degree of overall nonspecific pathogenicity (aggressiveness). DNA analysis using ribosomal (rDNA) and mitochondrial DNA (mtDNA) RFLPs and RAPDs indicated that there was significant genetic grouping according to host origin. However, several isolates also showed identical rDNA patterns and similar mtDNA indicating a common ancestry despite differential pathogenicity. The authors concluded that the pathogenic specialization coupled with the molecular analysis indicates that most infection occurs within rather than between crops so that within the *C. gloeosporioides* gene pool in these mixed gardens a degree of host specialization does occur. In another study of isolates (J.M. Waller and P.D. Bridge, unpublished) from coffee and interplanted spice crops in the same area, mtDNA profiles were also shown to be highly variable between host species, but there was little evidence of grouping according to host origin.

Studies of *C. gloeosporioides* isolates from subtropical and temperate fruits (3) also provided evidence of host specialization as indicated by lesion sizes in cross-inoculation studies. A comparison of isolates from almond, apple, avocado, mango, and pecan demonstrated that isolates from almond were most specialized for that host. Within the tropical fruit anthracnoses, there is evidence that mango anthracnose is caused primarily by a distinct genetic form of *C. gloeosporioides*. When DNA profiles using rDNA and mtDNA RFLPs from a collection of *C. gloeosporioides* isolated from fruit lesions of banana, avocado, mango, and papaya originating worldwide (11) were examined, considerable variation was found, and isolates could be distinguished in relation to their host source within geographic localities. This may indicate adaptation from a local *C. gloeosporioides* gene pool. However, isolates from different hosts never had the same banding patterns, and those from mango were specific to that host, suggesting a unique form pathogenic to mango. Further work (10) on Australian fruit isolates using RAPD molecular markers and cross-inoculation experiments also demonstrated the existence of a genetically and pathologically distinct mango biotype of *C. gloeosporioides.*

There are many other examples of host specialization within the general population of *C. gloeosporioides* on other crops and more recently within populations of *C. acutatum*. Molecular techniques have enabled genetic differences to be associated with this specialization, whereas morphological, cultural, and physiological differences have only proved useful as indicators of pathogenic variability where there are significant differences at the species level. Even here, however, the phenotypic plasticity of *Colletotrichum* makes species differentiation based solely on morphological criteria difficult.

Can the emergence of special pathogenic forms and their subsequent ecological separation and development eventually lead to the formation of new species? The evidence presented earlier indicates that the diversity within populations of *C. gloeosporioides* has enabled the fungus to adapt well to certain host species and that different degrees of adaptation exist. This specialization might represent stages along a sequential train of events leading to the development of quite separate and distinctive new populations of species rank.

Variability Within Hosts

Among the different forms or species of *Colletotrichum* frequently isolated from the same host species, some are apparently nonpathogenic members of the surface microbial population, others are apparently (harmless) endophytes, and others are more specialized pathogens. All are part of crop-associated microbial biodiversity and likely to perform different functions in relation to the host plant's "well-being." For example, on some crops such as coffee the different forms have been shown to compete (13), whereas on other crops, different forms have been associated with different disease manifestations. On mango, forms of *C. gloeosporioides* associated with leaf spotting, flower blight, and fruit anthracnose could be distinguished on the basis of pectic isozyme analysis (9). In some instances, different *Colletotrichum* species cause the same disease possibly under different ecological conditions. This relationship appears to occur with *C. gloeosporioides* and *C. acutatum* although confusion between these two species has made previous assertions dubious. Both species have been recorded from mango in Australia, and both can cause anthracnose. In a collection of isolates of mango anthracnose from across the world in Herb IMI at CABI Bioscience, *C. acutatum* was only found among Australian isolates. In Sri Lanka, *C. acutatum* and *C. gloeosporioides* both caused leaf fall of rubber and are equally virulent (5) although a recent report (12) suggests that *C. acutatum* is the major cause of the disease (9). Pathogenic variability of *Colletotrichum* isolates causing leaf fall has previously been reported from China and W. Africa (7) where previously resistant clones have succumbed to the disease; it is tempting to speculate whether *C. acutatum* is involved here as well.

CITRUS

Colletotrichum populations from citrus, on which the fungus is associated with a range of disease problems, vary greatly. Recent investigations of a range of citrus isolates (P.D. Bridge, M. Fortune, and P. Cannon in preparation) have demonstrated some distinct groupings based on morphology, isozyme profiles, and RAPD patterns that clearly distinguish *C. acutatum*. Generally, *C. gloeosporioides* is considered a weak pathogen or secondary invader of damaged or senescent citrus tissues. However, pathogenic strains of *C. acutatum* have recently been shown to cause anthracnose of key lime and postbloom fruit drop of oranges (6). The key lime pathogen, originally referred to as *Gloeosporium limetticola,* has long been considered distinct, but postbloom fruit drop of citrus in Belize was originally identified as being caused by *C. gloeosporioides.* Differentiation of *C. acutatum* and *C. gloeosporioides* is often difficult using morphological criteria alone, but recent molecular techniques have been established for differentiating the two species (18). On the basis of these techniques, other *C. gloeosporioides* forms reputedly causing specific diseases may need re-evaluation.

COFFEE

The variability of isolates of *Colletotrichum* from coffee obtained from Africa, Asia, and South America has been studied at CABI Bioscience for several years. In general, these studies show considerable heterogeneity in terms of mtDNA, but indicate that some diseases may be associated with clonal populations. For example, isolates obtained from leaves and fruit with oil spot disease in South America show different mtDNA RFLP patterns, but within these mtDNA populations there is little evidence of further heterogeneity, and each group produces homogenous patterns with PCR-fingerprinting techniques.

In studies to elucidate the epidemiology and to control coffee berry disease (CBD) in the 1950s and 60s, four main morphological *Colletotrichum* types could be readily distinguished from microflora on coffee in E. Africa. Two were forms of *C. gloeosporioides*, one was typical *C. acutatum*, and another was the pathogen (formerly referred to as a form of *Colletotrichum coffeanum*) causing coffee berry disease (CBD)—a specific anthracnose of young, expanding *Coffea arabica* berries, which continues to cause significant losses to coffee in most arabica-producing areas of Africa. Subsequent work clarified the distinctiveness of the CBD pathogen in terms of colony morphology, physiological properties, and ecology, and it was designated *Colletotrichum kahawae* (20). This species is very close to *C. gloeosporioides* as evidenced from the ITS sequence data (17) and presumably evolved from it in the recent past. However, it did not evolve in the center of diversity of *C. arabica* (Ethiopia) which the disease did not reach until 1970 when much of the wild *C. arabica* was found to be susceptible to the disease. The pathogen most likely originated from the general *C. gloeosporioides* gene pool in the forests of the great lakes region of central Africa on the diploid progenitors of the tetraploid *C. arabica* which have their centers of diversity in that area. When *C. arabica* was first planted in western Kenya in the 1920s, the crop encountered this pathogenic fungal population which spread relatively rapidly across the new areas of *C. arabica* in Africa (14). In doing so, it became an integral part of the *Colletotrichum* population on coffee but remained ecologically distinct; relative populations of the three species vary according to altitude (Table 1) and other factors.

Table 1. Relative frequencies of *Colletotrichum* species isolated (as percentage of total fungal isolates) from the microflora of arabica coffee stems from crops grown at different altitudes in Kenya (data sourced from [16])

Colletotrichum species	High altitude coffee	Medium altitude	Low altitude
C. acutatum	6.18	0.01	0.04
C. gloeosporioides	2.32	4.51	2.54
C. kahawae	0.77	0.12	0.05

Examination of a range of *Colletotrichum kahawae* isolates from across Africa showed a single unique mtDNA genotype despite the existence of some variation in vegetative compatibility groups (vcg) (2). Further analysis of *C. kahawae* isolates with RAPD primers, simple repetitive sequence fingerprinting primers, and AFLPs showed almost no variability within the group indicating a virtually clonal population. However, different pathogenic strains do occur (16) (L. Omondi, unpublished), populations resistant to MBC fungicides have developed (15), and vcg differences between geographically separated populations from East and West Africa have been confirmed (P.D. Bridge, J.M. Waller, and D. Davis, unpublished). Because a perfect state of *C. kahawae* is unknown, opportunities to develop genetic heterogeneity are limited. There has also been limited selection pressure for emergence of new genetic traits on this successful pathogen. Under these conditions, a seemingly clonal population can spread rapidly. Some evidence now suggests that variability at the functional level within the population is developing.

Conclusions

The examples given illustrate several stages in a process of speciation of *Colletotrichum* from initial host specialization within the general *C. gloeosporioides* population in the ecosystem as in Sri Lanka, through the emergence of specialized host specific strains, as seems to occur on mango, to the emergence of a new biological species, as exemplified by *C. kahawae* on coffee. In all cases, the intrinsic variability of the *C. gloeosporioides* population has enabled an 'opportunistic' response to exploit the sudden abundance of a susceptible host created by agricultural developments. The relatively rapid evolution of *C. kahawae* to a biological species status was driven by the encounter of a small variant population from the wild *Colletotrichum* gene pool with an expanding area of susceptible *Coffea arabica* plantations on which it had not co-evolved. The situation is similar in some ways to the relatively sudden emergence of the virulent Dutch elm pathogen *Ophiostoma novo-ulmi* where an initial clonal population developed into a genetically diverse species as it spread from western Asia to create a pandemic throughout susceptible elm populations (3,4). Evidence is also accumulating that other new pathogens of agricultural crops can emerge from the variable 'wild' populations of other *Colletotrichum* species such as *C. acutatum*.

Literature Cited

1. Alakahoon, P.W, Brown, A.E., and Sreenivasaprasad, S. 1994. Cross-infection potential of genetic groups of *Colletotrichum gloeosporioides* on tropical fruits. Physiol. Mol. Plant Pathol. 44:93-103.
2. Beynon, S.M., Coddington, A., Lewis B.G., and Varzea, V. 1995. Genetic variation in the coffee berry disease pathogen, *Colletotrichum kahawae*. Physiol. Mol. Plant Pathol. 46:457-470.
3. Brasier, C.M. 1988. Rapid changes in the genetic structure of epidemic populations of *Ophiostoma ulmi*. Nature 332:538-541.
4. Brasier, C.M. 1991. *Ophiostoma novo-ulmi* sp. Nov., the causative agent of current Dutch elm disease pandemics Mycopathologia 115:151-161.
5. Brown, A.E. and Soepena, H. 1994. Pathogenicity of *Colletotrichum acutatum* and *C. gloeosporioides* on leaves of *Hevea* spp. Mycol. Res. 98:264-266.
6. Brown, A.E., Sreenivasaprasad, S., and Timmer, L.W. 1996. Molecular characterisation of slow-growing orange and key lime anthracnose

strains of *Colletotrichum* from citrus as *C. acutatum*. Phytopathology 86:523-527.
7. Chee, K.H. 1990. The present state of rubber diseases and their control. Rev. Plant Pathol. 69:423-430.
8. Freeman, S. and Shabi, E. 1996. Cross-infection of subtropical and temperate fruits by *Colletotrichum* species from various hosts. Physiol. Mol. Plant Pathol. 49:395-404.
9. Gantotti, B.V. and Davis, M.J. 1993. Pectic zymogram analysis for characterizing genetic diversity of the mango anthracnose pathogen. Acta Hortic. 341:353-359.
10. Hayden, H.L., Pegg, K.G., Aitken, E.A.B., and Irwin, J.A.G. 1994. Genetic relationships as assessed by molecular markers and cross-infection among strains of *Colletotrichum gloeosporioides*. Austral. J. Bot. 42:9-18.
11. Hodson, A., Mills, P.R., and Brown, A.E. 1993. Ribosomal and mitochondrial DNA polymorphisms in *Colletotrichum gloeosporioides* isolated from tropical fruits. Mycol. Res. 97:329-335.
12. Jayasinghe, C.K., Fernando, T.H.P.S., and Priyanka, U.M.S. 1997. *Colletotrichum acutatum* is the main cause of *Colletotrichum* leaf disease of rubber in Sri Lanka. Mycopathologia 137:53-56.
13. Masaba, D.M. 1991. The role of saprophytic surface microflora in the development of coffee berry disease (*Colletotrichum coffeanum*) Ph.D. thesis. University of Reading, UK.
14. Masaba, D.M. and Waller, J.M. 1992. Coffee berry disease—the current status. Pages 237-249 in: *Colletotrichum*: Biology, Pathology and Control. J.A. Bailey and M.J. Jeger, eds. CAB International, Wallingford. UK.
15. Ramos, A.H. and Kamidid, R.E. 1982. Determination and significance of mutation rates for *Colletotrichum coffeanum* from benomyl sensitivity to benomyl tolerance. Phytopathology 72:181-185.
16. Rodrigues, C.J. Jr., Varzea, V.M., and Medeiros, E.F. 1992. Evidence for the existence of physiological races of *Colletotrichum coffeanum* Noack sensu Hindorf. Kenya Coffee 57:1417-1420.
17. Sreenivasaprasad, S., Brown, A.E., and Mills, R.R. 1993. Coffee berry disease pathogen in Africa: Genetic structure and relationship to the group species *Colletotrichum gloeosporioides*. Mycol. Res. 97:995-1000.
18. Sreenivasaprasad, S., Mills, P.R., and Brown, A.E. 1994. Nucleotide sequence of the rDNA spacer 1 enables identification of *Colletotrichum* as *C. acutatum*. Mycol. Res. 98:186-188.
19. Waller, J.M. 1992. *Colletotrichum* diseases of tropical perennial and cash crops. Pages 167-185 in: *Colletotrichum*: Biology, Pathology and Control. J.A. Bailey and M.J.Jeger, eds. CAB International, Wallingford, UK.

20. Waller, J.M., Bridge, P.D., Black, R., and Hakiza, G. 1993. Characterisation of the coffee berry disease pathogen *Colletotrichum kahawae* sp. nov. Mycol. Res. 97:989-994.

Chapter 21

Host-Pathogen Interaction and Variability of *Colletotrichum lindemuthianum*

Maeli Melotto, Ricardo S. Balardin, and James D. Kelly

Anthracnose, caused by the anamorphic fungus *Colletotrichum lindemuthianum* (Sacc. & Magnus) Briosi & Cav., is an important disease of common bean (*Phaseolus vulgaris* L.) in subtropical and temperate regions. The teleomorphic phase of this fungus, *Glomerella lindemuthiana* (Shear), however, is rarely found in culture or nature (51). Yield losses due to anthracnose have been reported as high as 95% in susceptible bean cultivars (29). Infected seeds of susceptible cultivars serve as a primary source of inoculum. Optimum conditions for disease development include high relative humidity, frequent precipitation, and temperature ranging from 18 to 22°C (29). Effective disease control is limited by efficient seed transmission of the fungus (66), the lack of cost-effective chemical control methods (51), the ability of *C. lindemuthianum* to survive up to 22 months in plant debris in northern latitudes (20), and the occasional development of sclerotia (64). Anthracnose can be controlled by quarantine, establishment of tolerance levels for infected seeds, and seed production in disease-free areas (51,65). Seed treatment, cultural measures, sanitation procedures, fungicide spraying, and the use of resistant cultivars can eradicate *C. lindemuthianum*. Host resistance has been the most appropriate method for control of anthracnose, particularly in countries where alternative methods are difficult to implement. The high variability present in *C. lindemuthianum* (51), however, has resulted in continuous breakdown of resistance in commercial cultivars. Therefore, studies on the variability of *C. lindemuthianum*, using standard procedures, are needed to direct breeding efforts towards long-term resistance to anthracnose.

Race Identification and Distribution

VIRULENCE ANALYSIS

Variability in *C. lindemuthianum* was first described by Barrus (10,11) when races α and β were identified based on their reaction on 139 bean cultivars. Since then, new races have been described in many parts of the world (Table 1). Research groups in many countries have used local codes rather than the original Greek letters to identify races (Table 2). Despite the tentative equivalency of described races in different countries, data collected using local hosts has limited the knowledge of the variability present within *C. lindemuthianum*, worldwide. A standard differential series of 12 bean cultivars of diverse origin and a binary system based on the position of each cultivar within this series were established to better understand the variability and structure in *C. lindemuthianum* populations (19,48,49) (Table 3).

Table 1. Races of *C. lindemuthianun* identified in different countries from 1911 to 1994 using Greek letters

Country	Race	Reference
USA	α, β, γ, δ	2,10,11,14,28
Chile	α, β, γ	44
France	α, β, γ, δ, ε, λ	9,13,16
Uganda	α, β, γ, δ, ε, λ	39
Malawi	α, β, γ, δ	3
Europe	αB, ι, λ, κ	23,24,30,31,32,38,59
Canada	ε, δ, λ, αB	65,67,68
Brazil	α, αB, β, γ, δ, ε, λ, κ, θ, ξ, μ, ζ	6,37,41,47

Table 2. Races of *C. lindemuthianum* classified by using local codes and their possible equivalency to Greek letters

Country	Race (Ref.)	Greek (Ref.)
Germany	A-E, G-N, X (54)	α, β, γ (60)
Australia	1 to 8 (17,70)	none
Mexico	Groups I to III (72)	none
Mexico	MA-11 to 13, MA-20 to 25 (25,45,71)	α (25,45,71)
France	PV6, D10, F86, I4, 1, 5 (9)	α, β, γ, δ, ε, race 5 = γ and δ combined (unpublished)
Brazil	BA-1 to 10 (46,55)	α, δ, Mexico I and II, groups I and II (46,55)
Guatemala	C-236 (61)	none

Table 3. Anthracnose differential series, resistance genes, host gene pool, and the binary number of each cultivar

Differential Cultivar	Host Genes[a]	Gene Pool[b]	Binary Number[c]
A. Michelite	--	MA	1
B. MDRK	*Co-1*	A	2
C. Perry Marrow	*Co-1*^{3}	A	4
D. Cornell 49242	*Co-2*	MA	8
E. Widusa	--	A	16
F. Kaboon	*Co-1*^{2}	A	32
G. Mexico 222	*Co-3*	MA	64
H. PI 207262	--	MA	128
I. TO	*Co-4*	MA	256
J. TU	*Co-5*	MA	512
K. AB 136	*Co-6, co-8*	MA	1024
L. G 2333	*Co-4*^{2}, *Co-5, Co-7*	MA	2048

[a]Host resistance genes (1,34,73,75). Not all resistance genes have been characterized.

[b]MA: Middle American gene pool; A: Andean gene pool (63).

[c]Binary number: 2^n, n is equivalent to the place of the cultivar within the series. The sum of cultivars with susceptible reaction will give the binary number of a specific race (49). E.g., race 17 = virulent on Michelite [1] + Widusa [16].

The adoption of this standard procedure allows comparison and compilation of data from different research groups and a fuller characterization of the wide variability present in *C. lindemuthianum*. The re-characterization of races previously assigned Greek letters or local codes suggests an overestimation of variability reported for this pathogen (Table 4).

Races 7, 64, 65, and 73 were identified in the United States (4,36) by using the binary code. In Mexico, 35 new races were reported (25,27,58) in addition to 11 races collected on wild *P. vulgaris* (62). Nine races were identified from 10 isolates in Nicaragua (56). In Colombia, 33 races were characterized from 178 isolates (57). Forty-one races, characterized from 138 isolates collected in Argentina, Brazil, the Dominican Republic, Honduras, Mexico, and the United States, fell into two categories, those found over a wide geographic area and those restricted to a single country. Only three races, 7, 65, and 73, were widely distributed (7) and isolated repeatedly in North, South, and Central America. These three races showed limited virulence on the differential cultivars. The remaining races were detected only in a single country, and about one-third of these localized races were isolated repeatedly within a country. Results of virulence

Table 4. Traditional races (Greek letters) of *Colletotrichum lindemuthianum* characterized using the standard set of differentials and named based on the binary nomenclature system

Race		Differential Cultivars[a]											
Binary	Greek	A	B	C	D	E	F	G	H	I	J	K	L
17	α	S	R	R	R	S	R	R	R	R	R	R	R
19	ε	S	S	R	R	S	R	R	R	R	R	R	R
23	δ	S	S	S	R	S	R	R	R	R	R	R	R
31	κ	S	S	S	S	S	R	R	R	R	R	R	R
55	λ	S	S	S	R	S	S	R	R	R	R	R	R
65	ε	S	R	R	R	R	R	S	R	R	R	R	R
81	ξ	S	R	R	R	S	R	S	R	R	R	R	R
87	μ	S	S	S	R	S	R	S	R	R	R	R	R
89	αB	S	R	R	S	S	R	S	R	R	R	R	R
99	θ	S	S	R	R	R	S	S	R	R	R	R	R
102	γ	R	S	S	R	R	S	S	R	R	R	R	R
130	β	R	S	R	R	R	R	R	S	R	R	R	R
141	C236	S	R	S	S	R	R	R	S	R	R	R	R
337	MexII	S	R	R	R	S	R	S	R	S	R	R	R
453	ζ	S	R	S	R	R	R	S	S	S	R	R	R

[a] Sequence of differential cultivars and associated binary codes are shown in Table 3. R: resistant reaction (rating 1-3); S: susceptible reaction (rating 4-9) (7).

diversity within 948 isolates (115 races) of *C. lindemuthianum* collected in 13 Latin American countries showed that 63% were unique to one country, whereas 17% occurred in two countries and less than 10% occurred in five or more countries (50). These results suggest that variability in *C. lindemuthianum* is higher within, rather than between, countries. In Central America, where the highest variability was observed (76%; i.e., number of detected races per number of isolates from a specific region), races were mostly characterized from a single isolate. Variability in *C. lindemuthianum* decreased either northwards (North America 7%) or southwards (South America 17%).

Genome Analysis

Combining virulence and molecular analyses has led to a better understanding of the variability present in *C. lindemuthianum*. DNA-DNA hybridization, restriction fragment length polymorphism (RFLP), polymerase chain reaction (PCR)-based markers, and isozymes are the most common molecular tools used in fungal systematics and population genetics. Molecular analysis of entire genomes can reveal the extent of variability in one species. Isolates of *C. lindemuthianum* collected in the Andean and northern regions of Colombia (57) and Brazil (42,69) have been identified as races using RAPD markers. No congruence was observed, however, between RAPD analysis and virulence data, geographic location, or host gene pool of 60 isolates of *C. lindemuthianum* primarily from Latin America (7). In a similar study conducted in South America, East Africa, and Europe, 12 isolates of *C. lindemuthianum* were grouped by using the percentage of shared amplicons as an index (22) and again, DNA polymorphism and the geographic origin of isolates were not correlated. Isolates collected from wild common bean populations in Mexico were the most polymorphic, and the most frequent isolates in Argentina and Ecuador were absent in Mexico (62). The majority of amplicons present in populations from Ecuador and Argentina were found in Mexico, suggesting that Andean populations have been derived from the Middle American center. It is still not clear how well RAPD data represent phylogenetic relationships in *C. lindemuthianum.*

RIBOSOMAL DNA POLYMORPHISM

Sequence analysis of specific regions of the genome, such as ribosomal DNA (rDNA), is a powerful approach and complements studies on intraspecific diversity. A specific set of primers, PN3 and PN10, was developed to amplify the rDNA region between the 18S and 28S genes in *C. lindemuthianum* (ITS 1, 5.8S rRNA gene, and ITS 2; 22). PCR amplification of this region produced a single reproducible 580-bp DNA fragment in all 57 isolates of *C. lindemuthianum* including *C. orbiculare* (8). Digestion with the endonucleases *Hae* III and *Msp* I generated two patterns (8,22,62). Isolates in group I had one additional *Hae* III and *Msp* I restriction site located adjacent to position 415 bp (GG/C/CGG) (8). One hundred and twenty-eight isolates (25 pathotypes) of *C. lindemuthianum,* collected from wild *Phaseolus* in Mexico, Ecuador and Argentina, were grouped according to the three host gene pools based on restriction of the two ITS regions (62). Three groups corresponded to three host gene pools, suggesting adaptation of races on host cultivars of the same geographic origin. Based on the same restriction analysis, 12 isolates (five avirulence spectra) of *C. lindemuthianum* from Africa, Europe, and South America were separated into two groups independent of geographical origin or host gene pool. Isolates in one group possessed the same avirulence spectrum, whereas no correlation between DNA polymorphism and pathogenicity was observed among isolates in the second group (22). Within a collection of 57 isolates of *C. lindemuthianum*, 85% of the Andean races belonged to one RFLP group, whereas 65% of Middle American races were present in the second group (8). The Middle American isolates, identified by Sicard et al. (62), also showed less association with RFLP groups or geographic origin than the Andean isolates. For instance, race 38 from the Dominican Republic clustered with races 457 from Mexico and 23 from the United States according to ITS sequence data, but no correlation was observed between these races based on virulence (8). Polymorphism in rDNA did not appear to be linked to a specific virulence phenotype or structured with the geographic origin of races or host gene pool. A variable level of genetic similarity among races and the lack of association of geographic origin with polymorphism in the rDNA region is clear evidence of the high level of molecular variability within *C. lindemuthianum*. In addition, insufficient support for parallel evolution of *C. lindemuthianum* and its host *P. vulgaris* is indicated by the high sequence similarity of the ITS region among Andean and Middle American races (8).

Intra-Race Variability

Intra-race polymorphism, determined using RAPD and AFLP markers, was observed within several isolates of different races of *C. lindemuthianum* collected in different countries (7,8,27,62). Among the 25 pathotypes of *C. lindemuthianum* isolated from wild beans, 75 different RAPD amplicons were observed (62). The most frequent pathotype (v1), displayed over 20 different RAPD amplicons. Isolates of race 73 from Honduras, Mexico, and the United States showed distinct RAPD patterns within and among countries. Isolates of race 65 characterized in the United States showed a different RAPD pattern, whereas isolates from Brazil were monomorphic (7). Isolates of race 0 collected in Mexico showed distinct AFLP patterns (27). Four of seven races compared using RFLP-ITS did not display intra-race variability, whereas isolates of races 7, 17, 31, and 73 from different countries showed polymorphism (8). ITS sequence data of isolates of race 73 from Mexico and the United States showed a large genetic distance (0.9%) based on pair-wise Jukes-Cantor distance estimates. The value exceeded that between distinct races from different countries, suggesting that more variability may exist within than between virulence phenotypes (8). The intra-race polymorphism observed using molecular markers suggested a high level of molecular variability within *C. lindemuthianum* and emphasized the limitation of virulence analysis for assessing variability. The significance of such intra-race variability on pathogen population structure is poorly understood, but it suggests independent evolution of specific virulence phenotypes, such as race 73, in different geographic regions.

Genetic Analysis of the *C. lindemuthianum-P. vulgaris* Pathosystem

None of the races collected on wild bean (62) were more virulent than those found on the cultivated bean (7). The most virulent wild Andean races attacked nine differentials, all except Mexico 222, AB136, and G2333, whereas the Middle American races from Mexico were avirulent on differentials with the *Co-1, Co-4*^{2}*, Co-5, Co-6, Co-7* genes, and the uncharacterized gene(s) in Widusa. The only genotype not to be defeated by races collected from cultivated beans was G2333 with the gene combination *Co-4*^{2}*/Co-5/Co-7* (75). A large number of isolates from wild beans were not pathogenic on the 12 differential hosts (pathotype v1) (62). Twenty-eight of the 29 isolates from Ecuador fell in the pathotype group v1 or race 0 (62). Given that Ecuador is the site of ancestral *Phaseolus* (33), the isolates pathogenic on these wild beans might represent less specialized races and may not demonstrate the same level of molecular polymorphism present in isolates collected from the cultivated species. Thirteen of 50 isolates collected by Sicard et al. (62) in Mexico fell in this same pathotype group (v1) and were more variable than the same pathotypes from Ecuador and Argentina and demonstrated two RFLP patterns. Mexico represents a Middle American domestication center where both wild and cultivated beans coexisted (26) and pathogens may have specialized more to exist in such an environment.

Morphological, biochemical, and molecular markers have been used to demonstrate the existence of a Middle American and an Andean gene pool within *P. vulgaris* (26,63). Unique molecular and biochemical markers detected within wild bean populations from northern Peru and Ecuador, suggest the presence of a third ancestral gene pool within *P. vulgaris* (33). Host resistance genes are classified as either Middle American or Andean depending on the gene pool origin of the host cultivar. Races of *C. lindemuthianum* are similarly classified as either Middle American or Andean depending on the gene pool of the host cultivar for which each was isolated. Races of *C. lindemuthianum* virulent to Middle American hosts show greater diversity in pathogenicity by attacking germplasm from both gene pools, whereas races virulent to Andean hosts are mostly pathogenic on Andean germplasm (5,7,50). The first suggestion of a possible association between variability in a bean pathogen and its host was made before common bean was classified by gene pool and was based on the virulence reaction of races of *C. lindemuthianum* (39). Races α and δ were found to be more pathogenic on small-seeded, indeterminate (Middle American gene pool) cultivars, whereas races β and γ were pathogenic on large-seeded, determinate cultivars (Andean gene pool). Reciprocal selection of resistance genes in *P. vulgaris* and virulence genes in *C. lindemuthianum* might occur (26). Since bean cultivars were first grouped according to their reaction to the races α and β of *C. lindemuthianum* (11), selecting for resistance to anthracnose has become a major objective in bean breeding programs worldwide.

The Andean *Co-1* (*A*) gene was the first major gene utilized to develop anthracnose resistant cultivars of common bean (14). Prior to 1978, the *Co-1* gene was used as the only source of resistance in navy beans grown in Michigan and Ontario. After the appearance of race δ in Ontario (65) and the United States (28), the Middle American *Co-2* (*Are*) gene, characterized in a black bean from Venezuela (40), became the main source of anthracnose resistance in the North America. The identification of race 73 in Michigan and race αB in Ontario, however, limited the further utilization of the *Co-2* gene (36,67). Combining the *Co-1* and *Co-2* genes in single cultivars was suggested as the best short-term protection against all known races currently present in North America (36,73).

The *Co-2* gene was the first predominant source of resistance to be used worldwide. The appearance of races 31, 63, and 89 virulent to the *Co-2* gene, however, emphasized the need for new sources of resistance (38,24). In France, new resistance genes *Co-3 (Mexique I)*, *Co-4 (Mexique II)* and *Co-5 (Mexique III)* were characterized from a collection of bean germplasm from Mexico (24). The *Co-4* and *Co-5* genes conferred resistance to current races virulent to the *Co-2* gene, whereas the *Co-3* gene was overcome by race αB. Bean cultivars TU, AB 136, G 2333, G 2338, G 3991, and G 4032 were identified as possessing resistance to nine races of *C. lindemuthianum* from Brazil (6). A collection of 13,000 bean accessions, maintained at the Centro Internacional de Agricultura Tropical (CIAT), was screened for resistance to Latin American and European isolates of *C. lindemuthianum* (61). Thirty Latin American isolates were more virulent than the four European races and only 0.25% of the accessions was resistant to all isolates. Among 20,144 bean accessions screened with a mixture of 14 isolates (10 races) of *C. lindemuthianum* from Middle American and Andean countries, 1.7% of the genotypes were resistant, including cultivar G 2333 (53). Virulence of *C. lindemuthianum* has been monitored in different countries, and evolving races have continually overcome resistant germplasm. Sources of resistance widely used in many breeding programs in Europe (24,28,40), North America (36,65) and South America (6,41) were defeated by emerging races in different countries. The cultivars TO, PI 207262, and Mexico 222, reported as resistant to European and Latin American isolates (24,38,61), were susceptible to races identified in Brazil (41). In Mexico, 32 different races were reported as virulent to the cultivars Cornell 49-242, Mexico 222, PI 207262, TO, TU, and AB 136 (25,58). In Nicaragua, isolates were identified that overcame resistance in the cultivars PI 207262, TO, TU, and AB 136 (56). Races isolated from Middle American hosts were virulent to 11 cultivars present in the differential series, except to cultivar G 2333 (7). These races were categorized into two groups. One group was virulent to specific Middle American resistance genes (*Co-2*, *Co-3*, *Co-4*, *Co-5*, and *Co-6*) and the second group was formed by races virulent to genotypes belonging to both gene pools. Durability of resistance depends on how efficiently new predominant virulence phenotypes are detected within a population, and how fast resistance against the newest race can be incorporated into commercial germplasm. Constant monitoring of the variability in *C. lindemuthianum* is needed.

Breeding for Resistance

The information on variability in *C. lindemuthianum* and the specialization of specific races on one host gene pool is invaluable in breeding for resistance (5). Numerous studies have indicated that resistance to *C. lindemuthianum* in common bean is controlled by major genes acting singly (Table 3) (40,73), as duplicate or complementary factors (15,18,43), or as members of an allelic series (24,75). The most resistant differential cultivar, G 2333, first thought to possess two dominant genes (53), was shown to possess three independent genes $Co\text{-}4^2$, *Co-5,* and *Co-7* (74,75). Pyramiding major resistance genes may be the most appropriate breeding strategy for long-term resistance in *P. vulgaris* (74). Knowledge of gene complementarity has been suggested to improve the efficiency of pyramiding genes for durable resistance (21). If pathogens were specialized on one of the host gene pools, pyramiding resistance genes from different gene pools may also provide a more durable anthracnose resistance (74). For instance, the incorporation of Andean resistance genes in bean breeding populations in Honduras where Middle American races predominate; or Middle American resistance genes in germplasm in the Dominican Republic where Andean races prevail could result in more durable resistance in each country (5). The efficient pyramiding of epistatic resistance genes with the aid of linked molecular markers has been demonstrated in common bean (35). The reduced virulence of isolates from Andean regions compared to isolates from Middle American regions suggests that deployment of resistance genes between gene pools should extend host resistance (5,26,74).

Despite the large number of studies indicating that resistance to anthracnose is controlled by either major dominant (Table 3) (34) or recessive resistance genes (1,15,43), other resistance mechanisms have been reported (12,52). The cultivar, ICA Llanogrande is susceptible to anthracnose in seedling assays, but is resistant to the same isolates under field conditions in widely different geographic regions of South America and Africa. Under controlled greenhouse testing, plants of ICA Llanogrande exhibit progressively higher resistance with age, but the mechanism or mode of inheritance is unknown (12). Under field conditions, the Brazilian cultivar Rio Negro, which carries the defeated *Co-2* gene, exhibits resistance against races of *C. lindemuthianum* to which it is known to be susceptible in seedling assays. The possibility of combining nonspecific resistance with major gene resistance for anthracnose may, therefore, exist (12).

Conclusions

A better resolution of the structure of the variability in *C. lindemuthianum* is obtained by combining virulence and molecular analyses. Unfortunately, virulence analysis based on local codes and different cultivar differential series has limited the understanding of the broad variability within this pathogen worldwide. The imbalance among cultivars from the two *P. vulgaris* gene pools in the new differential series might favor identification of races belonging to the Middle American reaction group. In addition, the multigenic resistance in some differentials such as G 2333 might select races with multiple avirulence genes. Therefore, information related with virulence may still be biased. Developing durable resistance to *C. lindemuthianum* must be based on reliable characterization of variability in this pathogen. Neutral molecular markers may sample a large portion of the genome and be more informative in assessing variability of organisms. No congruence was observed between data obtained by RAPD/AFLP and virulence analyses. Likewise, RFLP and sequence analyses of the variable ITS region in the rRNA genes in *C. lindemuthianum* have not supported the virulence data or shown a strong association with host gene pools or the geographic origin of isolates. The intra-specific variability shown by molecular analysis is evidence of limitations of virulence analysis. Combining virulence data and molecular markers for characterizing variability in *C. lindemuthianum* may have limited value in identifying new sources of resistance for local breeding programs. Pyramiding and deploying resistance genes according to host gene pools and specific geographic regions may result in the development of more durable resistance to *C. lindemuthianum.*

Literature Cited

1. Alzate-Marin, A.L., Baia, G.S., de Paula, T.J. Jr., de Carvalho, G.A., de Barros, E.G., and Moreira, M.A. 1997. Inheritance of anthracnose resistance in common bean differential cultivar AB 136. Plant Dis. 81:996-998.
2. Andrus, C.F. and Wade, B.L. 1942. The factorial interpretation of anthracnose resistance in beans. USDA Tech. Bull. No. 310:1-29.
3. Ayonoadu, U.W.U. 1974. Races of bean anthracnose in Malawi. Turrialba 24:311-314.

4. Balardin, R.S. and Kelly, J.D. 1996. Identification of Race 65-epsilon of bean anthracnose (*Colletotrichum lindemuthianum*) in Michigan. Plant Dis. 80:712.
5. Balardin, R.S. and Kelly, J.D. 1998. Interaction among races of *Colletotrichum lindemuthianum* and diversity in *Phaseolus vulgaris*. J. Amer. Soc. Hort. Sci. 123:1038-1047.
6. Balardin, R.S., Pastor-Corrales, M.A., and Otoya, M.M. 1990. Variabilidade patogênica de *Colletotrichum lindemuthianum* no Estado de Santa Catarina. Fitopatol. Bras. 15:243-245.
7. Balardin, R.S., Jarosz, A., and Kelly, J.D. 1997. Virulence and molecular diversity in *Colletotrichum lindemuthianum* from South, Central and North America. Phytopathology 87:1184-1191.
8. Balardin, R.S., Smith, J.S., and Kelly, J.D. 1999. Ribosomal DNA polymorphism in *Colletotrichum lindemuthianum*. Mycol. Res. 103:841-848.
9. Bannerot, H. 1965. Results de l'infection d'une collection de haricots par six races physiologiques d'anthracnose. Annales de Amelioration des Plantes (Paris) 15:201-222.
10. Barrus, M.F. 1911. Variation of cultivars of beans in their susceptibility to anthracnose. Phytopathology 1:190-195.
11. Barrus, M.F. 1918. Varietal susceptibility of beans to strains of *Colletotrichum lindemuthianum* (Sacc. & Magn.) B. & C. Phytopathology 8:589-605.
12. Beebe, S.E. and Pastor-Corrales, M.A 1991. Breeding for disease resistance. Pages 561-617 in: Common Beans: Research for Crop Improvement. A. van Schoonhoven and O. Voysest, eds. CAB International, CIAT, Colombia.
13. Blondet, A. 1963. L'anthracnose du haricot: Etudé des races physiologiques du *Colletotrichum lindemuthianum*. Ph.D. thèse, Faculté de Sciense, Paris, France.
14. Burkholder, W.H. 1923. The gamma strain of *Colletotrichum lindemuthianum* (Sacc. et Magn.) B. et C. Phytopathology 13:316-323.
15. Cardenas, F., Adams, M.W., and Andersen, A. 1964. The genetic system for reaction of field beans (*Phaseolus vulgaris* L.) to infection by three physiologic races of *Colletotrichum lindemuthianum*. Euphytica 13:178-186.
16. Charrier, A. and Bannerot, H. 1970. Contribution à l'étude des races physiologiques de l'anthracnose du haricot. Ann. Phytopathol. 2:489-506.
17. Cruickshank, I.A.M. 1966. Strains of *Colletotrichum lindemuthianum* (Sacc. and Magn.) in eastern Australia. J. Aust. Inst. Agric. Sci. 32:134-135.
18. Del Peloso, M.J., Cardoso, A.A., Vieira, C., Saraiva, L.S., and Zimmerman, M.J. de Oliveira. 1989. Genetic system for the reaction of

Phaseolus vulgaris to BA-2 (Alpha) race of *Colletotrichum lindemuthianum*. Brazil. J. Genet.12:313-318.

19. Drifhout, E. and Davis, J.H.C. 1989. Selection of a new set of homogeneously reacting bean (*Phaseolus vulgaris*) differentials to differentiate races of *Colletotrichum lindemuthianum*. Plant Pathol. 38:391-396.
20. Dillard, H.R. and Cobb, A.C. 1993. Survival of *Colletotrichum lindemuthianum* in bean debris in New York State. Plant Dis. 77:1233-1238.
21. Duvick, D.N. 1996. Plant breeding, an evolutionary concept. Crop Sci. 36:539-548.
22. Fabre, J.V., Julien, J., Parisot, D., and Dron, M. 1995. Analysis of diverse isolates of *Colletotrichum lindemuthianum* infecting common bean using molecular markers. Mycol. Res. 99:429-435.
23. Fouilloux, G. 1975. L'anthracnose du haricot: Étude des relations entre les pathotypes anciens et nouveaux; étude de nouvelles sources de résistance "totale". Pages 81-92 in: Réunion Eucarpia Haricot, Versailles, France. Centre National de Recherches Agronomiques, Paris, France.
24. Fouilloux, G. 1979. New races of bean anthracnose and consequences on our breeding programs. Pages 221-235 in: Disease of Tropical Food Crops. H. Maraite and J.A. Meyer, eds. Louvain-la-Neuve, Belgium.
25. Garrido, E.R. 1986. Identificación de razas fisiológicas de *Colletotrichum lindemuthianum* (Sacc. & Magn.) Scrib. en Mexico y busqueda de resistência genética a este hongo. M.Sc. thesis. Institución de Ensenanza y Investigación en Ciencias Agrícolas, Montecillos, Mexico.
26. Gepts, P. 1988. A Middle American and an Andean common bean gene pool. Pages 375-390 in: Genetic Resources of *Phaseolus* Beans; Their Maintenance, Domestication, and Utilization. P. Gepts, ed. Kluwer, London.
27. Gonzalez, M., Rodriguez, R., Zavala, M.E., Jacobo, J.L., Hernandez, F., Acosta, J., Martinez, O., and Simpson, J. 1998. Characterization of Mexican isolates of *Colletotrichum lindemuthianum* by using differential cultivars and molecular markers. Phytopathology 88:292-299.
28. Goth, R.W. and Zaumeyer, W.J. 1965. Reactions of bean cultivars to four races of anthracnose. Plant Dis. Reptr. 49:815-818.
29. Guzman, P., Donado, M.R., and Galvez, G.E. 1979. Perdidas economicas causadas por la antracnosis del frijol *Phaseolus vulgaris* en Colombia. Turrialba 29:65-67.
30. Hallard, J. and Trebuchet, G. 1976. Bean anthracnose in Western Europe. Annu. Rep. Bean Improv. Coop. 19:44-46.

31. Hubbeling, N. 1976. Selection for resistance to anthracnose, particularly in respect to the "ebnet" race of *Colletotrichum lindemuthianum*. Annu. Rep. Bean Improv. Coop. 19:49-50.
32. Hubbeling, N. 1977. The new iota race of *Colletotrichum lindemuthianum*. Annu. Rep. Bean Improv. Coop. 20:58.
33. Kami, J., Velasquez, V.B., Debouck, D.G., and Gepts, P. 1995. Identification of presumed ancestral DNA sequences of phaseolin in *Phaseolus vulgaris*. Proc. Natl. Acad. Sci. USA 92:1101-1104.
34. Kelly, J.D. and Young, R.A. 1996. Proposed symbols for anthracnose resistance genes. Annu. Rep. Bean Improv. Coop. 39:20-24.
35. Kelly, J.D. and Miklas, P.N. 1998. The role of RAPD markers in breeding for disease resistance in common bean. Mol. Breed. 4:1-11.
36. Kelly, J.D., Afanador, L., and Cameron, L.S. 1994. New races of *Colletotrichum lindemuthianum* in Michigan and implications in dry bean resistance breeding. Plant Dis. 78:892-894.
37. Kimati, H. 1966. Algumas raças fisiológicas de *Colletotrichum lindemuthianum* (Sac. et Magn.) Scrib. que ocorrem no Estado de São Paulo. An. Esc. Super. Agric. Luiz de Queiróz USP 23:411-437.
38. Kruger, J., Hoffmann, G.M., and Hubbeling, N. 1977. The kappa race of *Colletotrichum lindemuthianum* and sources of resistance to anthracnose in *Phaseolus* beans. Euphytica 26:23-25.
39. Leakey, C.L.A. and Simbwa-Bunnya, M. 1972. Races of *Colletotrichum lindemuthianum* and implications for bean breeding in Uganda. Ann. Appl. Biol. 70:25-34.
40. Mastenbroek, C. 1960. A breeding programme for resistance to anthracnose in dry shell haricot beans, based on a new gene. Euphytica 9:177-184.
41. Menezes, J.R. and Dianese, J.C. 1988. Race characterization of Brazilian isolates of *Colletotrichum lindemuthianum* and detection of resistance to anthracnose in *Phaseolus vulgaris*. Phytopathology 78:650-655.
42. Mesquita, A.G.G., Paula, T.J. Jr., Moreira, M.A., and Barros, E.G. 1998. Identification of races of *Colletotrichum lindemuthianum* with the aid of molecular markers. Annu. Rep. Bean Improv. Coop. 41:88-89.
43. Muhalet, C.S., Adams, M.W., Saettler, A.W., and Ghaderi, A. 1981. Genetic system for the reaction of field beans to beta, gamma, and delta races of *Colletotrichum lindemuthianum*. J. Amer. Soc. Hort. Sci. 106:601-604.
44. Mujica, R.F. 1952. Razas fisiológicas y susceptibilidad varietal de los frijoles chilenos a la antracnosis. Agri. Téc. (Santiago) 12:37-45.
45. Noyola, I.T.U., Lopez, G.F., and Muruaga, J. 1984. Determinación de razas fisiológicas de *Colletotrichum lindemuthianum* (Sacc. & Magn.)

Bri & Cav. Resúmenes del XI Cong. Nac. de Fitopatología Soc. Mexicana de Fitopatología, A.C. San Luis Potosí, Mexico.

46. Oliari, L., Vieira, C., and Wilkinson, R.E. 1973. Physiologic races of *Colletotrichum lindemuthianum* in the state of Minas Gerais, Brazil. Plant Dis. Rep. 57:870-872.
47. Oliveira, E.A., Antunes, I.F., Costa, J.G.C. da. 1973. Raças fisiológicas de *Colletotrichum lindemuthianum* identificadas no Rio Grande do Sul e em Santa Catarina de 1968 a 1972. Inst. Pesq. Agron. do Sul, Pelotas, 5 pp. 1973. Comunicado Técnico, No. 8.
48. Pastor-Corrales, M.A., ed. 1988. La antracnosis del frijol común, *Phaseolus vulgaris*, en America Latina. Documento de Trabajo No. 113. CIAT, Cali, Colombia.
49. Pastor-Corrales, M.A. 1991. Estandarización de variedades diferenciales y de designación de razas de *Colletotrichum lindemuthianum*. Phytopathology 81:694 (abstract).
50. Pastor-Corrales, M.A. 1996. Traditional and molecular confirmation of the coevolution of beans and pathogens in Latin America. Annu. Rep. Bean Improv. Coop. 39:46-47.
51. Pastor-Corrales, M.A. and Tu, J.C. 1989. Anthracnose. Pages 77-104 in: Bean Production Problems in the Tropics. H.F. Schwartz and M.A. Pastor-Corrales, eds. CIAT, Cali, Colombia.
52. Pastor-Corrales, M.A., Llano, G.A., Afanador, L., and Castellanos, G. 1985. Disease resistance mechanisms in beans to the anthracnose pathogen. Phytopathology 75:1176 (abstract).
53. Pastor-Corrales, M.A., Erazo, O.A., Estrada, E.I., and Singh, S.P. 1994. Inheritance of anthracnose resistance in common bean accession G 2333. Plant Dis. 78:959-962.
54. Peurer, H. 1931. Continued investigation on the occurrence of biological strains in *Colletotrichum lindemuthianum* (Sacc. & Magn.) Bri. et Cav. Phytopathol. Z. 4:83-112.
55. Pio-Ribeiro, G. and Chaves, G.M. 1975. Raças fisiológicas de *Colletotrichum lindemuthianum* (Sacc. et Magn.) Scrib. que ocorrem em alguns municípios de Minas Gerais, Espírito Santo e Rio de Janeiro. Experientiae 19:95-118.
56. Rava, C.A., Molina, J., Kauffmann, M., and Briones, I. 1993. Determinación de razas fisiológicas de *Colletotrichum lindemuthianum* en Nicaragua. Fitopatol. Bras. 18:388-391.
57. Restrepo, S. 1994. DNA polymorphism and virulence variation of *Colletotrichum lindemuthianum* in Colombia. M.Sc. thesis. Universite Paris IV, Paris-Grignon, France.
58. Rodriguez-Guerra, R. 1991. Identificación de razas patogenicas de *Colletotrichum lindemuthianum* (Sacc. y Magn.) Scrib. en el estado de Durango mediante un sistema propuesto internacionalmente y respuesta de genótipos de frijol tolerantes a sequia a razas del patógeno. M.Sc.,

Parasitologia Agricola, Universidad Autonoma Agraria "Antonio Narro", Buenavista, Mexico.

59. Schnock, M.G., Hoffmann, G.M., and Kruger, J. 1979. A new physiological strain of *Colletotrichum lindemuthianum* infecting *Phaseolus vulgaris* L. HortScience 10:140.
60. Schreiber, F. 1932. Resistenzzuchtung bei *Phaseolus vulgaris*. Phytopathol. Z. 4:415-454.
61. Schwartz, H.F., Pastor-Corrales, M.A., and Singh, S.P. 1982. New sources of resistance to anthracnose and angular leaf spot of beans (*Phaseolus vulgaris*). Euphytica 31:741-754.
62. Sicard, D., Michalakis, Y., Dron, M., and Neema, C. 1997. Genetic diversity and pathogenic variation of *Colletotrichum lindemuthianum* in the three centers of diversity of its host, *Phaseolus vulgaris*. Phytopathology 87:807-813.
63. Singh, S.P., Gepts, P., and Debouck, D.G. 1991. Races of common bean (*Phaseolus vulgaris*, Fabaceae). Econ. Bot. 45:379-396.
64. Sutton, B.C. 1992. The genus *Glomerella* and its anamorph *Colletotrichum*. Pages 1-26 in: *Colletotrichum:* Biology, Pathology and Control. J.A. Bailey and M.J. Jeger, eds. CAB International, Wallingford, UK.
65. Tu, J.C. 1988. Control of bean anthracnose caused by delta and lambda races of *Colletotrichum lindemuthianum* in Canada. Plant Dis. 72:5-7.
66. Tu, J.C. 1992. *Colletotrichum lindemuthianum* on bean. Population dynamics of the pathogen and breeding for resistance. Pages 203-224 in: *Colletotrichum:* Biology, Pathology and Control. J.A. Bailey and M.J. Jeger, eds. CAB International, Wallingford, UK.
67. Tu, J.C. 1994. Occurrence and characterization of the alpha-Brazil race of bean anthracnose (*Colletotrichum lindemuthianum*) in Ontario. Can. J. Plant Pathol. 16:129-131.
68. Tu, J.C., Sheppard, J.W., and Laidlaw, D.M. 1984. Occurrence and characterization of the epsilon race of bean anthracnose in Ontario. Plant Dis. 68:69-70.
69. Vilarinhos, A.D., Paula, T.J. Jr., de Barros, E.G., and Moreira, M.A. 1995. Characterization of races of *Colletotrichum lindemuthianum* by the random amplified polymorphic DNA technique. Fitopatol. Bras. 20:194-198.
70. Waterhouse, W.L. 1955. Studies of bean anthracnose in Australia. Proc. Linn. Soc. N.S.W. 80:71-83.
71. Yerkes, W.D. Jr. 1958. Additional new races of *Colletotrichum lindemuthianum* in Mexico. Plant Dis. Reptr. 42:329.
72. Yerkes, W.D. and Teliz-Ortiz, M. 1956. New races of *Colletotrichum lindemuthianum* in Mexico. Phytopathology 46:564-567.

73. Young, R.A. and Kelly, J.D. 1996. Characterization of the genetic resistance to *Colletotrichum lindemuthianum* in common bean differential cultivars. Plant Dis. 80:650-654.
74. Young, R.A. and Kelly, J.D. 1997. RAPD markers linked to three major anthracnose resistance genes in common bean. Crop Sci. 37:940-946.
75. Young, R.A., Melotto, M., Nodari, R.O., and Kelly, J.D. 1998. Marker assisted dissection of oligogenic anthracnose resistance in the common bean cultivar, G2333. Theor. Appl. Genet. 96:87-94.

Chapter 22

Colletotrichum coccodes on Potato

L. Tsror (Lahkim) and D.A. Johnson

Black dot of potato (*Solanum tuberosum* L.) caused by *Colletotrichum coccodes* (Wallr.) Hughes is common in many areas of the world. It is characterized by the development of small black sclerotia on senescent and dead plant tissue, on decaying roots and stems, and on stolons and daughter tubers (5,10,34).

Distribution of the Disease

C. coccodes has been reported worldwide on many different hosts. It is primarily found on the *Solanaceae* and *Cucurbitaceae* and is particularly destructive on potato and tomato. It has been reported on potatoes in the United States (4,13,21,30,40), Chile (12), Europe (35,2), Australia (14), South Africa (7), and Israel (43) and on tomatoes in Canada, the United States, and France (17,25). Several Solanaceous weed species, including velveltleaf and eastern black nightshade (1,32,45) are also hosts. Although *C. coccodes* is not a pathogen of strawberry in nature, several isolates were reported as moderately to highly virulent in artificial inoculations of stolons and fruits (27).

Until the early 1990s, the disease had been considered of minor importance throughout potato-growing areas worldwide (37,38,40). However, its importance has increased recently, and it now causes economic losses severe enough to necessitate control measures (2,7,8,36). Black dot of potato has been recognized as a potential economic problem in the United States (4,21) and in Israel, where *C. coccodes* recently caused economic damage in several commercial fields, reducing yields and infesting soil and tubers, especially in fields that had been fumigated with methyl bromide, vapam, or formalin prior to planting (43).

The Causal Pathogen

C. coccodes conidia are cylindrical with obtuse ends, hyaline, aseptate, and 16-24 x 2.5-4.5 µm in dimension. Colonies appear on various media as a white, superficial mycelium. Acervuli that produce conidia and setae are formed on stems, roots, and fruits. Sclerotia (100 µm to 0.5 mm in diameter) develop from acervuli and are produced in abundance in culture and on host tissues. Sclerotia survive for periods of one year or more in soil (11).

C. coccodes is a morphologically diverse fungus with isolates varying in sclerotium size, color, aerial mycelium, and frequency of sector formation (5,24). Some isolates may not produce the typically black sclerotia or may not even produce them at all (4). Variability in virulence among isolates of *C. coccodes* has been established on tomato and strawberry (26,27) and also on potato (4,41). However, variations in virulence among isolates need to be examined more thoroughly.

Disease Symptoms

Black dot describes abundant, dotlike black sclerotia on different parts of the plant. Disease symptoms include yellowing and wilting of the foliage, rot of roots, belowground stems, and stolons, and eventually early death of plants. It may be confused with symptoms caused by other wilt pathogens and is frequently a part of the potato early dying syndrome. As stems dry, small black sclerotia develop externally and internally on senescent and dead plant tissue, on decaying roots and stems, and on stolons and daughter tubers (5,34).

On tubers, the disease produces a brown-gray blemish which detracts from their appearance. Particularly noticeable on packaged potatoes washed before sale (19), it reduces the market value of affected tubers (24). In storage, grayish areas on tubers may closely resemble silver scurf (*Helminthosporium solani* Dur. and Mont.). These blemishes are usually superficial, but if they are extensive, the tubers may shrivel (16); deeper necrotic lesions may develop if diseased tubers are stored at -1°C (29).

Effect on Yield

C. coccodes can cause latent infections and severe disease symptoms in potato and reduce yields in experiments under controlled conditions (4,24). When *C. coccodes* sclerotia were added to field soil in the greenhouse, potato yield and plant growth were markedly reduced (40). Field inoculations reduced yields in the late-maturing cv. Russet Burbank, but not in the early-maturing cv. Norgold Russet (28). Other researchers confirmed this deleterious effect on yield in Russet Burbank in field and greenhouse experiments (8-12% and 29-43%, respectively) (4,20,28).

Yield loss in several cultivars was also demonstrated in field experiments over 3 years in Israel, after foliar sprays with conidia or soil amendments with sclerotia. Yield loss in Agria was 19-27%, in Desiree 3-27%, in Alpha 5-30%, in Cara 11-25%, and in Nicola 4-22% (43,44).

A significant difference in cultivar reaction to black dot disease was also reported in the UK in regard to disease incidence. Desiree, Maris piper, Maris peer and Record were severely affected, whereas Cara, Pentland Crown, and Romano were less affected (33).

Disease Pathogenesis

Inoculum sources for disease infection in the field may be either one or a combination of the following: tuber-, soil- and air-borne inoculum. When disease results from tuber- or soil-borne inoculum, the fungus may attack any belowground parts of potato plants (including roots, belowground main stems, stolons and tubers), and basal stems (9,31), and in some areas the disease develops to the potato foliage (4,20,28).

The number of colony-forming units of *C. coccodes* from belowground and aboveground stems varied over time, which may reflect distinct phases for the disease below- and aboveground (20,22). In four cultivars tested in field experiments in Israel, latent infections of plants were observed 90 days after planting. *C. coccodes* infected the surface parts of the stem more than the vascular system (L. Tsror et al, unpublished data).

Tuber-borne inoculum may increase infection (6,19,24), and frequency (6,8). However, variations in seed tuber infection can be masked by soil-borne inoculum (34). Infected seed tubers, either internally or externally on the skin, are an important source of inoculum, spreading the disease in potato-production areas (23). The pathogen was not isolated from nuclear seed tubers, and incidence of infection was higher with high seed generations. The fungus was isolated from tuber periderm and outer medulla tissues, and isolation frequency was greater from tuber stem ends than from either bud ends or lateral sections. Seed tuber infection may increase the incidence of early season plant infection (23). Infections from tubers, as well as from foliar lesions result in *C. coccodes* colonization of internal stem tissues and the vascular cylinder (20). Internal infection of the tuber in the stem end, however, was not correlated with surface infections of the tuber (L. Tsror, unpublished data).

In greenhouse and field experiments where plants were inoculated with conidia or sclerotia, dark brown to black lesions developed on leaves, petioles, and stems. When conidia were applied to the foliage, significantly more lesions developed on foliage that was sand blasted just before inoculation than on unwounded foliage (22). Lesion numbers on foliage of wounded and inoculated plants increased significantly as the duration of the post-inoculation wet period increased from 2 to 48 hr. Sclerotia from soil are most likely carried with blowing soil and, under wet conditions of rain or sprinkler irrigation, infection may occur.

Epidemiology of the Disease

The fungus, originally thought to be predominantly a tuber-borne pathogen, can be introduced into the soil through contaminated seed tubers, where it establishes itself and eventually serves as soil-borne inoculum for future crops (18,34). Tuber-borne inoculum may lead to early season plant infection (4,24). Komm and Stevenson (24) demonstrated stem infection that varied from 0 to 72%, whereas root infection ranged from 31 to 93%, depending on the severity of the fungal inoculum carried on the seed tubers. Therefore, seed-borne inoculum should be regarded as an important means for pathogen dissemination and introduction of the pathogen into soils where it has not been found previously and where soil fumigation treatments have been applied to reduce pathogen levels.

A high incidence of *C. coccodes* was found in certified potato seed tubers planted in Washington State, originating from various sources and geographic areas in western North America (up to 90% in 1994 and up to 53% in 1995) (22). Monitoring of imported potato seeds to Israel during the years 1994-1996 showed low incidences of *C. coccodes* in shipments from Germany, Scotland, and Ireland. However, shipments from Holland and France had higher levels, particularly shipments from Holland in 1995 and 1996. Progeny tubers planted from infected seed tubers in Israel had only low incidences of disease, except in 1996 when it was exceptionally high. This indicates correspondence between the health of imported seeds (for spring planting) and the quality of the daughter tubers that are, in effect, the local seeds planted in the autumn (44).

The fungus may exist at undetectable levels in soil until production of a host crop, such as potato, allows it to multiply to detectable levels. Foliage infection may also eventually lead to soil infestation. Conidia are known to survive for up to 52 weeks in soil (3,11) although most fail to survive longer than a few weeks (3).

Recent reports on foliar infections indicate that *C. coccodes* might also be classified as an air-borne pathogen (2,20,28) with the potential for foliar infection and disease spread via conidia under arid conditions, especially when aided by the use of sprinkler irrigation and windstorms. It may also be introduced to new areas during dust storms (4).

During storage, the severity of black dot increases, especially in humid conditions. However, harvesting early and storing the tubers dry can prevent or reduce disease expression (29). Incidence and severity of black dot increased more during storage at 15°C than at 5°C, especially on unwashed tubers (34). Tuber blemishes caused by *C. coccodes* may be extensive and cause shriveling, especially in cultivars with red skin (16).

Effect of External Factors on Disease Expression

The direct effects of *C. coccodes* infection on potato yield are not well understood. The fungus causes severe disease symptoms or latent infections in potato and reduces potato yields in controlled, replicated inoculation experiments. Abiotic and biotic factors may interact with *C. coccodes* to further reduce yields. Latent infections and disease development under conditions of stress make its relative importance difficult to determine (4,21,24,28,40,41).

ABIOTIC FACTORS

Black dot is most frequently associated with light sandy soils, low nitrogen, high temperatures, and poor water drainage. In Indiana, the disease occurred on crops stressed by excessive rain and low temperatures early in the season, followed by a prolonged drought (40). In England, seed tubers rotted more often when they were diseased than when they were not, irrigation decreased infection of stem bases, roots, and tubers up to 18 weeks after planting, but later rotting was increased by irrigation (34). In Israel where the crop is irrigated routinely, disease incidence and yield loss were observed when temperature was high and soil relatively dry (L. Tsror, unpublished data). High day temperatures of 23°C or 28°C in growth chamber studies decreased yields more than at 18°C on Russet Burbank and Norkotah Russet plants grown in soil infested with *C. coccodes* when compared with plants grown in non-infested soil (Geary and Johnson, unpublished data). Dry soils may inhibit conidial production, transport, and infection or decrease tuber susceptibility (34).

Growth conditions in different seasons in regions where potatoes are grown twice a year may affect disease expression. In field studies in Israel, average yield reduction of several cultivars appears to be more severe in autumn (19%) than in spring (13%), especially in Nicola (2.4 fold higher yield loss in autumn) and Cara (1.9 fold higher yield loss). Alpha's response was the opposite: a severe reduction of yield occurred in the spring (1.5 fold higher than in autumn). Agria and Desiree were not tested in both seasons; however, yield loss in Desiree (14%) which was tested only in spring was smaller than in Agria, which was tested only in autumn (24%) (44).

Because day length becomes shorter in autumn and longer in the spring, the effect of illumination on disease expression was tested in controlled growth chambers. Previous findings indicated that *Verticillium* wilt symptoms in potato were enhanced under short-day as compared with long-day conditions (42). When potato plants inoculated with *C. coccodes* were grown in growth chamber with a regime of 8:16 L:D, disease severity and fungal colonization increased more and sclerotia formed more rapidly than in 16:8 L:D conditions (44).

BIOTIC FACTORS

C. coccodes commonly acts in combination with one or more other pathogens. Infection by *C. coccodes* was increased by the presence of *V. dahliae,* and inoculations with a mixture of *V. dahliae* and *C. coccodes* increased death of potato plants beyond that observed with either pathogen alone (30). Tsror et al. (44) confirmed this synergism in the interactions between *C. coccodes* and *V. dahliae in vitro*, using aseptic plantlets of Nicola, Desiree, Alpha, and Cara which were inoculated with either the same concentrations of each pathogen or with a mixture of both. The two pathogens together in Nicola significantly reduced plant weight and height and caused more severe symptoms when compared with effects after inoculation with the individual pathogens. In Desiree, however, only disease symptoms were more severe with the combined inoculation. On the other hand, Alpha and Cara cultivars demonstrated no interaction. In summary, compared with the effects of infection with either *C. coccodes* or *V. dahliae*, infection with both pathogens can, in some cultivars, increase the incidence of potato early dying syndrome and thus decrease yields. These findings have practical field implications in the development of such synergistic infections, whether they result from conducive soil conditions, sowing tubers, or both (43,44). Synergistic infections may occur in the field, either from soilborne inoculum or seedborne inoculum or both, and understanding these interactions can have practical implications for disease management, and screening for resistance or tolerance to these pathogens.

However, other researchers reported that infection of potato by *C. coccodes* did not appear to depend on concurrent or preceding infection by *V. dahliae* (21). In another study, coincident infection by *C. coccodes* and *V. dahliae* was less than would be expected if the infections were associated (13). Yield loss by *V. dahliae* in cv. Amethyst almost doubled in the presence of *C. coccodes*, whereas *C. coccodes* alone caused damage only late in the growing season in weakened plants (39).

In wilted plants *C. coccodes* was often isolated in association with *Fusarium oxysporum* and *V. dahliae*. Infection with *C. coccodes* was also inversely correlated with potato virus X infection (3).

Treatment with oxamyl (nematicide) increased black dot in a 5-year rotation (15). Black dot on stems and tubers was not affected by previous cropping but was much less severe in a plot that had not grown potatoes during the 7 years of the experiment. A possible interaction of the herbicide metribuzin with the incidence of black dot was reported in Idaho (28).

Control of the Disease

Because both seed- and soil-borne inoculum can be important in the epidemiology of black dot, control of the disease on seed tubers is likely to be insufficient.

Seed Treatment

Fungicide treatment of seed tubers might be effective in preventing introduction of the pathogen into uninfested soil (34). Imazalil and tolclofos methyl severely affected *C. coccodes* growth *in vitro*, as did mancozeb and thiabendazole to a lesser extent (2). Part of these findings were confirmed by Tsror et al. (44), who observed that prochloraz, tebuconazol, and imazalil (1-10 ppm) suppressed *C. coccodes* growth *in vitro*, as did fenpiclonil (above 100 ppm). In contrast, agents usually used to control *Rhizoctonia solani* (pencycuron, flutolanil, tolclofos methyl and ioprodione) were totally ineffective (44).

Treating seeds either by dipping or by spraying with an ultra-low volume of some of the previously mentioned agents can considerably reduce infection levels on the daughter tuber surface. Dipping seed tubers in fenpiclonil (0.2% for 2 min) and prochloraz (0.4%) significantly reduced *C. coccodes* incidence in progeny tubers, whereas tebuconazol was less efficient and imazalil even less so (44, L. Tsror, unpublished data). Prochloraz has proved effective against *C. coccodes* and *H. solani* when applied by dusting or dipping infected tubers prior to planting, or as a pre-storage treatment combined with a second treatment just before planting (7). Pathogens carried within the tuber in the vascular bundles are not affected by these treatments.

Postharvest quality can be improved by drying tubers immediately after harvest before cool storage, especially for crops harvested from wet soil (15,34).

Soil Treatment

Because chemical seed treatments are unlikely to suppress disease if soil inoculum is prevalent, soil solarization or fumigation should be considered as an alternative method of control. However, soil fumigation has not consistently reduced black dot severity. Fumigation of naturally infested field soil in pot experiments with dichloropropane and other chlorinated hydrocarbons or in combination with methyl isothiocyanate failed to control

C. coccodes (40). In contrast, soil fumigation with methyl bromide reduced disease. The incidence of black dot on the progeny of infected seed planted in nonfumigated soil was twice that on progeny in soil fumigated with methyl bromide (126 g m^{-2}). In addition, progeny of uninfected seed had 68.5 times higher disease incidence in nonfumigated soil (8). Soil fumigation with formalin also reduced disease incidence on potato tubers grown in sandy soil (L. Tsror, unpublished data). However, if the soil is fumigated prior to planting, using clean seed is extremely important; if the pathogen is introduced on seed tubers, disease may be increased because competition by other organisms has been eliminated by fumigation (28; L. Tsror, unpublished data).

Foliage Treatment

Foliar sprays of fungicides may reduce disease levels, consequently minimizing yield loss. Thus, sclerotia formation on dying plants may be reduced and contamination of the soil limited. Reduction of disease symptoms and plant infection was observed in field experiments where plants were sprayed weekly with prochloraz (3 L/H) (L. Tsror, unpublished data). Further studies are required to determine the economic feasibility of such treatments.

Resistant Cultivars

Most commercial cultivars differ in susceptibility to *C. coccodes*. Severe losses in quality or yield may be reduced or avoided by growing the less susceptible cultivars (24,44). Resistance exists in certain *Solanum chacoense* clones and in some of its hybrids (10).

In addition to the above mentioned control means, routine control strategies, e.g. crop rotation, well-drained soil, elimination of weed hosts and use of tolerant cultivars, should also be implemented to reduce black dot disease on potato.

Summary

Black dot has been recognized as an economically important disease only recently. The disease can cause yield reduction and/or quality damage. Because *C. coccodes* is a tuber-, soil-, or air-borne pathogen, control measures should take into account these three means of dissemination. Further research is required to improve our understanding of the etiology and epidemiology of the disease and to optimize its control.

Literature Cited

1. Andersen, H.N. and Walker, H.L. 1985. *Colletotrichum coccodes*: A pathogen of eastern black nightshade (*Solanum ptycanthum*). Weed Sci. 33:902-905.
2. Andrivon, D., Ramage, K., Guerin, C., Lucas, J.M., and Jouan, B. 1997. Distribution and fungicide sensitivity of *Colletotrichum coccodes* in French potato-producing areas. Plant Pathol. 46:722-728.
3. Blakeman, J.P. and Hornby, D. 1966. The persistence of *Colletotrichum coccodes* and *Mycosphaerella ligulicola* in soil with special reference to sclerotia and conidia. Trans. Br. Mycol. Soc. 49:227-240.
4. Barkdoll, A.W. and Davis, J.R. 1992. Distribution of *Colletotrichum coccodes* in Idaho and variation in pathogenicity on potato. Plant Dis. 76:131-135.
5. Chesters, C.G.C. and Hornby, D. 1965. Studies on *Colletotrichum coccodes*. I. The taxonomic significance of variation in isolates from tomato roots. Trans. Br. Mycol. Soc. 48:573-581.
6. Dashewood, E.P., Fox, R.A., and Perry, D.A. 1992. Effect of inoculum source on root and tuber infections by potato blemish disease fungi. Plant Pathol. 41:215-223.
7. Denner, F.D.N., Millard, C.P., Geldenhuys, A., and Wehner, F.C. 1997. Treatment of seed potatoes with prochloras for simultaneous control of silver scurf and black dot on daughter tubers. Potato Res. 40:221-227.
8. Denner, F.D.N., Millard, C.P., and Wehner, F.C. 1998. The effect of seed- and soilborne inoculum of *Colletotrichum coccodes* on the incidence of black dot on potatoes. Potato Res. 41:51-56.
9. Dickson, B.T. 1926. The "black dot" disease of potato. Phytopathology 16:131-135.
10. Dillard, H.R. 1992. *Colletotrichum coccodes*: The pathogen and its hosts. Pages 225-236: *Colletotrichum*: Biology, Pathology and Control. J.A. Bailey and M.J. Jeger, eds. CAB International, Wallingford, UK.
11. Farley, J.D. 1976. Survival of *Colletotrichum coccodes* in soil. Phytopathol. 66:640-641.
12. Fernandez, M.C. 1987. Identificación de *Colletotrichum atramentium* (Berk et Br) Tamb. (syn. *C. coccodes* (Wallr.) Hughes) en papa. Agric. Tec. Santiago 47:184-186.
13. Goodell, J.J., Powelson, M.L., and Allen, T.C. 1982. Interactions between potato virus x, *Verticillium dahliae* and *Colletotrichum atramentium* in potato. Phytopathology 72:631-634.
14. Harrison, D.E. 1963. Black dot disease of potato. J. Agric. Victoria Aust. 61:573-576.

15. Hide, G.A. and Read, P.J. 1991. Effects of rotation length, fungicide treatment of seed tubers and nematicides on disease and quality of potato tubers. Ann. Appl. Biol. 119:77-87.
16. Hunger, R.M. and McIntyre, G.A. 1979. Occurrence, development and losses associated with silver scurf and black dot on Colorado potatoes. Am. Pot. J. 56:289-306.
17. Illman, W.I., Ludwig, R.A., and Farmer, J. 1959. Anthracnose of canning tomatoes in Ontario. Can. J. Bot. 37:1237-1246.
18. Jeger, M.J., Hide, G.A., van den Boogert, P.H.J.F., Termorshuizen, A.J., and van Baarlen, P. 1996. Pathology and control of soil-borne fungal pathogens of potato. Potato Res. 39:437-469.
19. Jellis, G.J. and Taylor, G.S. 1974. The relative importance of silver scurf and black dot: Two figuring diseases of potato tubers. ADAS Quarterly Rev. 14:53-61.
20. Johnson, D.A. 1992. Effect of foliar infection caused by *Colletotrichum coccodes* on yield of Russet Burbank potato. Plant Dis. 78:1075-1078.
21. Johnson, D.A. and Miliczky, E.R. 1993. Distribution and development of black dot, Verticillium wilt, and powdery scab on Russet Burbank potatoes in Washington State. Plant Dis. 77:13-17.
22. Johnson, D.A. and Miliczky, E.R. 1993. Effects of wounding and wetting duration on infection of potato foliage by *Colletotrichum coccodes*. Plant Dis. 77:13-17.
23. Johnson, D.A., Rowe, R.C., and Cummings, T.F. 1997. Incidence of *Colletotrichum coccodes* in certified potato seed tubers planted in Washington State. Plant Dis. 81:1199-1202.
24. Komm, D.A. and Stevenson, W.R. 1978. Tuber-borne infection of *Solanum tuberosum* 'Superior' by *Colletotrichum coccodes*. Plant Dis. Rep. 62:682-687.
25. Loprieno, N. 1961. Investigations on tomato anthracnose on the growth of *Colletotrichum coccodes* (Wallr.) Hughes. I. The influence of carbohydrates. Caryologia 14:219-229.
26. Loprieno, N. and Guglielminetti, R. 1962. Investigations on tomato anthracnose III. The influence of amino acids on the growth of *Colletotrichum coccodes* (Wallr.) Hughes. Phytopathol. Z. 45:312-320.
27. Mass, J.L and Howard, C.M. 1985. Variation of several anthracnose fungi in virulence to strawberry and apple. Plant Dis. 69:164-166.
28. Mohan, S.K., Davis, J.R., Sorensen, L.H., and Schneider, A.T. 1992. Infection of aerial parts of potato plants by *Colletotrichum coccodes* and its effects on premature vine death and yield. Am. Potato J. 69:547-559.
29. Mooi, J.C. 1959. A skin necrosis occurring on potato tubers affected by black dot (*Colletotrichum atramentarium*) after exposure to low temperatures. Eur. Potato J. 2:58-68.

30. Otazu, V., Gudmestad, N.C., and Zink, R.T. 1978. The role of *Colletotrichum atramentarium* in potato wilt complex in North Dakota. Plant Dis. Reptr. 62:847-851.
31. Pavlista, A. and Kerr, E.D. 1992. Black dot of potato caused by *Colletotrichum coccodes* in Nebraska. Plant Dis. 76:1077.
32. Raid, R.N. and Pennypacker, S.P. 1987. Weeds as hosts for *Colletotrichum coccodes.* Plant Dis. 71:643-646.
33. Read, P.J. 1991. The susceptibility of tubers of potato cultivars to black dot (*Colletotrichum coccodes* (Wallr.) Hughes). Ann. Appl. Biol. 119:475-482.
34. Read, P.J. and Hide, G.A. 1988. Effects of inoculum sources and irrigation on black dot (*Colletotrichum coccodes* (Wallr.) Hughes) and its development during storage. Potato Res. 31:493-500.
35. Read, P.J. and Hide, G.A. 1995. Development of black dot disease (*Colletotrichum coccodes* (Wallr.) Hughes) and its effect on the growth and yield of potato plants. Ann. Appl. Biol. 127:57-72.
36. Read, P.J. and Hide, G.A. 1995. Effects of fungicides on the growth and conidial germination of *Colletotrichum coccodes* and on the development of black dot disease of potatoes. Ann. Appl. Biol. 126:437-447.
37. Read, P.J., Storey, R.M., and Hudson, D.R. 1995. A survey of black dot and other fungal tuber blemishing diseases in British potato crops at harvest. Ann. Appl. Biol. 126:249-258.
38. Rowe, R.C., Davis, J.R., Powelson, M.L., and Rouse, D.I. 1987. Potato early dying: causal agents and management strategies. Plant Dis. 71:482-489.
39. Scholte, K., Veenbaas-Rijks, J.W., and Labruyere, R.E. 1985. Potato growing in short rotations and the effect of *Streptomyces* spp., *Colletotrichum coccodes, Fusarium tabacinum* and *Verticillium dahliae* on plant growth and tuber yield. Potato Res. 28:331-348.
40. Stevenson, W.R., Green, R.J., and Bergesen, G.B. 1976. Occurrence and control of black dot root rot in Indiana. Plant Dis. Reptr. 60:248-251.
41. Thirumalachar, M.S. 1967. Pathogenicity of *Colletotrichum atramentarium* on some potato cultivars. Am. Potato J. 44:241-244.
42. Tsror (Lahkim), L., Livescu, L., and Nachmias, A. 1990. Effect of light duration and season on Verticillium wilt in potato. Phytoparasitica 18:331-339.
43. Tsror (Lahkim), L., Erlich, O., Hazanovsky, M., and Peretz, I. 1994. *Colletotrichum* on potato in Israel, is it a new disease? Phytoparasitica 22:88.
44. Tsror (Lahkim), L. 1998. Etiology and control of *Colletotrichum coccodes* in potato. Proc. Europ. Assoc. Potato Res., Pathol. Sec. Mtg., Umei, Sweden.

45. Wymore, L.A., Poirier, C., Watson, A.K., and Gotlieb, A.R. 1988. *Colletotrichum coccodes,* a potential bioherbicide for control of velvetleaf (*Abutilon theophrasti*). Plant Dis. 72:534-538.

Chapter 23

The Biology of *Colletotrichum graminicola* and Maize Anthracnose

G.C. Bergstrom and R.L. Nicholson

Maize anthracnose, caused by the fungus *Colletotrichum graminicola* Ces. Wils., is a disease of worldwide importance (30,91). The disease affects all parts of the maize plant and can be found at any time during the growing season, but is observed most frequently in the form of anthracnose leaf blight (ALB) or anthracnose stalk rot (ASR). Once considered synonymous, *C. graminicola* is now considered distinct from *C. sublineolum* Henn., Kabat & Bubak, which causes the anthracnose disease of sorghum (76).

Prior to the early 1970s, anthracnose was not considered an important problem in North America. However, severe epidemics occurred in the early 1970s in the north central and eastern USA (30,31,72,86,88,91). Possible explanations for this sudden and widespread occurrence of anthracnose in diverse environments included the appearance of more virulent biotypes of the pathogen, a shift in host genotypes to greater susceptibility, the occurrence of more favorable weather patterns, and changes in cultural practices that increased fungal inoculum. Today the pathogen is endemic in many maize production areas of North America, but dramatic grain losses have been reduced by the wide-scale deployment of partially resistant maize genotypes (13).

The severity of the North American epidemics spurred considerable fundamental research on the biology of this interaction. This chapter is excerpted from a more complete treatise (13) and summarizes our current understanding of the maize anthracnose cycle and the biology of *C. graminicola* and its interactions with maize plants. It is our intention that such knowledge will provide new strategies for enhanced protection of plants against this and related pathogens. Fungal genetics is also important to a complete understanding of the biology of *C. graminicola*, and this is addressed in Chapter 3 by Vaillancourt et al.

The Maize Anthracnose Cycle

Colletotrichum graminicola is well adapted for survival in a maize-based agroecosystem because it is both an aggressive pathogen of living maize plants and a facultative saprophyte on maize residues. The maize anthracnose cycle can be characterized by five temporal phases: production and dissemination of primary inoculum, seedling blight from primary inoculum, leaf blight from repeating secondary inoculum, systemic colonization and stalk rot, and saprophytic survival.

Primary Inoculum Phase

The reservoir of primary inoculum for leaf blight is overwintered maize residues that remain on the soil surface. Thus, it is not surprising that ALB has increased coincident with the widespread adoption of conservation tillage in conjunction with continuous cropping of maize in the USA (13). Initial infection of seedling leaves occurs from spores produced in acervuli on crop debris. It is important to note that horizontal dispersal of spores from infested debris is limited to distances traversed by splashing and blowing raindrops. This short-distance rain dispersal has also been documented for *Colletotrichum acutatum* on strawberry (51). Steep gradients of maize anthracnose from point or area sources of primary inoculum have been recorded in the field where there was greater within-row than across-row spread of the leaf blight phase of the disease (47). This understanding of pathogen dispersal suggests that local cultivation practices are useful for disease control. But localized tillage and crop rotation may be insufficient to control late season leaf blight and stalk rot in locales where the pathogen is endemic (14).

Seedling Blight Phase

Leaf blight of juvenile plants may be so severe in highly susceptible genotypes that it restricts normal growth and development and can even result in seedling death. Typically, however, plants at the seedling stage of development grow so rapidly that as new leaves emerge from the whorl they appear to be resistant and often fail to exhibit disease. In the seedling stage of development, symptoms typically consist of oval lesions often with concentric zones of enlargement (68).

Leaf Blight Phase

Secondary inoculum for the continued development of the disease comes from lesions on lower leaves. Conidia are disseminated vertically in the crop canopy by splashing rain. Repeated cycles of secondary inoculum production and dispersal occur throughout crop development. Leaves of seedlings and post-anthesis plants develop more disease than do leaves that are expanding rapidly (10,43,61). Individual leaves age and become senescent at which time they become more susceptible (61). Because the maize plant has a determinate growth pattern, leaves emerge, mature, enter senescence, and eventually die one after another in succession (61). Conidia that serve as secondary inoculum for leaf infection also can serve as inoculum for stem infections. Isolates of *Colletotrichum graminicola* from nearby grasses may provide additional inoculum for infection of mature leaves. Isolates from several grasses induced susceptible-type lesions on senescing maize leaves, but induced only chlorotic flecks on juvenile maize leaves (33). Therefore, only isolates from maize are significant contributors to maize anthracnose epidemics.

SYSTEMIC COLONIZATION / STALK ROT PHASE

The stalk rind epidermis appears to become infected in a manner similar to that of the leaf epidermis (Bandyopadhyay and Bergstrom, unpublished); hence, the stalk rind infection may be an extension of the leaf blight phase of the disease. Conidia formed on leaves may be washed behind the leaf sheath and initiate rind infection. But deposition of conidial suspensions behind leaf sheaths has not proven a reliable method to induce systemic ASR (89). It is often considered that the predominant symptom of anthracnose stalk rot is the discoloration of the rind, often to the extent that it appears black. However, one must be cautious with this assumption as stalk internodes often show internal pith discoloration due to anthracnose with little or no discoloration of the external rind. Conversely, there can also be discoloration of the stalk rind yet the inner pith tissue is not rotted.

Stem ingress by the pathogen occurs primarily through wounds that breach the rind. The most frequently encountered wounds in maize are those produced by stalk-boring insects, especially by larvae of the European corn borer (*Ostrinia nubilalis*) (8,38,54,55,56,89). *Colletotrichum graminicola* is an aggressive vascular pathogen of plants at late vegetative and early reproductive stages of development (7,37,55,56). It is well adapted as a colonist of xylem as it readily utilizes sucrose (the primary transport sugar in maize) as a carbon source and constitutively produces invertase, the enzyme that hydrolyzes sucrose to glucose and fructose (11,66,74). Small, oval conidia are produced in the xylem vessels (70). These conidia appear to become trapped at the nodal plate (Bergstrom, unpublished), the site where vascular bundles come together (39). Systemic colonization during early reproductive stages can result in significant grain yield reductions (38). Under favorable conditions, systemic vascular colonization of corn plants results in a vascular wilt known as "top dieback" in which the upper leaves and stem internodes are killed during early stages of grain formation (79,91).

The term "stalk rot" is widely used to connote the premature senescence and deterioration of basal stalk tissues at or just prior to normal physiological maturity of the corn plant. It is characterized by premature plant death and soft, often lodged, lower stalk internodes (22). *C. graminicola*, unlike most stalk rot fungi, can attack stems of living plants. It is also a predominant contributor, along with *Fusarium* and *Stenocarpella* species, to senescence-pattern stalk rot in maize at late reproductive stages (1,31). Late season infections of basal stalk tissues by *C. graminicola* may progress from earlier root infections (presumably from contact of roots with infested maize residues), from mycelial invasion through senescent stalk rind tissues, or from vascular colonization from infected wounds.

SAPROPHYTIC PHASE

C. graminicola survives as a saprophyte on infected maize residues on the surface of the soil. If residue is buried, however, the fungus has less capacity to survive as it is a poor competitor (45,46,57,84,85). Overwinter survival of the fungus in stalk tissues is extended into the sporulation period in spring. Conidia are produced on the overwintered stalks in an extracellular mucilage that protects them from desiccation and other adverse environmental conditions until susceptible maize seedlings are available.

Environment Conducive to Disease

High light intensity increases the expression of resistance of the maize plant to anthracnose (26,34,75). Importantly, this occurs regardless of the level of resistance/susceptibility exhibited by a genotype (75). Conversely, a reduction in light intensity results in an increase in the relative level of susceptibility of maize genotypes (75). This phenomenon is believed to occur because the expression of resistance in maize involves phenolic compound biosynthesis (48), the expression of which is essentially light dependent (21,29,64). Consistent with this is the fact that epidemics in the field have been observed to develop rapidly under cloudy, overcast conditions without necessarily having a prolonged period of rain. However, anthracnose development is favored by extended periods of high humidity, a condition necessary for sporulation of the fungus. Of course, the disease is also favored by rain since conidia are most easily dispersed by splashing raindrops.

Optimal temperatures for anthracnose must be considered with regard to both pathogen morphogenesis and to the development of lesions on the plant. Although germination and appressorium formation by *C. graminicola* occurs over a broad temperature range (15 to 35°C) penetration of the host appears to occur only in the narrow range of 25 to 30°C (41,77).

Other conditions that contribute to anthracnose development include stresses on the host by other parasites. For example, leaf blight is more severe in plants stressed by the root lesion nematode (61). Stalk rot is more prevalent and severe in plants injured by stalk-boring insects, especially the European corn borer (8,38,54,55,56).

Anthracnose stalk rot development is enhanced by various post-anthesis stresses to the maize plant. However, development of anthracnose in living stems of at least some maize genotypes is not explained by the photosynthetic-stress translocation-balance hypothesis of Dodd (22) which describes stalk rots (especially those induced by *Fusarium* and *Stenocarpella* species) as opportunistic diseases of senescent stems with low sugar content. Removal of photosynthate sinks (ears) did not reduce systemic ASR development in plants inoculated with *C. graminicola*, whereas it nearly eliminated development of soft, rotted basal internodes regardless of whether plants had been inoculated with *C. graminicola*, *Fusarium graminearum*, or no pathogen (Bergstrom, unpublished).

Biological Processes in Pathogenesis

Spore Dissemination in an Extracellular Matrix

The onset of sporulation of the fungus begins within as little as 10 to 12 hr after exposure of infected plant tissues to 100% relative humidity. Mature, sporulating acervuli generally then form within an additional 12 to 18 hr. Scanning electron microscopy has shown that conidia, whether produced in culture or in acervuli on plant tissues, are produced in, and surrounded by, an extracellular mucilaginous matrix. The subject of considerable investigation, the matrix is now known to be essential to the fungus as it ensures survival and successful dissemination of conidia.

The matrix was first shown to act as an anti-desiccant and to aid in the survival of conidia during periods of drought. It was demonstrated that conidia could remain viable for weeks to months even at relative humidities as low as 45%, but, in the absence of the matrix, conidia lost viability in just a few hours even at high relative humidities (66). Upon drying, the mucilaginous matrix was shown to form a thin film that surrounds and appears to encase conidia (65). Because conidia can survive as dry spore masses, they can easily be disseminated as dry particles by wind currents (Nicholson, unpublished).

The primary component of the matrix is a mixture of very high molecular weight glycoproteins (composed of oxygen-linked oligomers of mannose, rhamnose, galactose, and glucose and high levels of hydrophobic and hydroxylic amino acids) (74), that account for the anti-desiccant properties of the matrix (66). These glycoproteins contain unusually high levels of proline, similar to proteins in animal mucins (74). Proline-rich proteins had been shown to exhibit selective binding to polyphenolic compounds, which can be very toxic because of their ability to bind to and precipitate proteins (25). Components of the mucilaginous matrix of *C. graminicola* also bind selectively to polyphenolic compounds (62,65), allowing the fungus a means of removing toxic phenolic metabolites from the infection court.

Another matrix component that ensures pathogen survival is a water-soluble, low molecular-weight compound that functions as a self-inhibitor of conidium germination. The inhibitor, mycosporine-alanine, was found in the conidial matrix at levels up to 4 mM (42). Thus, masses of conidia in an undisturbed acervulus are surrounded by the mucilagenous matrix which contains inhibitory levels of mycosporine-alanine, that prevents the germination of conidia within the acervulus. If splashing rain water hits an acervulus, conidia are dispersed from one another, and the water-soluble matrix which surrounds them is brought into solution. When the concentration of mycosporine-alanine is lowered below 0.5 mM, the inhibitor is no longer effective, and conidia can then germinate. This simple mechanism ensures that conidia will only germinate when they are dispersed from the acervulus and from one other and can be moved to an appropriate infection court on the host surface. The self-inhibitor and the anti-desiccant glycoproteins maintain the long-term survival of conidia that are left dry and undisturbed in the mucilaginous matrix of the acervulus. This is especially important as the pathogen is a poor competitor in the soil and survives primarily in or on plant debris on the soil surface (45,57,85).

In addition to protecting spores from adverse environments, the conidial matrix also promotes pathogenesis of corn (12). Pathogen establishment on maize plants is believed to be aided by the activity of enzymes in the conidial matrix, including invertase, β-glucosidase, non-specific esterase, DNAse, RNAse, alkaline phosphatase, cutinase, and laccase (2,11,66,71,74,80). Four different cutinase enzymes are present in the conidial mucilage. When cutinase activity was inhibited, the fungus failed to penetrate into the leaf suggesting that the role of the enzymes is associated with initial events in the infection process that allow for establishment, possibly including penetration of the fungus into the host (71). Laccase, a polyphenol oxidase, was also shown to be a major component of the mucilage and to act in the oxidation and detoxification of phenolic compounds produced by the plant in response to infection (2). Thus, this enzyme is believed to aid the spread of the fungus by detoxification of "hazardous" materials in the environment and is a mechanism for protection similar to that of the proline-rich proteins.

Establishment of the Fungus in the Infection Court

A significant obstacle in fungal plant disease interactions is that of establishing the actual site for infection, the infection court. When conidia germinate, they secrete materials that act as adhesives that bind the fungal germling to the plant surface (63). This adhesion prevents the germling from being displaced by wind or water from the infection court prior to its penetration into host tissue (23,60). Conidia of *C. graminicola* require 6 to 8 hr to germinate after coming into contact with a maize leaf (52,59). Conidia are dispersed from the acervulus and spread across the leaf surface in a film of water. As soon as the ungerminated conidia touch the hydrophobic leaf surface, they begin to release an adhesive material that binds them to the surface of the leaf. This adhesion allows the conidium to bind tenaciously to the leaf surface, a phenomenon that prevents displacement and enhances disease development in the field (52,53).

Germination of the adhered conidium generally begins by 6 to 8 hr after contact with the leaf surface. When conidia germinate, they often form a sessile appressorium, rather than a germ tube and then an appressorium. When the appressorium is forming, it also secretes an adhesive which causes it to bind to the plant surface. Generally, by 15 to 18 hr the wall of the appressorium becomes melanized, and the appressorium is then completely differentiated and referred to as mature. Penetration of the host only occurs after the appressorium matures and becomes melanized. The presence of melanin in the wall of the appressorium is believed to allow for the development of exceptionally high hydrostatic pressures within the appressorium and this facilitates penetration of the host cell wall (32).

The overall infection process is aided by the presence and/or production of various enzymes by the fungus, including cutinases that degrade the cutin component of the plant cuticle (71) and cellulases, pectinases, and polygalacturonases, that degrade host cell wall materials (67). As with many other *Colletotrichum* species, *C. graminicola*, upon penetration of the host cell, appears to grow first as a biotroph (6,59). Once established in the leaf tissue, the pathogen begins to parasitize the host as a necrotroph, producing a variety of cell wall-degrading enzymes and killing the host tissue in advance of the now intracellularly developing mycelium.

Penetration of the plant occurs through an appressorial pore, a thin circular zone at the base of the appressorium that is not melanized. An infection hypha emerges through the pore and penetrates through the host cell wall. An infection vesicle is then formed, which sets the stage for the fungus to grow biotrophically for a short period of time. The infection vesicle does not penetrate the host plasma membrane, rather the plasma membrane is invaginated around the vesicle. Next, the fungus forms primary hyphae that are also outside of the host plasma membrane. This form of growth provides a very large surface area of the host membrane relative to the fungus, allowing the fungus greater access to nutrients from the host. The fungus remains growing biotrophically for a relatively short time (24 to 36 hr) after which it penetrates host cells and grows as a necrotroph.

The epidermal cells of noninjured, living stem rind tissues appear to be invaded in a biotrophic manner from appressoria similar to infection of the foliar epidermis (Bandyopadhyay and Bergstrom, unpublished). Stems of plants at early vegetative stages as well as stems of mature plants following stalk senescence also can be directly invaded via appressoria penetration without wounds.

Pathogen infection of internal tissues of living stems occurs primarily through wounds that breach the stalk rind (54,55,56). Conidia germinate at wound sites producing extremely long germ tubes (54). Maize cells are usually penetrated from appressoria formed where the germ tubes come in contact with cell walls, but occasionally, penetration at wound sites occurs from hyphal strands without the formation of appressoria (54). Invasion rapidly becomes necrotrophic with tissue maceration evident. Once the xylem vessels are invaded, the fungus becomes systemic in stems (54).

The predisposition to fungal infection of wounded stalk pith is transitory. Wound sites aged as little as 1 to 2 hr before inoculation show significantly less invasion by *C. graminicola* than do sites inoculated immediately after wounding (54,55,56). The European corn borer (ECB) larva plays key roles in overcoming the rapid wound-healing defense. It vectors *C. graminicola* both externally and internally (the fungus remains viable from gut into frass) and places the pathogen into fresh wounds (54). As it feeds, it continually wounds stalk tissues. Interestingly, the association also benefits the ECB. Larvae develop 20% faster and with less mortality on maize tissues infected by *C. graminicola* than on healthy tissues (16). The stimulation of insect growth appears to be related to maceration of maize tissues and breakdown of complex carbohydrates by fungal enzymes (16).

Fungal Latency

Appressoria are believed to survive sometimes on the surface of the maize leaf or stem rind even if the tissue is not penetrated. This association with the host is similar to the manner in which other *Colletotrichum* species persist on plant tissues (73) and accounts for the delayed appearance of disease in inoculated plant tissues. Whether penetration occurs immediately from the appressorium, and the fungus remains dormant for a period in the epidermis of leaves and stem rind, or penetration occurs with the advent of favorable environmental conditions is unknown. Latency is especially prevalent on plant tissues inoculated during a period of rapid vegetative growth.

Host Resistance

Maize inbreds and hybrids with varying levels of resistance to anthracnose leaf blight and/or stalk rot have been described, and the resistance may be controlled by one or a few genes with major effects or by several genes with small, additive effects (3,4,5,17,18,19,36,44,81,87,91). Resistance to leaf blight and stalk rot phases of the disease appear to be inherited independently (5,91,93). Little is known about the specific mechanisms through which the expression of resistance in maize is activated. However, a substantial amount is known about the general physiological and biochemical responses to fungal infection that are characteristic of resistance expression in maize.

The general foliar resistance response of maize involves the stimulation of phenolic compound biosynthesis, especially synthesis of phenylpropanoids. One byproduct of phenylpropanoid biosynthesis is lignin, a complex polymer that has the important physical property of being very hard. The first evidence of a stimulation in phenylpropanoid synthesis is the formation of a lignified papilla in the outer wall of the epidermal cell directly beneath the fungal appressorium. It is important to point out that papilla formation occurs before the fungus penetrates the host cell. Papilla formation is a nonspecific, general response to attempted fungal penetration of Poaceae hosts (15,83). If papilla formation occurs rapidly enough and the papilla becomes heavily lignified, penetration is usually prevented. This occurs regardless of the resistance or susceptibility of the maize cultivar to *C. graminicola*. If penetration is successful, the expression of resistance is at first evident as the upregulation of genes for phenylpropanoid biosynthesis with the result that the plant synthesizes two caffeic acid esters, one an ester of glucose and the second an ester of an organic acid (50). These caffeic acid esters are not fungitoxic; however, they are believed to represent the diversion of carbon through the phenylpropanoid pathway to structural forms that can be esterified to carbohydrate polymers in the host cell wall. This esterification results in conversion of cell wall polymers to forms that are less easily broken down by wall-degrading enzymes produced by the fungus. Only two phenylpropanoids become esterified to the host cell wall, *p*-coumaric and ferulic acid (48).

As with other fungal diseases of maize, the expression of resistance also includes the formation of what has been referred to as stress lignin (48).

This lignin forms in cells immediately surrounding lesions and, as in the case of cell wall esterification, the lignin represents a physical barrier to the growth of the fungus as well as a chemical barrier to cell wall degradation and lesion expansion. An important feature of this stress lignin is that it contains high levels of syringaldehyde, a compound derived from the phenylpropanoid sinapic acid. This was a surprising discovery because sinapic acid cannot be isolated from the maize plant! We now believe that the plant synthesizes sinapic acid by a phenomenon known as channeling in which the substrate for a progressive series of enzymatic reactions is moved from one enzyme to another without being released. Such synthesis is thought to occur on multienzyme complexes. We believe that through such a mechanism the plant uses the phenylpropanoid 5-methoxyferulic acid to synthesize sinapic acid and to incorporate it into lignin.

In many maize cultivars, at the time the resistance response is terminated, anthocyanin pigments are synthesized in a zone immediately surrounding lesions (27). A similar accumulation of anthocyanin pigment occurs in maize lines with the recessive *rhm* gene for resistance to *Cochliobolus heterostrophus* (28). This pigment was identified as cyanidin 3-dimalonyl glucoside. The compound, which is also derived from phenylpropanoids, was suggested to act as a scavenger of cationic and anionic molecules that can be toxic to the plant itself. The same phenomenon of anthocyanin synthesis occurred in several maize cultivars inoculated with the anthracnose pathogen and grown under high light intensity (26). Thus, the active defense response in maize is based on an augmentation of phenolic compound metabolism, specifically phenylpropanoid biosynthesis.

Other compounds that have been proposed to be associated with resistance expression to *C. graminicola* are hydroxamic acids. The compound DIMBOA was once suggested to be an important preformed antifungal defense factor in maize. However, because levels of hydroxamic acids in leaf tissue decrease rapidly as the leaf blade expands and matures, these compounds are now thought to be of little importance as antifungal agents (49).

Resistance to *C. graminicola* in stalks has been associated with host phenological stage, maize genotype, and a rapid wound-healing type of resistance. These forms of resistance are additive (55,56). Stem tissues become more susceptible to ASR prior to anthesis (37,55,56). Genotypes such as MP305 and LB31 have been shown to express high levels of resistance to systemic development of ASR. This resistance is inherited simply as one or a few genes (3,17,81). Even in fully susceptible genotypes, wounded pith tissues become nearly as resistant to pathogen ingress within a few hours after wounding as a nonwounded stalk surface (55,56). Stem resistance involves, at least in part, a delay in fungal development at the early stages of the interaction. Conidial germination, germ-tube development, cell penetration, tissue maceration, vascular colonization, and systemic symptom development each are consistently delayed and reduced in resistant interactions (54). Reduced tissue discoloration in resistant plants was associated with reduced isolation of *C. graminicola* and with decreased fungal biomass as estimated by ergosterol content (54). Biochemical or other bases for stem resistance are not yet understood.

Pathogenic Variability in the Fungus

Isolates of *C. graminicola* from maize vary widely in aggressiveness on corn stalks and leaves, yet there is no convincing evidence to date for physiologic races (35,69,90). The potential for sexual recombination is illustrated by the success of laboratory matings (producing viable ascospores) that occurred between self-fertile and self-sterile strains and between certain self-sterile strains of the fungus (82).

Opportunities for Disease Management

Genetic Improvement of Maize

Most proprietary maize hybrids in the USA (and in other nations where commercial hybrids are grown) are significantly less susceptible to anthracnose leaf blight and stalk rot today than were commercial hybrids of 20 years ago (13). The current industry approach of selecting for moderate levels of general resistance is highly responsible and reduces the likelihood of strong directional selection within the *C. graminicola* population for specific virulence. Contemporary hybrids develop anthracnose when exposed to plentiful inoculum in conducive environments, but disease progress is slower and yield losses are less. Continued vigilance is needed in assessing genotypes for anthracnose resistance to avoid an inadvertent shift back to susceptible hybrids. Sporadic and damaging epidemics of anthracnose stalk rot still occur in the USA, primarily in conjunction with severe infestations of European corn borer.

The most accurate, established method for rating resistance to ASR requires cutting and splitting large numbers of plants at crop maturity for scoring the extent of pith discoloration. Pith tissue samples removed from living plants before anthesis provide an alternative, nondestructive method of assessment that can be performed on parental lines prior to making pollination decisions. The development of *C. graminicola* in inoculated, detached pith tissues is an excellent predictor of resistance to ASR in intact plants (54). Fungal maceration of detached pith tissues occurred more rapidly and more extensively in susceptible than resistant genotypes (54).

Genetic markers for quantitative trait loci conditioning resistance to anthracnose appear to be a useful tool for introgression of additive anthracnose stalk rot resistance genes into elite maize lines (36).

Future control strategies might utilize aspects of our current understanding of the role of phenolic synthesis in resistance. For example, genotypes might be selected for rapid phenolic compound responses including papilla formation, as well as esterification of phenylpropanoids to the cell wall and stress lignin formation. Such strategies could prove useful for the management of corn leaf blights in general.

Another approach to management is to disrupt the fungal infection process. For example, it is clear that because the fungus does not germinate until 6 hr after contact with the leaf, it is important that the ungerminated conidia can adhere to the leaf surface within minutes. This prevents displacement of the fungus from the infection court. Because this adhesion only occurs on a hydrophobic surface, it is possible that plants with non-waxy surfaces conditioned by glossy genes (58) may not support the required adhesion of the fungus.

Useful transgenes for transformation of maize could conceivably come from many sources including microorganisms, other plant species, and corn genotypes with otherwise undesirable agronomic traits. One potentially useful strategy would be to transform maize with genes for deoxyanthocyanidin (potent antifungal compounds) biosynthesis from sorghum. This would fit with the fact that the major response of the maize plant is phenolic biosynthesis. Host genetic alteration could be a mechanism to augment the speed and magnitude of the phenylpropanoid response.

Interruption of Insect Life Cycles

Injury by European corn borer larvae or other insects allows *C. graminicola* ingress into a maize stalk. Treatments that reduce corn borer populations might reduce anthracnose stalk rot in the field. This might be accomplished with insecticides, biocontrol agents, or plant resistance to the insect. Transgenic Bt corn hybrids expressing the CryIA (b) toxin, and exposed in the field to both *C. graminicola* and low natural populations of European corn borer, developed significantly less anthracnose stalk rot than did their corresponding non-Bt, near-isogenic hybrids (9). The large-scale deployment of transgenic Bt maize hybrids might have profound effects in reducing anthracnose stalk rot.

BIORATIONAL CHEMICAL CONTROL

Biorational chemicals, active at low concentration and having no or minimal toxicity to nontarget organisms, might find application in the protection of high value corn crops such as sweet maize, pop maize, and hybrid seed. A new trend in commercial fungicide development is the identification of novel biochemical target sites followed by subsequent directed synthesis of inhibitory chemicals (40). Candidate biochemical targets in *C. graminicola* include the enzymes for synthesis of melanin, necessary for functional appressoria. Other compounds might be sought that would either block or prevent the synthesis of the self-inhibitor mycosporine-alanine, a feature that would ensure that conidia germinate even when conditions are not favorable for disease development. Chemicals could also be developed that mimic mycosporine-alanine for use as protectants that prevent germination.

Chemical activators of induced resistance, e.g., benzothiadiazole ('Actigard') (24), turn on active resistance genes in advance of infection, and, thus, speed up the inherent defense response of the plant. Similar stimulants or activators of phenylpropanoid synthetic genes might result in induced resistance of maize genotypes normally susceptible to anthracnose leaf blight.

BIOLOGICAL CONTROL

A largely unexplored yet promising strategy for protecting maize against anthracnose is biological control on the phylloplane, especially on high value sweet maize and in hybrid seed production. Yeasts or cell-free extracts from yeast cultures applied to leaf surfaces have been shown to interfere with infection of maize leaves by *C. graminicola* and have reduced leaf blight severity (20,92). Bacteria and other microorganisms showing promise for biocontrol on the phylloplane of other plants could also be tested for efficacy against maize anthracnose.

A completely unexplored possibility is the use of antagonistic or competitive microorganisms applied to maize stubble for reducing the saprophytic survival of *C. graminicola* or, at least, for suppressing the production of primary conidial inoculum in the spring. This strategy is an especially attractive possibility for continuous corn production systems on erodible soils. Saprophytic survival of the fungus in maize stubble is the greatest factor favoring the regional increase of maize anthracnose, and its reduction may be the key to successful management of the disease.

Literature Cited

1. Anderson, B. and White, D.G. 1987. Fungi associated with corn stalks in Illinois in 1982 and 1983. Plant Dis.71:135-137.
2. Anderson, D.W. and Nicholson, R.L. 1996. Characterization of a laccase in the conidial mucilage of *Colletotrichum graminicola.* Mycologia. 88:996-1002.
3. Badu-Apraku, B., Gracen, V.E., and Bergstrom, G.C. 1987. A major gene for resistance to anthracnose stalk rot in maize. Phytopathology 77:957-959.
4. Badu-Apraku, B., Gracen, V.E., and Bergstrom, G.C. 1987. A major gene for resistance to anthracnose leaf blight in maize. Plant Breed. 98:194-199.
5. Badu-Apraku, B., Gracen,V.E., and Bergstrom, G.C. 1987. Inheritance of resistance to anthracnose stalk rot and leaf blight in a maize inbred derived from a temperate by tropical germplasm combination. Maydica 32:221-237.
6. Bailey, J.A., O'Connell, R.J., Pring, R.J., and Nash, C. 1992. Infection strategies of *Colletotrichum* species. Pages 88-120 in: *Colletotrichum*: Biology, Pathology and Control. J.A. Bailey and M.J. Jeger, eds. CAB International, Wallingford, UK.
7. Bergstrom, F.B. and Bergstrom, G.C. 1987. Influence of maize growth stage on fungal movement, viability, and rot induction in stalks inoculated with *Colletotrichum graminicola.* Phytopathology 77:115. (Abstr.)
8. Bergstrom, G.C., Croskey, B.S., and Carruthers, R.I. 1983. Interaction between European corn borer and anthracnose stalk rot in corn. Proc. Ann. Northeastern Corn Improvement Conf. 38:131-134.
9. Bergstrom, G.C., Davis, P.M., and Waldron, J.K. 1997. Management of anthracnose stalk rot/European corn borer pest complex with transgenic Bt corn hybrids for silage production. Biological and Cultural Tests for Control of Plant Diseases 12:14.
10. Bergstrom, G.C. and Nicholson, R.L. 1979. Influence of host developmental stage on the susceptibility of maize to *Colletotrichum graminicola.* Proc. Intl. Congress of Plant Protection, 9th, Washington, D.C. Abstract No. 116.
11. Bergstrom, G.C. and Nicholson, R.L. 1981. Invertase in the spore matrix of *Colletotrichum graminicola.* Phytopath. Z. 102:139-147.
12. Bergstrom, G.C. and Nicholson, R.L. 1983. Effect of the *Colletotrichum graminicola* conidial matrix on the development of anthracnose seedling blight in maize. Fitopathol. Brasil. 8:447-453.
13. Bergstrom, G.C. and Nicholson, R.L. 1999. The biology of corn anthracnose. Plant Dis. 83:596-608.

14. Byrnes, K.J. and Carroll, R.B. 1986. Fungi causing stalk rot of conventional-tillage and no-tillage corn in Delaware. Plant Dis. 70:238-239.
15. Cadena-Gomez, G. and Nicholson, R.L. 1987. Papilla formation and associated peroxidase activity: A non-specific response to attempted fungal penetration of maize. Physiol. Molec. Plant Pathol. 31:51-67.
16. Carruthers, R.I., Bergstrom, G.C., and Haynes, P.A. 1986. Accelerated development of the European corn borer *Ostrinia nubilalis* (Lepidoptera: Pyralidae), induced by interactions with *Colletotrichum graminicola* (Melanconiales: Melanconiaceae), the causal fungus of maize anthracnose. Ann. Entomol. Soc. Am. 79:385-389.
17. Carson, M.L. 1981. Sources and inheritance of resistance to anthracnose stalk rot of corn. Ph.D. thesis. University of Illinois, Urbana-Champaign, Illinois.
18. Carson, M.L. and Hooker, A.L. 1981. Inheritance of resistance to anthracnose leaf blight in five inbred lines of corn. Phytopathology 71:488-491.
19. Carson, M.L. and Hooker, A.L. 1981. Inheritance of resistance to anthracnose stalk rot of corn caused by *Colletotrichum graminicola.* Phytopathology 71:1190-1196.
20. da Silva, S.R. and Pasholati, S.F. 1992. *Saccharomyces cerevisiae* protects maize plants, under greenhouse conditions, against *Colletotrichum graminicola.* Z. Planzenkr. Plfanzenshutz 99:159-167.
21. Dixon, R.A. and Paiva, N.L. 1995. Stress-induced phenylpropanoid metabolism. Plant Cell 7:1085-1097.
22. Dodd, J.L. 1977. A photosynthetic stress-translocation balance concept of corn stalk rot. Proc. Ann. Corn and Sorghum Research Conf. 32:122-130.
23. Epstein, L. and Nicholson, R.L. 1997. Adhesion of spores and hyphae to plant surfaces. Pages 11-25 in: The Mycota, Vol. V, Part A. G. Carroll and P. Tudzynski, eds. Springer-Verlag, Berlin.
24. Gorlach, J., Volrath, S., Knauf-Beiter, G., Hengy, G., Beckhove, U., Kogel, K., Oostendorp, M., Staub, T., Ward, E., Kessman, H., and Ryals, J. 1996. Benzothiadiazole, a novel class of inducers of systemic acquired resistance, activates gene expression and disease resistance in wheat. Plant Cell 8:629-643.
25. Hagerman, A.E. and Butler, L.G. 1981. The specificity of proanthocyanidin-protein interactions. J. Biol. Chem. 256:4494-4497.
26. Hammerschmidt, R. and Nicholson, R.L. 1976. Resistance of maize to anthracnose: Effect of light intensity on lesion development. Phytopathology 67:247-250.

27. Hammerschmidt, R. and Nicholson, R.L. 1977. Resistance of maize to anthracnose: Changes in host phenols and pigments. Phytopathology 67:251-258.
28. Hipskind, J., Wood, K., and Nicholson, R.L. 1996. Stimulation of anthocyanin accumulation and delineation of pathogen ingress in maize genetically resistant to *Bipolaris maydis* race O. Physiol. Molec. Plant Pathol. 49:247-256.
29. Holton, T.A. and Cornish, E.C. 1995. Genetics and biochemistry of anthocyanin biosynthesis. Plant Cell 7:1071-1083.
30. Hooker, A.L. 1976. Corn anthracnose leaf blight and stalk rot. Proc. Ann. Corn and Sorghum Res. Conf. 31:167-182.
31. Hooker, A.L. and White, D.G. 1976. Prevalence of corn stalk rot fungi in Illinois. Plant Dis. Reptr. 60:1032-1034.
32. Howard, R.J. 1997. Breaching the outer barriers-Cuticle and cell wall penetration. Pages 43-60 in: The Mycota, Vol. V, Part A. G. Carroll and P. Tudzynski, eds.Springer-Verlag, Berlin.
33. Jamil, F.F. and Nicholson, R.L. 1987. Susceptibility of corn to isolates of *Colletotrichum graminicola* pathogenic to other grasses. Plant Dis. 71:809-810.
34. Jenns, A.E., and Leonard, K.J. 1985. Effects of illuminance on the resistance of inbred lines of corn to isolates of *Colletotrichum graminicola*. Phytopathology 75:281-286.
35. Jenns, A.E., Leonard, K.J., and Moll, R.H. 1982. Variation in the expression of specificity in two maize diseases. Euphytica 31:269-279.
36. Jung, M., Weldekidan, T., Schaff, D., Patterson, A., Tingey, S., and Hawk, J. 1994. Generation means analysis and quantitative trait locus mapping of anthracnose stalk rot genes in maize. Theor. Appl. Genet. 89:413-418.
37. Keller, N.P. and Bergstrom, G.C. 1988. Developmental predisposition of maize to anthracnose stalk rot. Plant Dis. 72:977-980.
38. Keller, N.P., Bergstrom, G.C., and Carruthers, R.I. 1986. Potential yield reductions in maize associated with an anthracnose/European corn borer pest complex in New York. Phytopathology 76:586-589.
38. Kieselbach, T.A. 1980. The structure and reproduction of corn. The University of Nebraska Press, Lincoln.
39. Knight, S.C., Anthony, V.M., Brady, A.M., Greenland, A.J., Heaney, S.P., Murray, D.C., Powell, K.A, Schulz, M.A., Spinks, C.A., Worthington, P.A., and Youle, D. 1997. Rationale and perspectives on the development of fungicides. Annu. Rev. Phytopathol. 35:349-372.
40. Kollo, I.A. 1988. Influence of leaf age on anthracnose development in *Zea mays*. M.S. thesis, Purdue University, Indiana. 92 pp.

41. Leite, B. and Nicholson, R.L. 1992. Mycosporine-alanine: A self-inhibitor of germination from the conidial mucilage of *Colletotrichum graminicola*. Exper. Mycol. 16: 76-86.
42. Leonard, K.J. and Thompson, D.L. 1976. Effects of temperature and host maturity on lesion development of *Colletotrichum graminicola* on corn. Phytopathology 66:635-639.
43. Lim, S.M. and White, D.G. 1978. Estimates of heterosis and combining ability for resistance of maize to *Colletotrichum graminicola*. Phytopathology 68:1336-1342.
44. Lipps, P.E. 1983. Survival of *Colletotrichum graminicola* in infested corn residues in Ohio. Plant Dis. 67:102-104.
45. Lipps, P.E. 1985. Influence of inoculum from buried and surface corn residues on the incidence of corn anthracnose. Phytopathology 75:1212-1216.
46. Lipps, P.E. 1988. Spread of corn anthracnose from surface residues in continuous corn and corn-soybean rotation plots. Phytopathology 75:756-761.
47. Lyons, P.C., Hipskind, J., Vincent, J.R., and Nicholson, R.L. 1993. Phenylpropanoid dissemination in maize resistant or susceptible to *Helminthosporium maydis*. Maydica 38:175-181.
48. Lyons, P.C. and Nicholson, R.L. 1989. Evidence that cyclic hydroxamate concentrations are not related to resistance of corn leaves to anthracnose. Can. J. Plant Pathol. 11:215-220.
49. Lyons, P.C., Wood, K.V., and Nicholson, R.L. 1990. Caffeoyl ester accumulation in corn leaves inoculated with fungal pathogens. Phytochemistry 29:97-101.
50. Madden, L.V., Yang, X., and Wilson, L.L. 1996. Effects of rain intensity on splash dispersal of *Colletotrichum acutatum*. Phytopathology 86:864-874.
51. Mercure, E.W., Kunoh, H., and Nicholson, R.L. 1994. Adhesion of *Colletotrichum graminicola* to corn leaves: A requirement for disease development. Physiol. Molec. Plant Pathol. 45:407-420.
52. Mercure, E.W., Kunoh, H., and Nicholson, R.L. 1995. Visualization of materials released from adhered, ungerminated conidia of *Colletotrichum graminicola*. Physiol. Molec. Plant Pathol. 46:121-135.
53. Muimba-Kankolongo, A. 1991. Nature of resistance to anthracnose stalk rot associated with maizegenotype, ontogeny, and wound response. Ph.D. thesis. Cornell University, New York.
54. Muimba-Kankolongo, A. and Bergstrom, G.C. 1990. Transitory wound predisposition of maize to anthracnose stalk rot. Can. J. Plant Pathol. 12:1-10.
55. Muimba-Kankolongo, A. and Bergstrom, G.C. 1992. Wound predisposition of maize to anthracnose stalk rot as affected by internode

position and inoculum concentration of *Colletotrichum graminicola*. Plant Dis.76:188-195.

56. Naylor, V.D. and Leonard, K.J. 1977. Survival of *Colletotrichum graminicola* in infected corn stalks in North Carolina. Plant Dis. Reptr. 61:382-383.
57. Neuffer, M.G., Coe, E.H., and Wessler, S.R. 1997. Mutants of Maize. Cold Spring Harbor Laboratory Press, Plainview, NY. 468 pp.
58. Nicholson, R.L. 1992. *Colletotrichum graminicola* and the anthracnose disease of corn and sorghum. Pages 186-202 in: Colletotrichum: Biology, Pathology and Control. J.A. Bailey and M.J. Jeger, eds. CAB International, Wallingford, UK.
59. Nicholson, R.L. 1996. Adhesion of fungal propagules: Significance to the success of the fungal infection process. Pages 117-134 in: Histology, Ultrastructure and Molecular Cytology of Plant-Microorganism Interactions. M. Nicole and V. Gianinazzi-Pearson, eds. Kluwer Academic PublishersDordrecht, The Netherlands.
60. Nicholson, R.L., Bergeson, G.B., DeGennaro, F.P., and Viveiros, D.M. 1985.Single and combined effects of the lesion nematode and *Colletotrichum graminicola* on growth and anthracnose leaf blight on corn. Phytopathology 75:654-661
61. Nicholson, R.L., Butler, L.G., and Asquith, T.N. 1986. Glycoproteins from *Colletotrichum graminicola* that bind phenols: Implicatons for survival and virulence of phytopathogenic fungi. Phytopathology 76:1315-1318.
62. Nicholson, R.L. and Epstein, L. 1991. Adhesion of fungi to the plant surface: Prerequisite for pathogenesis. Pages 3-23 in: The Fungal Spore and Disease Initiation in Plants and Animals. H. Hoch and G. Cole, eds. Plenum, New York.
63. Nicholson, R.L. and Hammerschmidt, R. 1992. Phenolic compounds and their role in disease resistance. Annu. Rev. Phytopathol. 30:369-389.
64. Nicholson, R.L., Hipskind, J., and Hanau, R.M. 1989. Protection against phenol toxicity by the spore mucilage of *Colletotrichum graminicola*. Physiol. Mol. Plant Pathol. 35:243-252.
65. Nicholson, R.L. and Moraes, W.B.C. 1980. Survival of *Colletotrichum graminicola*: Importance of the spore matrix. Phytopathology 70:255-261.
66. Nicholson, R.L., Turpin, C.A., and Warren, H.L. 1976. Role of pectic enzymes in susceptibility of living maize pith to *Colletotrichum graminicola*. Phytopathol. Z. 87:324-336.
67. Nicholson, R.L. and Warren, H.L. 1976. Criteria for evaluation of resistance to maize anthracnose. Phytopathology 66:86-90.

68. Nicholson, R.L. and Warren, H.L. 1981.The issue of races of *Colletotrichum graminicola* pathogenic to corn. Plant Dis. 65:143-145.
69. Panaccione, D.G., Vaillancourt, L., and Hanau, R.M. 1989. Conidial dimorphism in *Colletotrichum graminicola.* Mycologia 81:876-883.
70. Pascholati, S.F., Deising, H., Leite, B., Anderson, D., and Nicholson, R.L. 1993. Cutinase and non-specific esterase activities in the conidial mucilage of *Colletotrichum graminicola.* Physiol. Molec. Plant Pathol. 42:37-51.
71. Perkins, J.M. and Hooker, A.L. 1979. The effects of anthracnose stalk rot on corn yields in Illinois. Plant Dis. Reptr. 63:26-30.
72. Prusky, D. and Plumbley, R.A. 1992. Quiescent infections of *Colletotrichum* in tropical and subtropical fruits. Pages 289-307 in: *Colletotrichum*: Biology, Pathology and Control. J.A. Bailey and M.J. Jeger, eds. CAB International, Wallingford, UK.
73. Ramados, C.S., Uhlig, J., Carlson, D.M., Butler, L.G., and Nicholson, R.L. 1985. Composition of the mucilaginous spore matrix of *Colletotrichum graminicola,* a pathogen of corn, sorghum, and other grasses. J. Agric. Food Chem. 33:728-732.
74. Schall, R.A., Nicholson, R.L., and Warren, H.L. 1980. Influence of light on maize anthracnose in the greenhouse. Phytopathology70:1023-1026.
75. Sherriff, C., Whelan, M.J., Arnold, G.M., and Bailey, J.A. 1995. rDNA sequence analysis confirms the distinction between *Colletotrichum graminicola* and *C. sublineolum.* Mycol. Res. 99:475-478.
76. Skoropad, W.P. 1967. Effect of temperature on the ability of *Colletotrichum graminicola* to form appressoria and penetrate barley leaves. Can. J. Plant Sci. 47:431-434.
77. Smith, D.R. 1976. Yield reduction in dent corn caused by *Colletotrichum graminicola.* Plant Dis. Reptr. 60:967-970.
78. Smith, D.R. and White, D.G. 1988. Diseases of corn. Pages 687-766 in: Corn and Corn Development, 3rd ed., Agronomy Series No. 18. G.F. Sprague and J.W. Dudley, eds. Am. Soc. Agron., Madison, Wisconsin.
79. Snyder, B.A. 1990. The infection process of *Colletotrichum graminicola* (Ces.) Wils. on *Sorghum bicolor* L. and analysis of an extracelluar DNAse produced by the fungus. Ph.D. thesis, Purdue University, Indiana.
80. Toman, J. Jr. and White, D.G. 1993. Inheritance of resistance to anthracnose stalk rot of corn. Phytopathology 83:981-986.
81. Vaillancourt, L.J. and Hanau, R.M. 1991. A method for genetic analysis of *Glomerella graminicola* (*Colletotrichum graminicola*) from maize. Phytopathology 81:530-534.
82. Vance, C.R., Kirk, T.P., and Sherwood, R.T. 1980. Lignification as a mechanism of disease resistance. Annu. Rev. Phytopathol. 18:259-288.

83. Vizvary, M.A. and Warren, H.L. 1974. Saprophytic survival of *Colletotrichum graminicola* in sweet and dent corn stalk tissues. Proc. Am. Phytopathol. Soc. 1:130-131. (Abstr.)
84. Vizvary, M.A. and Warren, H.L. 1982. Survival of *Colletotrichum graminicola* in soil. Phytopathology 72:522-525.
85. Warren, H.L., Nicholson, R.L., Ullstrup, A.J., and Sharvelle, E.G. 1973. Observations of *Colletotrichum graminicola* on sweet corn in Indiana. Plant Dis. Reptr. 57:143-144.
86. Weldekidan, T. and Hawk, J.A. 1993. Inheritance of anthracnose stalk rot resistance in maize. Maydica 38:189-192.
87. Wheeler, H., Politis, D.J., and Poneleit, C.G. 1974. Pathogenicity, host range, and distribution of *Colletotrichum graminicola* on corn. Phytopathology 64:293-296.
88. White, D.G. and Humy, C. 1976. Methods for inoculation of corn stalks with *Colletotrichum graminicola.* Plant Dis. Reptr. 60:898-899.
89. White, D.G., Yanney, J., and Anderson, B. 1987. Variation in pathogenicity, virulence, and aggressiveness of *Colletotrichum graminicola* on corn. Phytopathology 77:999-1001.
90. White, D.G., Yanney, J., and Natti, T.A. 1979. Anthracnose stalk rot. Proc. Ann. Corn and Sorghum Res. Conf. 34:1-15.
91. Williamson, M.A. and Fokkema, N.J. 1985. Phyllosphere yeasts antagonize penetration from appressoria and subsequent infection of maize leaves by *Colletotrichum graminicola.* Neth. J. Pl. Path. 91:265-276.
92. Zuber, M.S., Ainsworth, T.C., Blanco, M.H., and Darrah, L.L. 1981. Effect of anthracnose leaf blight on stalk rind strength and yield in F single crosses in maize. Plant Dis. 65:719-722.